SUPERSYMMETRIC GAUGE FIELD THEORY AND STRING THEORY

Graduate Student Series in Physics
Other books in the series

Gauge Theories in Particle Physics
J J R AITCHISON and A J G HEY

Introduction to Gauge Field Theory
D BAILIN and A LOVE

Mechanics of Deformable Media
A B BHATIA and R N SINGH

Hadron Interactions
P D B COLLINS and A D MARTIN

The Physics of Structurally Disordered Matter: An Introduction
N E CUSACK

Collective Effects in Solids and Liquids
N H MARCH and M PARRINELLO

Geometry, Topology and Physics
M NAKAHARA

Supersymmetry, Superfields and Supergravity: An Introduction
P P SRIVASTAVA

Superfluidity and Superconductivity
D R TILLEY and J TILLEY

GRADUATE STUDENT SERIES IN PHYSICS

Series Editor: Professor Douglas F Brewer, MA, DPhil
Professor of Experimental Physics, University of Sussex

SUPERSYMMETRIC GAUGE FIELD THEORY AND STRING THEORY

DAVID BALIN
Physics and Astronomy Subject Group
School of Mathematical and Physical Sciences
University of Sussex

ALEXANDER LOVE
Department of Physics
Royal Holloway and Bedford New College
University of London

Taylor & Francis
Taylor & Francis Group
New York London

Published in 1994 by
Taylor & Francis Group
270 Madison Avenue
New York, NY 10016

Published in Great Britain by
Taylor & Francis Group
2 Park Square
Milton Park, Abingdon
Oxon OX14 4RN

Reprinted with corrections 1996

No claim to original U.S. Government works
Printed in the United States of America on acid-free paper
10 9 8 7 6 5 4 3 2

International Standard Book Number-10: 0-7503-0267-4 (Softcover)
International Standard Book Number-13: 978-7503-0267-8 (Softcover)
DOI: 10.1201/9780367805807

Library of Congress Cataloging-in-Publication Data

Catalog record is available from the Library of Congress

informa
Taylor & Francis Group
is the Academic Division of Informa plc.

Visit the Taylor & Francis Web site at
http://www.taylorandfrancis.com

To Anjali
and
To Christine

PREFACE

There has been considerable enthusiasm in recent years for the idea that the gauge field theories describing elementary particle interactions should possess global supersymmetry. This enthusiasm has been generated to a considerable extent by the capacity of supersymmetry to solve the gauge hierarchy problem of grand unified theories, fine tuning at each order of perturbation theory to preserve the electroweak scale being avoided in supersymmetric theories as a consequence of non-renormalization theorems.

Once one is committed to global supersymmetry the commitment does not end there. Potentially realistic supersymmetric gauge theories have a supersymmetry-breaking scale sufficiently large that effects of gravity cannot be neglected so that one must derive the globally supersymmetric theory from a theory of supergravity in which the supersymmetry is local. However, supergravity, like any (point particle) field theory containing gravity is non-renormalizable and cannot be the fundamental theory of interactions. There is, at the time of writing, only one known renormalizable theory that can describe quantum gravity in the presence of matter, namely the theory of superstrings. Thus, once embarked on a study of supersymmetry we are led almost inevitably to string theory.

This book introduces the reader to supersymmetry, supergravity and superstring theory in a single volume. In view of the fact that there is potentially enough material to fill five or six volumes, we have been very selective. In particular, the discussion of supersymmetry is entirely in terms of component fields, the discussion of supergravity entirely from the Noether procedure standpoint, and the development of string theory does not go beyond interactions at tree level. However, we have thought it appropriate to include two chapters on the construction of string theories in four dimensions so as to make contact between string theory and low-energy supergravity.

We are grateful to many colleagues, including D R T Jones, G G Ross, B Sendhoff and especially D C Dunbar and S Thomas for the supersymmetric physics that we have learned from them, and to Miss A S Clark for her very careful and speedy typing of the manuscript. Finally, we wish to thank our wives, to whom this book is dedicated, for their invaluable encouragement throughout the writing of the book.

David Bailin
Alexander Love

CONTENTS

1

SUPERSYMMETRY ALGEBRA AND MULTIPLETS

1.1 Introduction

At the time of writing particle accelerators are beginning to probe the 100 GeV to 1 TeV energy scale at which many theorists expect the first direct evidence of supersymmetry to appear. We shall see shortly that supersymmetry implies that all particles possess supersymmetric partners having opposite statistics. This is because supersymmetry multiplets consist of equal-mass particles whose spins differ by $\frac{1}{2}$. So far no supersymmetric partners of any known particles have been discovered, so supersymmetry if it exists is not only broken, but broken at an energy scale beyond the reach of accelerators to date. Nevertheless, supersymmetry remains attractive, at least to particle theorists, for several reasons.

First, it provides the only known solution of the 'technical hierarchy problem'. This will be described in detail in Chapter 6. For the present we merely remark that in a grand unified theory (of strong and electroweak interactions) the unification scale is at least 10^{15} GeV, and the (hierarchy) problem is to understand how the electroweak scalars remain massless way below this scale when they are not protected by any symmetry that would guarantee it. This problem is solved in a supersymmetric theory by a cancellation of Feynman diagrams that separately would generate the undesired mass scale. This illustrates the second attractive feature of supersymmetry, which is that supersymmetric theories have better high-energy behaviour than non-supersymmetric theories. In fact some (extended) supersymmetric theories are so 'well-behaved' that they are completely finite(1). (Thus these theories meet Dirac's criticism(2) of (non-supersymmetric) theories, such as quantum electrodynamics, that they cannot be considered as complete physical theories if one has to calculate and manipulate infinite quantities.) The third, and most recent, reason for supersymmetry being so well regarded is that it appears to be an indispensable ingredient of the 'superstring' theories which we shall discuss in Chapter 9, and the succeeding chapters. These theories have some quite remarkable properties, as we shall see, and at the moment they are the best candidates we have for 'theories of everything', i.e. *quantum* theories of the strong, electroweak and gravitational interactions.

The cancellation that solves the hierarchy problem arises because of the negative sign associated with closed fermion loops, as compared to bosonic

DOI: 10.1201/9780367805807-1

loops. The fact (if it is one) that each boson has a fermionic partner of equal mass suggests the enlargement of the Poincaré algebra by the inclusion of a *spinor generator* Q. Then since Q commutes with the mass operator, but not with the spin operator, we obtain irreducible representations of the enlarged algebra that have a definite mass, but different spin values. We can then see how this might, in principle, solve the hierarchy problem: since we know how to arrange that fermions stay massless, by having a chiral theory, we can ensure that their bosonic partners are also kept massless by making such a theory supersymmetric.

In the first instance, supersymmetry is introduced as a global symmetry of the Lagrangian and this will be the standpoint adopted in Chapters 1–3. However, it is attractive to think that supersymmetry, like gauge symmetries, may occur as a local symmetry. In that case, because the supersymmetry algebra contains the generators P_μ of translations we will be considering translations that vary from point to point in space-time. Thus, a theory of local supersymmetry will contain general coordinate transformations of space-time, and so, amongst other things, will be a theory of gravity. Such supergravity theories will be discussed in Chapter 4.

One might imagine that it will only be necessary to take account of supergravity, as distinct from global supersymmetry, at very high energies, close to the Planck scale. However, it turns out, as will be discussed further in §4.1, that potentially realistic supersymmetric theories have supersymmetry breaking scales of 10^{10}–10^{11} GeV. As will be seen in Chapter 5, this supersymmetry breaking feeds through into the low-energy theory as masses for scalar partners of fermionic states of the order of 10^2–10^3 GeV. Thus, it is not possible to neglect the effects of supergravity even at low energies. Supergravity will be developed in Chapters 4 and 5.

Like gravity itself, supergravity is a non-renormalizable theory and cannot therefore be an acceptable final theory of everything. The only known theory containing gravity that is renormalizable is the relativistic string and, as will be discussed further at the end of Chapter 12, supergravity should be regarded as an effective low-energy theory derived from the fundamental string theory. The development of string theory will take up the second half of the book. In string theory, supersymmetry enters in two distinct ways. First, it occurs in the formulation of the superstring in Chapter 8 as a symmetry associated with the two-dimensional world sheet of the string. Second, it can also occur as a space-time symmetry such as is studied in this chapter. Although space-time supersymmetry need not necessarily arise from a superstring theory (with world sheet supersymmetry) theories that do not enjoy space-time supersymmetry usually (if not always) lack a stable ground state when the question of stability of the ground state is studied beyond string tree level.

At the time of writing the first fairly direct evidence of supersymmetry may already have been seen in the running of gauge coupling constants from their measured low-energy values to high energies. When non-supersymmetric renormalization group equations for the standard model are employed nothing special happens, but when supersymmetric renormalization group equations, taking account of the supersymmetry partners of the standard model particles, are run instead, the $SU(3) \times SU(2) \times U(1)$ gauge coupling constants reach a common value at around 10^{16} GeV. This may be regarded as evidence for supersymmetric grand unification. This empirical observation is a little more tricky to interpret in the context of superstring theory where there is a natural unification of gauge coupling constants at tree level at around 10^{18} GeV, regardless of whether there is a grand unified group or not, and the observed unification at about 10^{16} GeV may require string loop threshold corrections to the renormalization group equations to move the unification scale down in energy.

1.2 Dirac, Weyl and Majorana spinors

We start by reviewing the Poincaré algebra. A Poincaré transformation P is a proper Lorentz transformation Λ followed by a translation a. Let x^μ ($\mu = 0, 1, 2, 3$) denote the coordinates of a space-time point. Then the Poincaré transformed coordinates are given by

$$x'^\mu = \Lambda^\mu{}_\nu x^\nu + a^\mu \tag{1.1}$$

where Λ is the restricted Lorentz transformation. So

$$\det \Lambda = +1 \qquad \Lambda^0{}_0 > 1 \tag{1.2}$$

and all such transformations are continuously connected to the identity. We denote such a Poincaré transformation by

$$P = \{\Lambda, a\}. \tag{1.3}$$

The generators of the Poincaré group are evidently the six generators $M^{\mu\nu}$ of the Lorentz group plus the four generators P^λ of the translation group. We use Hermitian generators so that P^λ is the energy–momentum operator, and $M^{\mu\nu}$ is the angular momentum tensor. By considering infinitesimal translations

$$x'_\mu = x_\mu + a_\mu \equiv x_\mu - \mathrm{i}\, a_\lambda (P^\lambda)_\mu \tag{1.4}$$

we find

$$(P^\lambda)_\mu = \mathrm{i}\, \delta^\lambda_\mu . \tag{1.5}$$

Similarly for an infinitesimal Lorentz transformation

$$x'^\rho = x^\rho + \omega^\rho{}_\sigma x^\sigma \equiv x^\rho - \tfrac{1}{2}\mathrm{i}\, \omega_{\mu\nu} (M^{\mu\nu})^\rho{}_\sigma x^\sigma \tag{1.6}$$

where $\omega_{\rho\sigma} = -\omega_{\rho\sigma}$, it follows that

$$(M^{\mu\nu})_{\rho\sigma} = \mathrm{i}(\delta^{\mu}_{\rho}\delta^{\nu}_{\sigma} - \delta^{\mu}_{\sigma}\delta^{\nu}_{\rho}). \tag{1.7}$$

It is then easy to verify that the Poincaré algebra is

$$[P^{\lambda}, P^{\mu}] = 0 \tag{1.8a}$$

$$[M^{\mu\nu}, P^{\lambda}] = \mathrm{i}(\eta^{\nu\lambda} P^{\mu} - \eta^{\mu\lambda} P^{\nu}) \tag{1.8b}$$

$$[M^{\mu\nu}, M^{\rho\sigma}] = \mathrm{i}(\eta^{\nu\rho} M^{\mu\sigma} + \eta^{\mu\sigma} M^{\nu\rho} - \eta^{\mu\rho} M^{\nu\sigma} - \eta^{\nu\sigma} M^{\mu\rho}) \tag{1.8c}$$

and we are using the 'Bjorken and Drell' Minkowski space-time metric

$$\eta_{\mu\nu} = \eta^{\mu\nu} = \mathrm{diag}(1, -1, -1, -1). \tag{1.9}$$

We have already declared our intention to enlarge this algebra by the introduction of a spinor generator, but to avoid confusion we need to be quite precise about the various spinors that arise. First we have the familiar Dirac spinor, discussed in §3.4 of Bailin and Love I. This is defined in terms of the 4×4 matrices γ^{μ} which satisfy

$$\{\gamma^{\mu}, \gamma^{\nu}\} \equiv \gamma^{\mu}\gamma^{\nu} + \gamma^{\nu}\gamma^{\mu} = 2\eta^{\mu\nu} I_4. \tag{1.10}$$

It is easy to verify that the matrices

$$\tfrac{1}{2}\Sigma^{\mu\nu} \equiv \frac{\mathrm{i}}{4}(\gamma^{\mu}\gamma^{\nu} - \gamma^{\nu}\gamma^{\mu}) \tag{1.11}$$

satisfy the Lorentz algebra (1.8c), and in fact on the Dirac spinor the Poincaré generators are given by

$$P^{\lambda} = \mathrm{i}\,\partial^{\lambda} \tag{1.12a}$$

$$M^{\mu\nu} = x^{\mu}P^{\nu} - x^{\nu}P^{\mu} + \tfrac{1}{2}\Sigma^{\mu\nu}. \tag{1.12b}$$

When discussing massless solutions of Dirac equation it is particularly useful to use the Weyl representation for the gamma matrices. In the Weyl representation

$$\gamma^{\mu} = \begin{pmatrix} 0 & \sigma^{\mu} \\ \bar{\sigma}^{\mu} & 0 \end{pmatrix} \qquad (\mu = 0, 1, 2, 3) \tag{1.13}$$

where

$$\sigma^{\mu} \equiv (I_2, \boldsymbol{\sigma}) \tag{1.14a}$$

$$\bar{\sigma}^{\mu} \equiv (I_2, -\boldsymbol{\sigma}) = \sigma_{\mu}. \tag{1.14b}$$

Then

$$\gamma_5 \equiv \mathrm{i}\,\gamma^0\gamma^1\gamma^2\gamma^3 = \begin{pmatrix} -I_2 & 0 \\ 0 & I_2 \end{pmatrix} \tag{1.15}$$

and we see that in this representation the upper two components of the Dirac spinor Ψ_D have left chirality, while the bottom two components have right chirality. In other words, we can write

$$\Psi_D = \Psi_L + \Psi_R \tag{1.16}$$

where

$$\Psi_L = \tfrac{1}{2}(1 - \gamma_5)\Psi_D \tag{1.17a}$$

$$\Psi_R = \tfrac{1}{2}(1 + \gamma_5)\Psi_D \tag{1.17b}$$

and then Ψ_L has two non-zero components, denoted ψ_α $(\alpha = 1, 2)$, in the upper two components. The two non-zero components of Ψ_R are denoted $\bar{\chi}^{\dot{\alpha}}$ $(\alpha = 1, 2)$. These two component spinors are called Weyl spinors. We use a dotted label $\dot{\alpha}$ for the right-handed spinors, since the two types of spinor transform differently under Lorentz transformations. This is easily seen from (1.11), for example. Both spinors transform identically under rotations, since

$$\tfrac{1}{2}\Sigma^{ij} = \tfrac{1}{2}\epsilon^{ijk}\begin{pmatrix} \sigma^k & 0 \\ 0 & \sigma^k \end{pmatrix}. \tag{1.18}$$

The difference is in their behaviour under Lorentz boost transformations, since

$$\tfrac{1}{2}\Sigma^{0i} = \tfrac{1}{2}\begin{pmatrix} -\mathrm{i}\,\sigma^i & 0 \\ 0 & \mathrm{i}\,\sigma^i \end{pmatrix}. \tag{1.19}$$

Thus we write (in the Weyl representation)

$$\Psi_L = \begin{pmatrix} \psi_\alpha \\ 0 \end{pmatrix} \qquad (\alpha = 1, 2) \tag{1.20}$$

$$\Psi_R = \begin{pmatrix} 0 \\ \bar{\chi}^{\dot{\alpha}} \end{pmatrix} \qquad (\alpha = 1, 2). \tag{1.21}$$

Now suppose the Dirac spinor Ψ_D has 'charge' e, and satisfies the Dirac equation

$$\mathrm{i}\,\gamma^\mu(\partial_\mu - \mathrm{i}\,eA_\mu)\Psi_D = 0 \tag{1.22}$$

where A_μ is the vector potential associated with some external electromagnetic field. Then using $\bar{\Psi} \equiv \Psi^\dagger\gamma_0$, we find

$$-\mathrm{i}\,\gamma^{\mu\mathrm{T}}(\partial_\mu + \mathrm{i}\,eA_\mu)\bar{\Psi}_D{}^{\mathrm{T}} = 0. \tag{1.23}$$

The matrices $-\gamma_\mu{}^{\mathrm{T}}$ also satisfy (the Clifford algebra) (1.10), and (in four dimensions) there is a non-singular matrix C such that

$$C^{-1}\gamma^\mu C = -\gamma^{\mu\mathrm{T}}. \tag{1.24}$$

Thus if we define the 'charge-conjugate spinor' Ψ_D^c by putting

$$\Psi_D^c \equiv C\bar{\Psi}_D{}^T \tag{1.25}$$

we see that it has 'charge' $-e$ and satisfies

$$i\gamma^\mu(\partial_\mu + i\,eA_\mu)\Psi_D^c. \tag{1.26}$$

It is easy to show that C is always anti-symmetric, and in the Weyl representation (1.13) we may choose C to be (proportional to) $\gamma^0\gamma^2$. So

$$C = \omega\gamma^0\gamma^2 = \omega\begin{pmatrix} -\sigma^2 & 0 \\ 0 & \sigma^2 \end{pmatrix}. \tag{1.27}$$

Writing Ψ_D in terms of two-component spinors

$$\Psi_D = \begin{pmatrix} \psi \\ \bar{\chi} \end{pmatrix} \tag{1.28}$$

we find

$$\Psi_D^c = \omega\begin{pmatrix} -\sigma^2\bar{\chi}^* \\ \sigma^2\psi^* \end{pmatrix}. \tag{1.29}$$

We require that

$$(\psi_D^c)^c = \Psi_D \tag{1.30}$$

which implies

$$|\omega| = 1. \tag{1.31}$$

It is easy to verify that $\sigma^2\psi^*$ transforms in the same way as $\bar{\chi}$ does under Lorentz boosts, and that $\sigma^2\bar{\chi}^*$ transforms like ψ. We therefore introduce the following notation: first we *define*

$$\bar{\psi}_{\dot{\alpha}} \equiv (\psi_\alpha)^* \qquad \chi^\alpha \equiv (\bar{\chi}^{\dot{\alpha}})^* \tag{1.32}$$

and then use the matrix $\omega\sigma^2$ to raise dotted indices, and $-\omega\sigma^2$ to lower undotted indices. It is convenient to choose

$$\omega = -i \tag{1.33}$$

and then the two matrices are the inverses of each other:

$$(-\omega\sigma^2)_{\alpha\beta} = \epsilon_{\alpha\beta} = \begin{pmatrix} 0 & 1 \\ -1 & 0 \end{pmatrix} \tag{1.34}$$

$$(\omega\sigma^2)^{\dot{\alpha}\dot{\beta}} = \epsilon^{\dot{\alpha}\dot{\beta}} = \begin{pmatrix} 0 & -1 \\ 1 & 0 \end{pmatrix}. \tag{1.35}$$

Then

$$\chi_\alpha \equiv \epsilon_{\alpha\beta}\chi^\beta \tag{1.36}$$

$$\bar{\psi}^{\dot{\alpha}} \equiv \epsilon^{\dot{\alpha}\dot{\beta}}\bar{\psi}_{\dot{\beta}} \tag{1.37}$$

and we have

$$\Psi_{\mathrm{D}} = \begin{pmatrix}\psi_\alpha \\ \bar{\chi}^{\dot{\alpha}}\end{pmatrix} \qquad \Psi_{\mathrm{D}}^{\mathrm{c}} = \begin{pmatrix}\chi_\alpha \\ \bar{\psi}^{\dot{\alpha}}\end{pmatrix}. \tag{1.38}$$

The above definitions specify how to raise undotted indices and to lower dotted indices:

$$\chi^\alpha = \epsilon^{\alpha\beta}\chi_\beta \tag{1.39}$$

$$\bar{\psi}_{\dot{\alpha}} = \epsilon_{\dot{\alpha}\dot{\beta}}\bar{\psi}^{\dot{\beta}} \tag{1.40}$$

where

$$\epsilon^{\alpha\beta} = \begin{pmatrix}0 & -1 \\ 1 & 0\end{pmatrix} \tag{1.41}$$

$$\epsilon_{\dot{\alpha}\dot{\beta}} = \begin{pmatrix}0 & 1 \\ -1 & 0\end{pmatrix}. \tag{1.42}$$

Thus

$$\epsilon_{\alpha\beta} = \epsilon_{\dot{\alpha}\dot{\beta}} = \mathrm{i}\,\sigma^2 \tag{1.43}$$

$$\epsilon^{\alpha\beta} = \epsilon^{\dot{\alpha}\dot{\beta}} = -\mathrm{i}\,\sigma^2. \tag{1.44}$$

Evidently a Dirac spinor in general has four independent components, two for each Weyl spinor. A Majorana spinor Ψ_{M} is defined as one that is equal to its charge-conjugate spinor

$$\Psi_{\mathrm{M}}^{\mathrm{c}} = \Psi_{\mathrm{M}}. \tag{1.45}$$

It follows from (1.38) that this occurs if, and only if,

$$\psi_\alpha = \chi_\alpha \tag{1.46}$$

which implies

$$\bar{\chi}^{\dot{\alpha}} = \bar{\psi}^{\dot{\alpha}}. \tag{1.47}$$

Clearly a Weyl spinor cannot be a Majorana spinor, and vice versa. However, given a Weyl spinor ψ_α we can always construct a Majorana spinor from it:

$$\Psi_{\mathrm{M}} = \begin{pmatrix}\psi_\alpha \\ \bar{\psi}^{\dot{\alpha}}\end{pmatrix}. \tag{1.48}$$

A general Dirac spinor Ψ_D can always be written in terms of *two* Majorana spinors:

$$\Psi_D = \Psi_{M1} + i\,\Psi_{M2} \tag{1.49}$$

where

$$\Psi_{M1} = \frac{1}{2}(\Psi_D + \Psi_D^c) \tag{1.50a}$$

$$\Psi_{M2} = \frac{1}{2i}(\Psi_D - \Psi_D^c). \tag{1.50b}$$

The reason for introducing raised and lowered indices on the spinors is to facilitate the construction of Lorentz-invariant (and covariant) quantities in terms of Weyl spinors. (It is analogous to the definition of covariant and contravariant vectors from which we construct Lorentz scalars.) Consider first the behaviour of a Dirac spinor under the Lorentz transformations (1.1) with $a^\mu = 0$. The invariance of the Dirac equation requires that the wave function $\Psi'_D(x')$ describing a spin-$\frac{1}{2}$ particle in the Poincaré transformed coordinates is related to the wave function in the original frame by

$$\Psi'_D(x') = S(\Lambda)\Psi_D(x) \tag{1.51}$$

where $S(\Lambda)$ satisfies

$$S(\Lambda)^{-1}\gamma^\mu S(\Lambda) = \Lambda^\mu{}_\nu\gamma^\nu. \tag{1.52}$$

We may write the general (proper) Lorentz transformation Λ in the form

$$\Lambda = \exp\left[-\frac{i}{2}\omega_{\mu\nu}M^{\mu\nu}\right] \tag{1.53}$$

with $M^{\mu\nu}$ given in (1.6), and $\omega_{\mu\nu} = -\omega_{\nu\mu}$, and then

$$S(\Lambda) = \exp\left[-\frac{i}{2}\omega_{\mu\nu}\frac{1}{2}\Sigma^{\mu\nu}\right] \tag{1.54}$$

with $\Sigma^{\mu\nu}$ given in (1.11). In the Weyl representation given in (1.13) we may write

$$\frac{1}{2}\Sigma^{\mu\nu} = \begin{pmatrix} i\sigma^{\mu\nu} & 0 \\ 0 & i\bar{\sigma}^{\mu\nu} \end{pmatrix} \tag{1.55}$$

where

$$\sigma^{\mu\nu} \equiv \tfrac{1}{4}(\sigma^\mu\bar{\sigma}^\nu - \sigma^\nu\bar{\sigma}^\mu) \tag{1.56a}$$

$$\bar{\sigma}^{\mu\nu} \equiv \tfrac{1}{4}(\bar{\sigma}^\mu\sigma^\nu - \bar{\sigma}^\nu\sigma^\mu). \tag{1.56b}$$

Evidently the matrices $\sigma^{\mu\nu}$ and $\bar{\sigma}^{\mu\nu}$ control the transformation properties of the dotted and undotted spinors, and clearly have indices

$$(\sigma^{\mu\nu})_\alpha{}^\beta \qquad (\bar{\sigma}^{\mu\nu})^{\dot{\alpha}}{}_{\dot{\beta}} \tag{1.57}$$

which is consistent with σ^μ and $\bar{\sigma}^\mu$ having indices

$$(\sigma^\mu)_{\alpha\dot{\alpha}} \qquad (\bar{\sigma}^\mu)^{\dot{\alpha}\alpha}. \tag{1.58}$$

Then under the Lorentz transformation (1.53) the undotted spinor ψ_α transforms to

$$\psi'_\alpha = S_1(\Lambda)_\alpha{}^\beta \psi_\beta \tag{1.59}$$

where

$$S_1(\Lambda) = \exp\left(\tfrac{1}{2}\omega_{\mu\nu}\sigma^{\mu\nu}\right). \tag{1.60}$$

Similarly the dotted spinor $\bar{\chi}^{\dot{\alpha}}$ transforms to

$$\bar{\chi}^{\dot{\alpha}\prime} = S_2(\Lambda)^{\dot{\alpha}}{}_{\dot{\beta}}\bar{\chi}^{\dot{\beta}} \tag{1.61}$$

where

$$S_2(\Lambda) = \exp\left(\tfrac{1}{2}\omega_{\mu\nu}\bar{\sigma}^{\mu\nu}\right). \tag{1.62}$$

It is easy to verify that $\sigma^{\mu\nu\dagger} = -\bar{\sigma}^{\mu\nu}$ and hence that

$$S_1(\Lambda)^\dagger = S_2(\Lambda)^{-1}. \tag{1.63}$$

The transformation properties of the undotted raised spinor ψ^α follow from its definition (1.39):

$$\psi'^\alpha = \epsilon^{\alpha\beta}\psi'_\beta = \epsilon^{\alpha\beta}S_1(\Lambda)_\beta{}^\gamma\psi_\gamma = \epsilon^{\alpha\beta}S_1(\Lambda)_\beta{}^\gamma\epsilon_{\gamma\delta}\psi^\delta \equiv S_3(\Lambda)^\alpha{}_\delta\,\psi^\delta. \tag{1.64}$$

We leave it as an exercise to check that

$$\begin{aligned}
\sigma^2\sigma^\mu\sigma^2 &= \bar{\sigma}^{\mu\mathrm{T}} \\
\sigma^2\bar{\sigma}^\mu\sigma^2 &= \sigma^{\mu\mathrm{T}} \\
\sigma^2\sigma^{\mu\nu}\sigma^2 &= -\sigma^{\mu\nu\mathrm{T}} \\
\sigma^2\bar{\sigma}^{\mu\nu}\sigma^2 &= -\bar{\sigma}^{\mu\nu\mathrm{T}}
\end{aligned} \tag{1.65}$$

and from these it follows that

$$S_3(\Lambda) = S_1(\Lambda)^{-1\mathrm{T}}. \tag{1.66}$$

Similarly

$$\bar{\chi}'_{\dot{\alpha}} = S_4(\Lambda)_{\dot{\alpha}}{}^{\dot{\beta}}\,\bar{\chi}_{\dot{\beta}} \tag{1.67}$$

where

$$S_4(\Lambda) = S_2(\Lambda)^{-1\mathrm{T}} = S_1(\Lambda)^* \tag{1.68}$$

which is just as well, since this is certainly required by (1.32) and (1.59). Finally we may check that the quantities

$$\chi^\alpha \psi_\alpha = -\chi_\alpha \psi^\alpha \tag{1.69a}$$

$$\bar{\chi}_{\dot{\alpha}} \bar{\psi}^{\dot{\alpha}} = -\bar{\chi}^{\dot{\alpha}} \bar{\psi}_{\dot{\alpha}} \tag{1.69b}$$

are Lorentz invariant, as the notation suggests. We may also construct covariant four-vectors from the Weyl spinors using the matrices σ^μ, $\bar{\sigma}^\mu$. Thus

$$\bar{\chi}_{\dot{\alpha}} (\bar{\sigma}^\mu)^{\dot{\alpha}\alpha} \psi = \bar{\chi}^{\dot{\alpha}} (\sigma^\mu)_{\alpha\dot{\alpha}} \psi^\alpha \tag{1.70a}$$

$$\chi^\alpha (\sigma^\mu)_{\alpha\dot{\alpha}} \bar{\psi}^{\dot{\alpha}} = \chi_\alpha (\bar{\sigma}^\mu)^{\dot{\alpha}\alpha} \bar{\psi}_{\dot{\alpha}} \tag{1.70b}$$

both transform as vectors. Similarly

$$\chi^\alpha (\sigma^{\mu\nu})_\alpha{}^\beta \psi_\beta = \chi_\alpha (\sigma^{\mu\nu})_\beta{}^\alpha \psi^\beta \tag{1.71a}$$

$$\bar{\chi}_{\dot{\alpha}} (\bar{\sigma}^{\mu\nu})^{\dot{\alpha}}{}_{\dot{\beta}} \bar{\psi}^{\dot{\beta}} = \bar{\chi}^{\dot{\alpha}} (\bar{\sigma}^{\mu\nu})^{\dot{\beta}}{}_{\dot{\alpha}} \bar{\psi}_{\dot{\beta}} \tag{1.71b}$$

transform as tensors. We can make the indices occur in the natural order on the right-hand sides of (1.70), (1.71) by interchanging the two spinors. However, in doing this we must remember that the spinors are all *Grassmann variables*[3]. That is to say all spinors are anti-commuting (c-numbers). Then

$$\{\psi, \chi\} = \{\bar{\psi}, \bar{\chi}\} = \{\psi, \bar{\chi}\} = 0 \tag{1.72}$$

for upper or lower indices. It follows that

$$\chi^\alpha \psi_\alpha = \psi^\alpha \chi_\alpha \tag{1.73a}$$

$$\bar{\chi}_{\dot{\alpha}} \bar{\psi}^{\dot{\alpha}} = \bar{\psi}_{\dot{\alpha}} \bar{\chi}^{\dot{\alpha}} \tag{1.73b}$$

$$\bar{\chi}_{\dot{\alpha}} (\bar{\sigma}^\mu)^{\dot{\alpha}\alpha} \psi_\alpha = -\psi^\alpha (\sigma^\mu)_{\alpha\dot{\alpha}} \bar{\chi}^{\dot{\alpha}} \tag{1.74a}$$

$$\chi^\alpha (\sigma^{\mu\nu})_\alpha{}^\beta \psi^\beta = -\psi^\alpha (\sigma^{\mu\nu})_\alpha{}^\beta \chi_\beta \tag{1.74b}$$

$$\bar{\chi}_{\dot{\alpha}} (\bar{\sigma}^{\mu\nu})^{\dot{\alpha}}{}_{\dot{\beta}} \bar{\psi}^{\dot{\beta}} = -\bar{\psi}_{\dot{\alpha}} (\bar{\sigma}^{\mu\nu})^{\dot{\alpha}}{}_{\dot{\beta}} \bar{\psi}^{\dot{\beta}}. \tag{1.74c}$$

It is often useful to use an abbreviated notation and omit the summed spinor indices. Thus we define

$$\chi\psi \equiv \chi^\alpha \psi_\alpha = \psi\chi \tag{1.75a}$$

$$\bar{\chi}\bar{\psi} = \bar{\chi}_{\dot{\alpha}} \bar{\psi}^{\dot{\alpha}} = \bar{\psi}\bar{\chi} \tag{1.75b}$$

where we have used (1.73) to establish the right-handed sides. The reason for the definition $\bar{\chi}\bar{\psi}$ as $\bar{\chi}_{\dot{\alpha}}\bar{\psi}^{\dot{\alpha}}$ rather than $\bar{\chi}^{\dot{\alpha}}\bar{\psi}_{\dot{\alpha}}$ is so that we may *define* the Hermitian conjugate of a product

$$(\chi\psi)^{\dagger} \equiv \psi^{\dagger}\chi^{\dagger}. \tag{1.76}$$

For single spinors Hermitian conjugation is just complex conjugation, so

$$(\psi_{\alpha})^{\dagger} \equiv (\psi_{\alpha})^{*} = \bar{\psi}_{\dot{\alpha}} \tag{1.77a}$$

$$(\chi^{\alpha})^{\dagger} \equiv (\chi^{\alpha})^{*} = \bar{\chi}^{\dot{\alpha}} \tag{1.77b}$$

from (1.32). Then because of the reversal of order in the definition

$$(\chi\psi)^{\dagger} = \bar{\psi}\bar{\chi} = \bar{\chi}\bar{\psi}. \tag{1.78}$$

Also if we define

$$\chi\sigma^{\mu}\bar{\psi} \equiv \chi^{\alpha}(\sigma^{\mu})_{\alpha\dot{\beta}}\bar{\psi}^{\dot{\beta}} \tag{1.79}$$

then

$$(\chi\sigma^{\mu}\bar{\psi})^{\dagger} = \psi\sigma^{\mu}\bar{\chi} = -\bar{\chi}\bar{\sigma}^{\mu}\psi = -(\bar{\psi}\,\bar{\sigma}^{\mu}\,\chi)^{\dagger}. \tag{1.80}$$

Similarly defining

$$\chi\sigma^{\mu\nu}\psi \equiv \chi^{\alpha}(\sigma^{\mu\nu})_{\alpha}{}^{\beta}\,\psi_{\beta} \tag{1.81}$$

implies

$$(\chi\sigma^{\mu\nu}\psi)^{\dagger} = -(\bar{\psi}\bar{\sigma}^{\mu\nu}\bar{\chi}) = \bar{\chi}\bar{\sigma}^{\mu\nu}\,\bar{\psi} = -(\psi\sigma^{\mu\nu}\chi)^{\dagger}. \tag{1.82}$$

We may now use this notation to express the usual Dirac covariant bilinears in terms of the Weyl spinors which appear in the Dirac spinors. We write

$$\Psi = \begin{pmatrix}\psi_{\alpha}\\ \bar{\chi}^{\dot{\alpha}}\end{pmatrix} \qquad \Phi = \begin{pmatrix}\varphi_{\alpha}\\ \bar{\eta}^{\dot{\alpha}}\end{pmatrix} \tag{1.83}$$

in the Weyl representation, and then

$$\bar{\Psi}\Phi = \bar{\psi}\bar{\eta} + \chi\varphi = (\bar{\Phi}\Psi)^{\dagger} \tag{1.84a}$$

$$\bar{\Psi}\gamma_5\Phi = \bar{\psi}\bar{\eta} - \chi\varphi = -(\bar{\Phi}\gamma_5\Psi)^{\dagger} \tag{1.84b}$$

$$\bar{\Psi}\gamma^{\mu}\Phi = \chi\sigma^{\mu}\bar{\eta} + \bar{\psi}\bar{\sigma}^{\mu}\varphi = (\bar{\Phi}\gamma^{\mu}\Psi)^{\dagger} \tag{1.84c}$$

$$\bar{\Psi}\gamma^{\mu}\gamma_5\Phi = \chi\sigma^{\mu}\bar{\eta} - \bar{\psi}\bar{\sigma}^{\mu}\varphi = (\bar{\Phi}\gamma^{\mu}\gamma_5\Psi)^{\dagger} \tag{1.84d}$$

$$\bar{\Psi}\Sigma^{\mu\nu}\Phi = \mathrm{i}\,\chi\sigma^{\mu\nu}\varphi + \mathrm{i}\,\bar{\psi}\bar{\sigma}^{\mu\nu}\bar{\eta} = (\bar{\Phi}\Sigma^{\mu\nu}\Psi)^{\dagger}. \tag{1.84e}$$

In the same way we may also express the Majorana bilinear covariants in terms of the various Weyl spinor covariants. Defining the Majorana spinors

$$\Psi_M \equiv \begin{pmatrix} \psi_\alpha \\ \bar{\psi}^{\dot{\alpha}} \end{pmatrix} \qquad \Phi_M = \begin{pmatrix} \varphi_\alpha \\ \bar{\varphi}^{\dot{\alpha}} \end{pmatrix} \tag{1.85}$$

it follows from (1.84) that

$$\bar{\Psi}_M \Phi_M = \bar{\psi}\bar{\varphi} + \psi\varphi = \bar{\Phi}_M \Psi_M = (\bar{\Psi}_M \Phi_M)^\dagger \tag{1.86a}$$

$$\bar{\Psi}_M \gamma_5 \Phi_M = \bar{\psi}\bar{\varphi} - \psi\varphi = \bar{\Phi}_M \gamma_5 \Psi_M = -(\bar{\Psi}_M \gamma_5 \Phi_M)^\dagger \tag{1.86b}$$

$$\bar{\Psi}_M \gamma^\mu \Phi_M = \psi\sigma^\mu\bar{\varphi} + \bar{\psi}\bar{\sigma}^\mu\varphi = -\bar{\Phi}_M \gamma^\mu \Psi_M = -(\bar{\Psi}_M \gamma^\mu \Phi_M)^\dagger \tag{1.86c}$$

$$\bar{\Psi}_M \gamma^\mu \gamma_5 \Phi_M = \psi\sigma^\mu\bar{\varphi} - \bar{\psi}\bar{\sigma}^\mu\varphi = \bar{\Phi}_M \gamma^\mu \gamma_5 \Psi_M = (\bar{\Psi}_M \gamma^\mu \gamma_5 \Phi_M)^\dagger \tag{1.86d}$$

$$\bar{\Psi}_M \Sigma^{\mu\nu} \Phi_M = \mathrm{i}\,\psi\sigma^{\mu\nu}\varphi + \mathrm{i}\,\bar{\psi}\bar{\sigma}^{\mu\nu}\varphi = -\bar{\Phi}_M \Sigma^{\mu\nu} \Psi_M = -(\bar{\Psi}_M \Sigma^{\mu\nu} \Phi_M)^\dagger\,. \tag{1.86e}$$

Just like Dirac spinors, the Weyl spinors also satisfy various Fierz identities. All of these may be derived from the basic identity

$$\delta_{\alpha\beta}\delta_{\gamma\delta} = \tfrac{1}{2}[\delta_{\alpha\delta}\delta_{\gamma\beta} + \sigma^i_{\alpha\delta}\sigma^i_{\gamma\beta}]$$

which expresses the completeness of the set I_2, σ^i as a set of 2×2 matrices. Then, for instance, it follows that

$$\delta_\alpha{}^\beta\, \delta^{\dot{\gamma}}{}_{\dot{\delta}} = \tfrac{1}{2}(\sigma^\mu)_{\alpha\dot{\delta}}(\bar{\sigma}_\mu)^{\dot{\gamma}\beta} \tag{1.87a}$$

using the definitions (1.14). Hence

$$(\theta\varphi)(\bar{\chi}\bar{\eta}) = -\tfrac{1}{2}(\theta\sigma^\mu\bar{\eta})(\bar{\chi}\bar{\sigma}_\mu\,\varphi) \tag{1.87b}$$

with the minus sign arising from the anti-commutation of the Grassmann variables. Similarly we can write

$$\delta_\alpha{}^\beta\, \delta_\gamma{}^\delta = \tfrac{1}{2}[\delta_\alpha{}^\delta\, \delta_\gamma{}^\beta - (\sigma^{\mu\nu})_\alpha{}^\delta(\sigma_{\mu\nu})_\gamma{}^\beta] \tag{1.88}$$

from which we may deduce

$$(\theta\varphi)(\chi\eta) = -\tfrac{1}{2}[(\theta\eta)(\chi\varphi) - (\theta\sigma^{\mu\nu}\eta)(\chi\sigma_{\mu\nu}\varphi)] \tag{1.89}$$

and

$$(\bar{\theta}\bar{\varphi})(\bar{\chi}\bar{\eta}) = -\tfrac{1}{2}[(\bar{\theta}\bar{\eta})(\bar{\chi}\bar{\varphi}) - (\bar{\theta}\,\bar{\sigma}^{\mu\nu}\,\bar{\eta})(\bar{\chi}\bar{\sigma}_{\mu\nu}\,\bar{\varphi})] \tag{1.90}$$

follows using Hermitian conjugation and (1.82). In the special case $\theta = \eta$ we find

$$(\theta\varphi)(\chi\theta) = -\tfrac{1}{2}(\theta\theta)(\chi\varphi) = (\theta\varphi)(\theta\chi) \tag{1.91}$$

since

$$\theta\sigma_{\mu\nu}\theta = 0 \tag{1.92}$$

using (1.74*b*). Another useful identity, from which (1.91) also follows, is

$$\theta^\beta \theta_\gamma = \tfrac{1}{2}(\theta\theta)\delta^\beta{}_\gamma. \tag{1.93}$$

This too follows from (1.88) using (1.92). A complete set of Fierz identities is given in Appendix A together with some useful identities involving the matrices σ^μ, $\sigma^{\lambda\nu}$ etc.

1.3 Simple supersymmetry algebra[4]

We have already noted that supersymmetry involves the introduction of a spinor generator to supplement the usual (bosonic) generators of the Poincaré group. The simplest way to do this, and the one that we shall use, is to introduce a (two-component) Weyl spinor generator Q_α. Of course, given Q_α we can always construct a (four-component) Majorana spinor, as observed in (1.48), and we can then express the various commutation and anti-commutation relations satisfied by Q_α in terms of this Majorana spinor.

First, since Q_α is a Weyl spinor its transformation properties with respect to the Poincaré group are already determined:

$$[P^\mu, Q_\alpha] = 0. \tag{1.94}$$

This follows from (1.51), for example, where it is apparent that translations act only on the argument of a spinor wave function. Alternatively we can derive it using the Jacobi identity

$$[P^\mu, [P^\nu, Q_\alpha]] + [P^\nu, [Q_\alpha, P^\mu]] + [Q_\alpha, [P^\mu, P^\nu]] = 0. \tag{1.95}$$

Clearly the right-hand side of (1.94) must be a spinor quantity and the only possibility is

$$[P^\mu, Q_\alpha] = c\sigma^\mu_{\alpha\dot\beta}\, \bar{Q}^{\dot\beta} \tag{1.96}$$

with $\bar{Q}^{\dot\beta}$ defined by (1.32) and (1.37). It follows that

$$[P^\mu, \bar{Q}^{\dot\beta}] = -c^*\bar{\sigma}^{\mu\dot\beta\gamma}Q_\gamma \tag{1.97}$$

and then the Jacobi identity yields

$$|c|^2(\sigma^\mu\bar{\sigma}^\nu + \sigma^\nu\bar{\sigma}^\mu) = 0 \tag{1.98}$$

using (1.8*a*). Hence $c = 0$ and (1.94) follows. The Majorana spinor Q_{M} *constructed from* Q_α and $\bar{Q}^{\dot\alpha}$ as in (1.48) also commutes with P^μ:

$$[P^\mu, Q_{\mathrm{M}}] = 0. \tag{1.99}$$

Similarly, under an infinitesimal Lorentz transformation (1.53) we have from (1.59)

$$Q'_\alpha = (1 + \tfrac{1}{2}\omega_{\mu\nu}\sigma^{\mu\nu})_\alpha{}^\beta Q_\beta = U(\Lambda)^\dagger Q_\alpha U(\Lambda)$$

$$= Q_\alpha + \frac{i}{2}\omega_{\mu\nu}[M^{\mu\nu}, Q_\alpha]. \tag{1.100}$$

Thus

$$[M^{\mu\nu}, Q_\alpha] = -i\,(\sigma^{\mu\nu})_\alpha{}^\beta Q_\beta \tag{1.101a}$$

and likewise

$$[M^{\mu\nu}, \bar{Q}^{\dot\alpha}] = -i\,(\bar\sigma^{\mu\nu})^{\dot\alpha}{}_{\dot\beta}\,\bar{Q}^{\dot\beta}. \tag{1.101b}$$

In terms of the (four-component) Majorana spinor Q_M these may be combined to give

$$[M^{\mu\nu}, Q_M] = -\tfrac{1}{2}\Sigma^{\mu\nu}Q_M \tag{1.102}$$

with $\Sigma^{\mu\nu}$ defined as in (1.11) and (1.55).

To close the algebra we need to specify the anti-commutators $\{Q_\alpha, Q_\beta\}$ and $\{Q_\alpha, \bar{Q}^{\dot\beta}\}$. Evidently both of these are bosonic, rather than fermionic, so we require them to be linear in P^μ and $M^{\mu\nu}$. The only possibilities are then

$$\{Q_\alpha, Q^\beta\} = s(\sigma^{\mu\nu})_\alpha{}^\beta M_{\mu\nu} \tag{1.103a}$$

and

$$\{Q_\alpha, \bar{Q}_{\dot\beta}\} = t\sigma^\mu_{\alpha\dot\beta}P_\mu. \tag{1.103b}$$

Since Q_α, Q_β and $\bar{Q}_{\dot\beta}$ all commute with P^μ, by virtue of (1.94), both of the anti-commutators (1.103) also commute with P^μ. This requires

$$s = 0 \tag{1.104}$$

so

$$\{Q_\alpha, Q_\beta\} = 0 = \{\bar{Q}_{\dot\alpha}, \bar{Q}_{\dot\beta}\} \tag{1.105}$$

but does not restrict t. In fact the value of t must be positive (see below), and evidently depends upon the normalization of the generators Q_α, which we have not so far specified. We are therefore free to fix $t = 2$, and adopt the convention

$$\{Q_\alpha, \bar{Q}_{\dot\beta}\} = 2\sigma^\mu_{\alpha\dot\beta}P_\mu. \tag{1.106}$$

As before we may rewrite (1.105) and (1.106) in terms of the Majorana spinor Q_M (and its adjoint $\bar{Q}_M \equiv Q_M{}^\dagger\gamma_0$) and the γ-matrices (1.13):

$$\{Q_M, \bar{Q}_M\} = 2\gamma^\mu P_\mu. \tag{1.107}$$

There is an immediate and important consequence of (1.106). Since

$$\sigma^\mu\bar\sigma^\nu = \eta^{\mu\nu} + 2\sigma^{\mu\nu} \tag{1.108}$$

it follows that

$$\mathrm{tr}(\sigma^\mu \bar{\sigma}^\nu) = 2\eta^{\mu\nu} \tag{1.109}$$

Applying this to (1.106) yields

$$(\bar{\sigma}^\nu)^{\dot{\beta}\alpha}\{Q_\alpha, \bar{Q}_{\dot{\beta}}\} = 4P^\nu . \tag{1.110}$$

Now take $\nu = 0$ and take the matrix element of (1.110). Then

$$\begin{aligned} 4\langle\psi|P^0|\psi\rangle &= \langle\psi|Q_1\bar{Q}_{\dot{1}} + \bar{Q}_{\dot{1}}Q_1 + Q_2\bar{Q}_{\dot{2}} + \bar{Q}_{\dot{2}}Q_2|\psi\rangle \\ &= \langle\psi|Q_\alpha(Q_\alpha)^* + (Q_\alpha)^*Q_\alpha|\psi\rangle \geqslant 0 \end{aligned} \tag{1.111}$$

where we have used the defining property (1.32). Thus in a supersymmetric theory the energy of any non-vacuum state is positive definite, and, in fact, the vanishing of the vacuum energy is a necessary and sufficient condition for the existence of a unique vacuum:

$$\langle 0|P^0|0\rangle = 0 \Leftrightarrow Q_\alpha|0\rangle = 0 . \tag{1.112}$$

Another consequence of (1.107) is that in a supersymmetric theory every representation has an equal number of equal-mass bosonic and fermionic states. First, the mass-squared operator $P^2 \equiv P_\mu P^\mu$, which is a Casimir operator of the Poincaré algebra, is also a Casimir operator of the supersymmetry algebra, since from (1.94)

$$[P^2, Q_\alpha] = 0 = [P^2, \bar{Q}_{\dot{\alpha}}] \tag{1.113}$$

Next, the Pauli–Lubanski spin vector

$$W^\mu = \tfrac{1}{2}\epsilon^{\mu\nu\rho\sigma}P_\nu M_{\rho\sigma} \tag{1.114}$$

gives a Poincaré group Casimir

$$W^2 = -\, m^2 J^2 \tag{1.115}$$

where m^2 is the mass-squared eigenvalue, and $J^2 = j(j+1)$ is the angular momentum eigenvalue. Evidently

$$[W^2, Q_\alpha] \neq 0 \tag{1.116}$$

by virtue of (1.101). Thus the (massive) irreducible representations of the supersymmetry algebra will certainly contain different spins. To see that these spin states are split equally between the bosonic and fermionic sectors, we note first that Q_α and $\bar{Q}_{\dot{\beta}}$ each change the fermion number by *one* unit, and thus change a bosonic state into a fermionic one, and a fermionic state into a bosonic state. The anti-commutator $\{Q_\alpha, \bar{Q}_{\dot{\beta}}\}$ therefore maps the fermionic sector into itself, and the bosonic sector into itself. However, equation (1.107) shows that this same mapping is accomplished (essentially) by P_μ which *in most representations* is a one-to-one operator. It follows that Q_α (and $\bar{Q}_{\dot{\beta}}$) are also one-to-one operators and therefore that the bosonic sector has the same dimension as the fermionic sector. More formally (and perhaps less generally) we note that, since Q_α changes fermion number by one unit, we may write

$$(-1)^{N_F} Q_\alpha = -Q_\alpha (-1)^{N_F} \tag{1.117}$$

where N_F is the fermion number operator. Now consider a finite-dimensional representation R of the algebra. Then

$$\begin{aligned} \mathrm{tr}[(-1)^{N_F}\{Q_\alpha, \bar{Q}_{\dot\beta}\}] &= \mathrm{tr}[-Q_\alpha(-1)^{N_F}\bar{Q}_{\dot\beta} + (-1)^{N_F}\bar{Q}_{\dot\beta}Q_\alpha] \\ &= \mathrm{tr}[-Q_\alpha(-1)^{N_F}\bar{Q}_{\dot\beta} + Q_\alpha(-1)^{N_F}\bar{Q}_{\dot\beta}] = 0 \end{aligned} \tag{1.118}$$

where we have used the cyclic property of the trace to rewrite the second term. It follows from (1.107) that

$$2\sigma^\mu_{\alpha\dot\beta}\, \mathrm{tr}[(-1)^{N_F}P_\mu] = 0 \tag{1.119}$$

and so

$$\mathrm{tr}(-1)^{N_F} = 0 \tag{1.120}$$

for fixed non-zero P_μ. Since $(-1)^{N_F}$ has value $+1$ on a bosonic state and -1 on a fermionic state, this means that

$$n_B(R) - n_F(R) = 0 \tag{1.121}$$

where $n_{B\,(F)}(R)$ are respectively the number of bosons (fermions) in the representation R of the supersymmetry algebra.

1.4 Supersymmetry multiplets

Before discussing the supersymmetric field theory which is our primary interest, it is instructive to consider the representations of the supersymmetry algebra that can be realized by one-particle states. We start with the massless case, since in most of the phenomenologically interesting scenarios the non-zero masses of the particles that we observe are generated by supersymmetry-breaking effects. For massless particles $W^2 = P^2 = 0$, and in fact the spin vector W^μ and the energy–momentum vector P^μ are parallel:

$$W^\mu = \lambda P^\mu. \tag{1.122}$$

It is easy to see that (for positive energy representations) λ is just the helicity:

$$\lambda = (\boldsymbol{J} \cdot \boldsymbol{P})P_0^{-1} \tag{1.123}$$

where

$$J^i = \tfrac{1}{2}\epsilon^{ijk}M_{jk} \tag{1.124}$$

is the total angular momentum. Now consider the (normalized) massless state $|p, \lambda\rangle$ with momentum p. Then

$$P^\mu|p, \lambda\rangle = p^\mu|p, \lambda\rangle \tag{1.125a}$$

where

$$p^\mu = (E, 0, 0, E) \tag{1.125b}$$

and λ is the helicity:

$$W^\mu|p, \lambda\rangle = \lambda p^\mu|p, \lambda\rangle. \tag{1.126}$$

We may choose $|p, \lambda\rangle$ in such a way that

$$Q_\alpha|p, \lambda\rangle = 0 \qquad (\alpha = 1, 2). \tag{1.127}$$

To see this we note that (1.106) shows that

$$Q_\alpha Q_\alpha = 0 \qquad \text{(no summation)} \tag{1.128}$$

so, if (1.127) is *not* satisfied, we may define

$$|p, \lambda'\rangle \equiv Q_\alpha|p, \lambda\rangle \tag{1.129}$$

and then

$$Q_\alpha|p, \lambda'\rangle = 0. \tag{1.130}$$

(Note that (1.94) implies that $|p, \lambda'\rangle$ also has momentum p^μ, as the notation implies.) Thus we can always choose $|p, \lambda\rangle$ in such a way that (1.127) is satisfied. It follows that the only other possible states in the same supersymmetric representation as $|p, \lambda\rangle$ are $\bar{Q}_{\dot{\alpha}}|p, \lambda\rangle$ ($\dot{\alpha} = 1, 2$). However, $\bar{Q}_{\dot{1}}|p, \lambda\rangle$ is a state of zero norm. For, applying (1.106) to $|p, \lambda\rangle$ gives

$$\{Q_\alpha, \bar{Q}_{\dot{\beta}}\}|p, \lambda\rangle = 2(\sigma^\mu)_{\alpha\dot{\beta}}p_\mu|p, \lambda\rangle \tag{1.131}$$

and with p^μ given by (1.125)

$$\sigma^\mu p_\mu = E(\sigma^0 - \sigma^3) = 2E\begin{pmatrix} 0 & 0 \\ 0 & 1 \end{pmatrix}. \tag{1.132}$$

Then, using (1.127) we see that

$$\langle p, \lambda|Q_1\bar{Q}_{\dot{1}}|p, \lambda\rangle = 0 \tag{1.133}$$

which shows that

$$\bar{Q}_{\dot{1}}|p, \lambda\rangle = 0. \tag{1.134}$$

Thus the only other state is

$$|\psi\rangle \equiv (4E)^{-1/2}\bar{Q}_{\dot{2}}|p, \lambda\rangle = -(4E)^{-1/2}\bar{Q}^{\dot{1}}|p, \lambda\rangle \tag{1.135}$$

where the factor $(4E)^{-1/2}$ is included so that $|\psi\rangle$ is a normalized state. As before, equation (1.94) shows that $|\psi\rangle$ has momentum p^μ, and so is also a massless state. Now, from the definition (1.114), and the commutators (1.94) and (1.101), it follows that

$$[W_\mu, \bar{Q}^{\dot{\alpha}}] = -\frac{\mathrm{i}}{2}\epsilon_{\mu\nu\rho\sigma}P^\nu(\bar{\sigma}^{\rho\sigma})^{\dot{\alpha}}{}_{\dot{\beta}}\bar{Q}^{\dot{\beta}}. \tag{1.136}$$

Applying this to the state $|p, \lambda\rangle$, with p as in (1.125b), gives

$$[W_0, \bar{Q}^{\dot{\alpha}}]|p, \lambda\rangle = -\tfrac{1}{2}p_0(\sigma^3\bar{Q})^{\dot{\alpha}}|p, \lambda\rangle \tag{1.137}$$

Thus from (1.126)

$$W_0(\bar{Q}^{\dot{1}}|p, \lambda\rangle) = (\lambda - \tfrac{1}{2})p_0(\bar{Q}^{\dot{1}}|p, \lambda\rangle) \tag{1.138}$$

so $\bar{Q}^{\dot{1}}|p, \lambda\rangle$ has helicity $\lambda - \frac{1}{2}$, and

$$(4E)^{-1/2}\bar{Q}_{\dot{2}}|p, \lambda\rangle \equiv |p, \lambda - \tfrac{1}{2}\rangle \tag{1.139}$$

is the normalized state in the same supersymmetric representation as $|p, \lambda\rangle$. It is easy to see that there are no other states in this representation; for instance, $Q_2|p, \lambda - \frac{1}{2}\rangle$ is proportional to $|p, \lambda\rangle$, and (1.127) shows that $\bar{Q}_{\dot{2}}|p, \lambda - \frac{1}{2}\rangle$ is zero. The fact that there are just these two states in the supersymmetry representation is consistent with our previous observation (1.121) that any such representation has equal numbers of bosons and fermions—one in this case.

The most common of these representations that we shall encounter are those with $\lambda = \frac{1}{2}, 1, 2$ (together with their *TCP*-conjugate representations). The $\lambda = \frac{1}{2}$ supermultiplet consists of a Weyl spinor with helicity $\frac{1}{2}$ and a scalar particle. To construct a Lorentz invariant field theory it is necessary to include also the *TCP*-conjugate representation, which has a Weyl fermion with helicity $-\frac{1}{2}$ and another scalar. Together, these two representations constitute a Majorana fermion and a complex scalar field. These 'chiral supermultiplets' arise in the applications that we consider for all matter fields (quarks and leptons), as well as for the Higgs particles. The scalar partners of the quarks are called 'squarks', and the scalar partners of the leptons are 'sleptons'. The fermionic partners of Higgs particles are usually called 'Higgsinos', but occasionally 'shigges' appears in the literature. The boson in the $\lambda = 1$ representation has helicity 1, so together with its *TCP*-conjugate which has helicity -1, it describes a (massless) vector particle, such as the gauge bosons that arise whenever we have a (locally) gauge-invariant theory. $\lambda = 1$ supermultiplets are therefore called 'vector supermultiplets'. The fermion partners of the gauge particles (having helicity $\pm\frac{1}{2}$) are called 'gauginos' generically, and 'photinos', 'Winos', 'Zinos' and 'gluinos' in particular. Evidently the $\lambda = 1$ representation and its *TCP*-conjugate together constitute a vector (gauge) field and a Majorana (gaugino) field. In the same way the helicity $\pm\frac{3}{2}$ particle, which is the partner of

Table 1.1 $N = 1$ supermultiplet examples.

Particle	λ	Helicity	Degeneracy	*TCP*-conjugate helicity	Super-multiplet
Quark, lepton Higgsino	$\frac{1}{2}$	$\frac{1}{2}$	1	$-\frac{1}{2}$	Chiral
Squark, slepton Higgs		0	1	0	
Gauge boson	1	1	1	-1	Vector
Gaugino		$\frac{1}{2}$	1	$-\frac{1}{2}$	
Graviton	2	2	1	-2	Gravity
Gravitino		$\frac{3}{2}$	1	$-\frac{3}{2}$	

the helicity ± 2 graviton field which mediates gravitational interactions is called the 'gravitino'. These examples are summarized in table 1.1.

The treatment of the massive representations of supersymmetry proceeds similarly, but is a little more involved. We consider a particle of mass m in a normalized state $|p, s, s_3\rangle$, where p is the momentum, s is the spin and s_3 its third component. In the rest frame we have

$$P^\mu |p, s, s_3\rangle = p^\mu |p, s, s_3\rangle \tag{1.140a}$$

with

$$p^\mu = (m, 0, 0, 0). \tag{1.140b}$$

Also

$$W_\mu W^\mu |p, s, s_3\rangle = -m^2 J^2 |p, s, s_3\rangle = -m^2 s(s+1) |p, s, s_3\rangle \tag{1.141}$$

and

$$J^3 |p, s, s_3\rangle = s_3 |p, s, s_3\rangle. \tag{1.142}$$

As before we can always choose $|p, s, s_3\rangle$ in such a way that

$$Q_\alpha |p, s, s_3\rangle = 0 \qquad (\alpha = 1, 2). \tag{1.143}$$

From each of the $2s + 1$ states $|p, s, s_3\rangle$ we can construct *two* more normalized states also having momentum p:

$$|p, \dot{\alpha}\rangle \equiv \frac{1}{\sqrt{2m}} \bar{Q}_{\dot{\alpha}} |p, s, s_3\rangle \qquad (\dot{\alpha} = \dot{1}, \dot{2}). \tag{1.144}$$

To determine the angular momentum quantum numbers of these states we use (1.101) again. This shows that

$$[J^i, \bar{Q}^{\dot{\alpha}}] = -\tfrac{1}{2}(\sigma^i \bar{Q})^{\dot{\alpha}} \tag{1.145}$$

from which it follows that

$$J^3|p, \dot{1}\rangle = (s_3 + \tfrac{1}{2})|p, \dot{1}\rangle \tag{1.146a}$$

$$J^3|p, \dot{2}\rangle = (s_3 - \tfrac{1}{2})|p, \dot{2}\rangle . \tag{1.146b}$$

In the case where the original state $|p, s, s_3\rangle$ has spin *zero* ($s = s_3 = 0$), it is easy to see that these two states form a spin-$\frac{1}{2}$ doublet. From (1.145) we can see that

$$[J^1 + \mathrm{i}J^2, \bar{Q}_{\dot{1}}] = 0 = [J^1 - \mathrm{i}J^2, \bar{Q}_{\dot{2}}] \tag{1.147}$$

from which it follows that

$$\frac{1}{\sqrt{2m}}\bar{Q}_{\dot{1}}|p, 0, 0\rangle = |p, \tfrac{1}{2}, \tfrac{1}{2}\rangle \tag{1.148a}$$

$$\frac{1}{\sqrt{2m}}\bar{Q}_{\dot{2}}|p, 0, 0\rangle = |p, \tfrac{1}{2}, -\tfrac{1}{2}\rangle . \tag{1.148b}$$

The only other independent state in this system is obtained by applying $\bar{Q}_{\dot{2}}$ to the first of these (or $\bar{Q}_{\dot{1}}$ to the second). Proceeding as above it is easy to verify that

$$\frac{1}{\sqrt{2m}}\bar{Q}_{\dot{2}}\bar{Q}_{\dot{1}}|p, 0, 0\rangle = |p, 0, 0\rangle' . \tag{1.149}$$

The prime is to distinguish it from the original state, which has the same quantum numbers. So in this case we have two spin-zero states and a spin-$\frac{1}{2}$ doublet, and

$$n_{\mathrm{F}} = n_{\mathrm{B}} = 2 . \tag{1.150}$$

A similar analysis works for the initial states $|p, \frac{1}{2}, \pm\frac{1}{2}\rangle$. We find

$$\frac{1}{\sqrt{2m}}\bar{Q}_{\dot{1}}|p, \tfrac{1}{2}, \tfrac{1}{2}\rangle = |p, 1, 1\rangle \tag{1.151a}$$

$$\frac{1}{\sqrt{2m}}\bar{Q}_{\dot{2}}|p, \tfrac{1}{2}, \tfrac{1}{2}\rangle = \frac{1}{\sqrt{2}}[|p, 1, 0\rangle + |p, 0, 0\rangle] \tag{1.151b}$$

$$\frac{1}{\sqrt{2m}}\bar{Q}_{\dot{1}}|p, \tfrac{1}{2}, -\tfrac{1}{2}\rangle = \frac{1}{\sqrt{2}}[|p, 1, 0\rangle - |p, 0, 0\rangle] \tag{1.151c}$$

$$\frac{1}{\sqrt{2m}}\bar{Q}_{\dot{2}}|p, \tfrac{1}{2}, -\tfrac{1}{2}\rangle = |p, 1, -1\rangle \qquad (1.151d)$$

$$\frac{1}{\sqrt{2m}}\bar{Q}_{\dot{2}}\bar{Q}_{\dot{1}}|p, \tfrac{1}{2}, \tfrac{1}{2}\rangle = |p, \tfrac{1}{2}, \tfrac{1}{2}\rangle \qquad (1.151e)$$

$$\frac{1}{\sqrt{2m}}\bar{Q}_{\dot{1}}\bar{Q}_{\dot{2}}|p, \tfrac{1}{2}, -\tfrac{1}{2}\rangle = |p, \tfrac{1}{2}, -\tfrac{1}{2}\rangle. \qquad (1.151f)$$

So we have two (fermion) doublets having $s = \frac{1}{2}$ and a (boson) $s = 1$ triplet and $s = 0$ singlet. In general, starting with a $(2s + 1)$-component multiplet of spin $s > 0$, we generate a spin-$(s + \frac{1}{2})$ multiplet, a spin-$(s - \frac{1}{2})$ multiplet and spin-s multiplets in this way. Thus the general massive representation has

$$n_{\mathrm{B}} = n_{\mathrm{F}} = 2(2s + 1). \qquad (1.152)$$

The operators Q_α $(\alpha = 1, 2)$, $\bar{Q}_{\dot{\alpha}}$ $(\dot{\alpha} = \dot{1}, \dot{2})$ generate an SO(4) algebra when acting on these states, since

$$\{Q_\alpha, \bar{Q}_{\dot{\beta}}\} = 2m\delta_{\alpha\beta}. \qquad (1.153)$$

If we define

$$\Gamma_\alpha \equiv \frac{1}{\sqrt{2m}}(Q_\alpha + \bar{Q}_{\dot{\alpha}}) \qquad (1.154a)$$

$$\Gamma_{2+\alpha} \equiv \frac{\mathrm{i}}{\sqrt{2m}}(Q_\alpha - \bar{Q}_{\dot{\alpha}}) \qquad (1.154b)$$

then the four gamma matrices generate the Clifford algebra

$$\{\Gamma_a, \Gamma_b\} = 2\delta_{ab} \qquad (a, b = 1, \ldots, 4) \qquad (1.155)$$

with the SO(4) invariance group, whose generators are

$$R_{ab} \equiv -\frac{\mathrm{i}}{4}[\Gamma_a, \Gamma_b]. \qquad (1.156)$$

The representation that we have constructed is the four-dimensional spinor representation with the bosonic and fermionic states each transforming as a two-dimensional 'Weyl' representation: if we define

$$\Gamma_5 = \Gamma_1\Gamma_2\Gamma_3\Gamma_4 \qquad (1.157)$$

then both the original and the generated spin-s states have $\Gamma_5 = +1$ while the spin-$(s \pm \frac{1}{2})$ states all have $\Gamma_5 = -1$.

1.5 Supersymmetric free-field theory[5]

The most important realization of supersymmetry is in quantum field theory. As we shall see, supersymmetric field theories permit the solution of the 'hierarchy problem', and at the time of writing this is the only known solution. To study supersymmetry in this context we need to determine how a field operator $\hat{\varphi}(x)$ transforms under a general symmetry. This is fixed by the connection

$$\psi(x) = \langle\psi|\hat{\varphi}(x)|0\rangle \tag{1.158}$$

which gives the wave function $\psi(x)$ describing the (one-particle) state $|\psi\rangle$. If the transformation properties of the (c-number) wave function are known, the above relation determines how $\hat{\varphi}(x)$ transforms. The wave function describing the transformed state

$$|\psi'\rangle = U|\psi\rangle \tag{1.159}$$

is given by

$$\psi'(x') = \langle\psi'|\hat{\varphi}(x')|0\rangle = \langle\psi|U^{\dagger}\hat{\varphi}(x')UU^{\dagger}|0\rangle = \langle\psi|U^{\dagger}\hat{\varphi}(x')U|0\rangle \tag{1.160}$$

assuming that the vacuum is invariant, so

$$U^{\dagger}|0\rangle = |0\rangle. \tag{1.161}$$

Now if the transformed wave function is related to the original wave function by

$$\psi'(x') = S\psi(x) = S\langle\psi|\hat{\varphi}(x)|0\rangle \tag{1.162}$$

we deduce that the field operator transforms according to

$$\hat{\varphi}'(x') \equiv U^{\dagger}\hat{\varphi}(x')U = S\hat{\varphi}(x). \tag{1.163}$$

For example, we may consider the transformation properties of a field $\hat{\varphi}(x)$ under translations

$$x' = x + a. \tag{1.164}$$

Then

$$\psi'(x') = \psi(x) \tag{1.165}$$

and

$$\hat{\varphi}'(x') = U^{\dagger}(a)\hat{\varphi}(x')U(a) = \hat{\varphi}(x' - a). \tag{1.166}$$

For an infinitesimal translation

$$U(a) = 1 - \mathrm{i}\,P.a \tag{1.167}$$

so

$$\hat{\varphi}'(x') = \hat{\varphi}(x') + \mathrm{i}\,[P.a, \hat{\varphi}(x')] = \hat{\varphi}(x') - a.\partial\hat{\varphi}(x'). \tag{1.168}$$

It follows that

$$\hat{\varphi}'(x) - \hat{\varphi}(x) \equiv \delta\hat{\varphi}(x) = [\mathrm{i}\, P.a, \hat{\varphi}(x)] = -\, a\, \partial\hat{\varphi}(x) \tag{1.169}$$

and

$$[P_\mu, \hat{\varphi}(x)] = \mathrm{i}\, \partial_\mu \hat{\varphi}(x). \tag{1.170}$$

An infinitesimal supersymmetry transformation is characterized by (constant) anti-commuting Grassmann parameters ξ^α and $\bar{\xi}_{\dot{\alpha}}$. Then

$$U(\xi) = 1 - \mathrm{i}\,(\xi^\alpha Q_\alpha + \bar{\xi}_{\dot{\alpha}} \bar{Q}^{\dot{\alpha}}) = 1 - \mathrm{i}\,(\xi Q + \bar{\xi}\bar{Q}) \tag{1.171}$$

using the notation (1.75), and

$$\begin{aligned} &\{\xi^\alpha, \xi^\beta\} = \{\xi^\alpha, \bar{\xi}_{\dot{\beta}}\} = 0 \\ &\{\xi^\alpha, Q_\beta\} = \{\xi^\alpha, \bar{Q}^{\dot{\beta}}\} = 0 \\ &\{\bar{\xi}_{\dot{\alpha}}, Q_\beta\} = \{\bar{\xi}_{\dot{\alpha}}, \bar{Q}^{\dot{\beta}}\} = 0 \\ &[\xi^\alpha, P_\mu] = [\xi^\alpha, M_{\mu\nu}] = 0 \text{ etc.} \end{aligned} \tag{1.172}$$

Using such Grassmann parameters the supersymmetry algebra can be rewritten entirely in terms of commutators:

$$\begin{aligned} &[P^\mu, \xi Q] = 0 = [P^\mu, \bar{\xi}\bar{Q}] \\ &[M^{\mu\nu}, \xi Q] = -\,\mathrm{i}\,(\xi\sigma^{\mu\nu} Q) \\ &[M^{\mu\nu}, \bar{\xi}\bar{Q}] = -\,\mathrm{i}\,(\bar{\xi}\bar{\sigma}^{\mu\nu}\bar{Q}) \\ &[\xi Q, \eta Q] = 0 = [\bar{\xi}\bar{Q}, \bar{\eta}\bar{Q}] \\ &[\xi Q, \bar{\eta}\bar{Q}] = 2(\xi\sigma^\mu\bar{\eta})P_\mu \end{aligned} \tag{1.173}$$

where $\eta_\alpha, \bar{\eta}^{\dot{\alpha}}$ are a second set of Grassmann parameters. (Note the minus signs in (1.69).) To characterize the supersymmetry transformation in field theory we are therefore required to specify the quantities

$$\delta_\xi \varphi(x) \equiv [\mathrm{i}\,(\xi Q + \bar{\xi}\bar{Q}), \varphi(x)] \tag{1.174}$$

for a general field $\varphi(x)$ in a way that is consistent with the supersymmetry algebra (1.173). The first three commutators merely specify the Poincaré transformation properties of $\delta_\xi \varphi$. The last two constrain these possibilities by showing how (the difference between) two successive supersymmetry transformations must close the algebra. We have already seen in §1.4 that supersymmetry has its simplest realization in the massless (chiral) supermultiplet that has just scalar and spinor particles. Thus we might anticipate a field theory realization involving *just* scalar and spinor fields.

Suppose then we start with a *complex* scalar field $\varphi(x)$. The (mass) dimension of $\varphi(x)$ is

$$[\varphi(x)] = 1 \tag{1.175}$$

and that of Q is

$$[Q] = \tfrac{1}{2} = [\bar{Q}] \tag{1.176}$$

from (1.173). Thus

$$[\xi] = -\tfrac{1}{2} = [\bar{\xi}] \tag{1.177}$$

so that ξQ is dimensionless. So the simplest possibility linear in ξ and having the right dimensions is

$$\delta_\xi \varphi(x) = a\xi\psi(x) + b\bar{\xi}\bar{\psi}(x) \tag{1.178}$$

where $\psi_\alpha(x)$ is a (Weyl) spinor field, and therefore has

$$[\psi] = \tfrac{3}{2} = [\bar{\psi}] \tag{1.179}$$

and a is a constant. Obviously $\delta_\xi \psi_\alpha(x)$ has dimension 2, so if we want to close the algebra using just these fields we must take out one dimension with a derivative:

$$\delta_\xi \psi_\alpha(x) = c\sigma^\mu_{\alpha\dot{\beta}}\, \bar{\xi}^{\dot{\beta}}\, \partial_\mu \varphi \tag{1.180}$$

which implies

$$\delta_\xi \bar{\psi}^{\dot{\alpha}} = -c^*(\bar{\sigma}^\mu)^{\dot{\alpha}\beta}\xi_\beta\, \partial_\mu \varphi^* \tag{1.181}$$

Thus

$$\delta_\eta \delta_\xi \varphi(x) = a\xi\, \delta_\eta \psi + b\bar{\xi}\, \delta_\eta \bar{\psi} = ac(\xi\sigma^\mu\bar{\eta})\, \partial_\mu \varphi - bc^*(\bar{\xi}\bar{\sigma}^\mu\eta)\, \partial_\mu \varphi^*. \tag{1.182}$$

Now

$$\delta_\eta \delta_\xi \varphi(x) \equiv [\mathrm{i}(\eta Q + \bar{\eta}\bar{Q}), [\mathrm{i}(\xi Q + \bar{\xi}\bar{Q}), \varphi(x)]] \tag{1.183}$$

so the Jacobi identity implies that

$$\begin{aligned}(\delta_\eta\delta_\xi - \delta_\xi\delta_\eta)\varphi(x) &= [[\mathrm{i}(\eta Q + \bar{\eta}\bar{Q}), \mathrm{i}(\xi Q + \bar{\xi}\bar{Q})], \varphi(x)] \\ &= \{-2(\eta\sigma^\mu\bar{\xi}) + 2(\xi\sigma^\mu\bar{\eta})\}[P_\mu, \varphi(x)] \\ &= 2(\xi\sigma^\mu\bar{\eta} - \eta\sigma^\mu\bar{\xi})\, \mathrm{i}\, \partial_\mu\varphi(x)\end{aligned} \tag{1.184}$$

using the supersymmetry algebra (1.173), and (1.170). Comparison with (1.182) gives

$$ac = 2\mathrm{i} \qquad b = 0 \tag{1.185}$$

unless $\varphi(x)$ is a constant field. For future reference it is useful to rewrite

(1.184) in terms of the Majorana spinors constructed from ξ and η, using (1.86*c*)

$$(\delta_\eta \delta_\xi - \delta_\xi \delta_\eta)\varphi(x) \equiv [\delta_\eta, \delta_\xi]\varphi(x) = 2(\bar{\xi}_M \gamma^\mu \eta_M)\, \mathrm{i}\, \partial_\mu \varphi. \quad (1.186)$$

In the same way we apply two successive supersymmetry transformations to ψ_α. This gives

$$\begin{aligned}\delta_\eta \delta_\xi \psi_\alpha &= ca\sigma^\mu_{\alpha\dot{\alpha}} \bar{\xi}^{\dot{\alpha}} (\eta\, \partial_\mu \psi) \\ &= -\mathrm{i}[(\partial_\mu \psi_\alpha)(\eta \sigma^\mu \bar{\xi}) + 2(\sigma^{\mu\nu})_\alpha{}^\beta (\partial_\mu \psi_\beta)(\eta \sigma_\nu \bar{\xi})] \\ &= -2\mathrm{i}[(\partial_\mu \psi_\alpha)(\eta\sigma^\mu \bar{\xi}) - (\sigma^\nu \bar{\sigma}^\mu\, \partial_\mu \psi)_\alpha (\eta \sigma_\nu \bar{\xi})]\end{aligned}$$

using (1.185) and the Fierz identity (A5). This gives

$$\begin{aligned}(\delta_\eta \delta_\xi - \delta_\xi \delta_\eta)\psi_\alpha(x) &= 2(\xi \sigma^\mu \bar{\eta} - \eta \sigma^\mu \bar{\xi})\, \mathrm{i}\, \partial_\mu \psi_\alpha \\ &= 2(\bar{\xi}_M \gamma^\mu \eta_M)\, \mathrm{i}\, \partial_\mu \psi_\alpha \end{aligned} \quad (1.187)$$

(again as required by (1.173) and (1.170)) *provided*

$$\bar{\sigma}^\mu\, \partial_\mu \psi = 0. \quad (1.188)$$

Thus the transformations (1.178), (1.181) are representations of the supersymmetry algebra provided ψ is a massless non-interacting Weyl field. In fact in these circumstances it is straightforward to verify that the action can be invariant under the supersymmetry transformations. Taking

$$\mathscr{L} = (\partial_\mu \varphi^*)(\partial^\mu \varphi) + \mathrm{i}\, \bar{\psi} \bar{\sigma}^\mu\, \partial_\mu \psi \quad (1.189)$$

which describes a free massless complex scalar field, and a free massless Weyl spinor field, we find

$$\begin{aligned}\delta_\xi \mathscr{L} &= (\partial_\mu \varphi^*) a(\xi\, \partial^\mu \psi) - a^*(\bar{\xi}\, \partial_\mu \bar{\psi})(\partial^\mu \varphi) \\ &\quad - \mathrm{i}\, c^*(\xi \sigma^\nu \bar{\sigma}^\mu\, \partial_\mu \psi)(\partial_\nu \varphi^*) + \mathrm{i}\, c(\bar{\psi} \bar{\sigma}^\mu \sigma^\nu \bar{\xi})(\partial_\mu\, \partial_\nu \varphi) \\ &= (a - \mathrm{i}\, c^*)(\partial_\mu \varphi^*)(\xi\, \partial^\mu \psi) - (\mathrm{i}\, c + a^*)(\bar{\xi}\, \partial_\mu \bar{\psi})(\partial^\mu \varphi) \\ &\quad + \mathrm{i}\, \partial_\mu [c(\bar{\xi} \bar{\psi})(\partial^\mu \varphi) - 2c^*(\xi \sigma^{\nu\mu} \psi)(\partial_\nu \varphi^*)]. \end{aligned} \quad (1.190)$$

The total divergence does not contribute to the action, so the action is supersymmetric if

$$a = \mathrm{i}\, c^*. \quad (1.191)$$

Combining with (1.185) gives

$$|c|^2 = 2$$

and a consistent choice is

$$a = \sqrt{2} \qquad c = \sqrt{2}\,\mathrm{i} \tag{1.192}$$

giving

$$\delta_\xi \varphi = \sqrt{2}\, \xi\psi \tag{1.193a}$$

$$\delta_\xi \psi = \mathrm{i}\sqrt{2}\, \sigma^\mu \bar{\xi}\, \partial_\mu \varphi \tag{1.193b}$$

when substituted into (1.178) and (1.180).

Thus we have an *on-shell* realization of the supersymmetry algebra in the field theory given by (1.189), but we have to use the equation of motion (1.188). In fact this is a reflection of the fact that the bosonic and fermionic degrees of freedom are *not* equal in the present context (unless we impose the on-shell condition (1.188)). Off-shell a (Weyl) spinor field has two complex components, whereas we have only introduced one (bosonic) complex scalar field. To match the degrees of freedom we have to introduce another complex scalar field $F(x)$, having mass dimension

$$[F] = 2. \tag{1.194}$$

We start with (1.193*a*), as before, but modify (1.193*b*):

$$\delta_\xi \psi = \mathrm{i}\sqrt{2}\, \sigma^\mu \bar{\xi}\, \partial_\mu \varphi + d\xi F. \tag{1.195}$$

This closes the algebra on $\varphi(x)$, as in (1.184), for any value of the constant d. The extra term contributes to $\delta_\eta\, \delta_\xi \psi$: it generates the extra term $d\xi\, \delta_\eta F$. Then we can close the algebra on ψ *without* using the equation of motion provided that we take

$$\delta_\xi F = e\bar{\xi}\bar{\sigma}^\mu\, \partial_\mu \psi \tag{1.196}$$

and

$$de = 2\,\mathrm{i}. \tag{1.197}$$

(As before, the highest-dimension field must transform by a total derivative.) Finally we require that the algebra closes on F, which it does without further constraint. The introduction of F has enabled us to realize the supersymmetry algebra *without* recourse to the field equations, but if we *do* use them we know that we can achieve this with $F = 0$. This suggests that F is an 'auxiliary' field, i.e. one that can be eliminated by using its equation of motion. We can easily modify (1.189) to accomplish this. We take

$$\mathcal{L} = (\partial_\mu \varphi^*)(\partial^\mu \varphi) + \mathrm{i}\, \bar{\psi}\bar{\sigma}^\mu\, \partial_\mu \psi + F^* F \tag{1.198}$$

and then the Euler–Lagrange equation for F gives $F = 0$ immediately. It is straightforward to verify that $\delta_\xi \mathcal{L}$ is a total derivative, so the action is supersymmetric. This technique of introducing auxiliary (non-propagating) fields will be used extensively in later chapters to formulate (off-shell) supersymmetric field theories *with* interactions.

1.6 Extended supersymmetry[6]

We have confined ourselves hitherto to the simplest extension of the Poincaré algebra, by the inclusion of a single spinor generator Q_α. For reasons that will become apparent, this 'simple' supersymmetry is likely to be the one with most relevance to the spectrum and interactions that can be explored by the current, and foreseeable, generation of accelerators. Nevertheless it is natural to wonder what would be the consequences of the introduction of more than one supersymmetry generator $Q_\alpha{}^A$ $(A = 1, \ldots, N)$ with A labelling some internal symmetry. Obviously the index A is a spectator in all of the commutation relations with the generators of the Poincaré algebra. Thus, as in (1.94), (1.101*a*)

$$[P^\mu, Q_\alpha{}^A] = 0 \tag{1.199}$$

$$[M^{\mu\nu}, Q_\alpha{}^A] = -\mathrm{i}(\sigma^{\mu\nu})_\alpha{}^\beta Q_\beta{}^A \tag{1.200}$$

and defining

$$\bar{Q}_{\dot{\alpha}A} \equiv (Q_\alpha{}^A)^\dagger \tag{1.201}$$

$$[M^{\mu\nu}, \bar{Q}^{\dot{\alpha}}{}_A] = -\mathrm{i}(\bar{\sigma}^{\mu\nu})^{\dot{\alpha}}{}_{\dot{\beta}}\bar{Q}^{\dot{\beta}}{}_A \tag{1.202}$$

as in (1.101*b*).

The indices A label the representation of the internal symmetry group to which $Q_\alpha{}^A$ belongs. We denote the Hermitian generators of the group by B^r, so

$$[B^r, B^s] = \mathrm{i}\, c^{rst} B^t \tag{1.203}$$

where c^{rst} are the structure constants. A representation $(b^r)^A{}_B$ satisfies

$$[b^r, b^s] = \mathrm{i}\, c^{rst} b^t \tag{1.204}$$

and

$$[B^r, Q_\alpha{}^A] = -(b^r)^A{}_C Q_\alpha{}^C \tag{1.205}$$

$$[B^r, \bar{Q}_{\dot{\alpha}A}] = \bar{Q}_{\dot{\alpha}C}(b^r)^C{}_A. \tag{1.206}$$

Finally we have to close the algebra of the supersymmetry generators. On general grounds, as in (1.103*b*), we should expect

$$\{Q_\alpha{}^A, \bar{Q}_{\dot{\beta}B}\} = \Delta^A{}_B \sigma^\mu_{\alpha\dot{\beta}} P_\mu. \tag{1.207}$$

Then taking the adjoint shows that the matrix Δ is Hermitian, and in fact positive definite, as before. So we can diagonalize Δ and rescale the Qs and $\bar{Q}$s so that

$$\{Q^A{}_\alpha, \bar{Q}_{\dot{\beta}B}\} = 2\delta^A_B \sigma^\mu{}_{\alpha\dot{\beta}} P_\mu. \tag{1.208}$$

The main difference in principle arises in the anti-commutator $\{Q_\alpha{}^A, Q_\beta{}^B\}$. We can exclude the Poincaré generators, as in (1.105), but, because of the internal symmetry, we are allowed

$$\{Q_\alpha{}^A, Q_\beta{}^B\} = \epsilon_{\alpha\beta} Z^{AB} \tag{1.209}$$

with $\epsilon_{\alpha\beta}$ defined in (1.34) and

$$Z^{AB} = -Z^{BA} \tag{1.210}$$

a linear combination of the internal symmetry generators:

$$Z^{AB} = (q^r)^{AB} B^r \tag{1.211}$$

where the quantities q have dimension

$$[q] = 1 \tag{1.212}$$

in view of (1.176). With a bit of work(7) we can show that

$$[Z^{AB}, Q_\alpha{}^C] = [Z^{AB}, \bar{Q}_{\dot{\alpha}C}] = [Z^{AB}, B^r] = [Z^{AB}, Z^{CD}] = 0. \tag{1.213}$$

That is to say, Z^{AB} commutes with everything: it belongs to the abelian invariant subalgebra of the internal symmetry group; hence the name 'central charges'(8).

In the absence of central charges the internal symmetry group is U(N), since the algebra is invariant under the substitutions

$$Q_\alpha{}^{A'} = U^{AB} Q_\alpha{}^B \qquad \bar{Q}_{\dot{\alpha}A} = U^{AB*} \bar{Q}_{\dot{\alpha}}{}^B \tag{1.214}$$

provided that U is unitary. The effect of the central charges is to reduce this symmetry. Note that the anti-symmetry (1.210) shows that central charges cannot occur in simple ($N = 1$) supersymmetry, and the U(1) invariance is called R-symmetry in this case.

We shall not explore the representations of extended supersymmetry in very much detail. Clearly the existence of extra supersymmetry generators has the effect of enlarging the number of fields/states that constitute a supermultiplet. In the absence of central charges the analysis preceding (1.139) holds for *each* supersymmetry generator $\bar{Q}_{\dot{2}A}$. Thus if we start with a massless single-particle state of helicity λ, there are N states with helicity $\lambda - \frac{1}{2}$:

$$(4E)^{-1/2} \bar{Q}_{\dot{2}A} |p, \lambda\rangle = |p, \lambda - \tfrac{1}{2}, A\rangle \qquad (A = 1, \ldots, N). \tag{1.215}$$

The states with helicity $\lambda - 1$ are obtained by applying $\bar{Q}_{\dot{2}B}$, with $B \neq A$, so there are ${}^N C_2 \equiv N!/2!(N-2)!$ of them. Proceeding in this way, the (massless) irreducible representation with highest helicity λ will close with a state

Table 1.2 $N = 2$ supermultiplet examples.

λ	Helicity	Degeneracy	*TCP*-conjugate helicity	Supermultiplet
1	1	1	-1	Vector
	$\frac{1}{2}$	2	$-\frac{1}{2}$	
	0	1	0	
$\frac{1}{2}$	$\frac{1}{2}$	1	—	Hypermultiplet
	0	2	—	
	$-\frac{1}{2}$	1	—	
2	2	1	-2	Gravity
	$\frac{3}{2}$	2	$-\frac{3}{2}$	
	1	1	-1	

having helicity $\lambda - N/2$, since each $\bar{Q}_{\dot{2}C}$ lowers the helicity by $\frac{1}{2}$ and only totally antisymmetric states ($A \neq B \neq C \neq \ldots$) are allowed; the supermultiplet closes when all N lowering operators have been applied. The dimensionality of the supermultiplet is evidently 2^N, since for each $\bar{Q}_{\dot{2}A}$ there are two possibilities: to apply it, or not to apply it, to the original state $|p, \lambda\rangle$. Together these states comprise a (possibly reducible) representation of the internal symmetry group U(N). If we wish to construct a supersymmetric Yang–Mills theory, then we shall certainly wish to include one-particle (gauge boson) states with helicity $|\lambda| = 1$, and to preserve the renormalizability we shall also require that there are no states with helicity $|\lambda| > 1$. It follows that the largest extended supersymmetry compatible with these requirements must have $N \leqslant 4$. In the same way, a theory involving the graviton with $|\lambda| = 2$, and that does not contain states with $|\lambda| > 2$, must have $N \leqslant 8$.

The simplest extension is, of course, to take $N = 2$. The above argument shows that the supermultiplets are four-dimensional (real) representations of U(2), in the absence of central charges. Thus the vector supermultiplet starts with a highest helicity $\lambda = 1$. It has *two* helicity $+\frac{1}{2}$ states and a single helicity-0 state. As for the $N = 1$ case, we shall need to include the *TCP*-conjugate states in order to construct a Lorentz invariant field theory. However, if we start with highest helicity $\lambda = \frac{1}{2}$, then the supermultiplet has two helicity-0 states and a single helicity $-\frac{1}{2}$ state. Thus *this* supermultiplet, is *TCP*-self-conjugate: it includes a Majorana fermion and a complex scalar

Table 1.3 $N = 4$ supermultiplets.

λ	Helicity	Degeneracy	*TCP*-conjugate helicity	Supermultiplet
1	1	1	—	Vector
	$\frac{1}{2}$	4	—	
	0	6	—	
	$-\frac{1}{2}$	4	—	
	-1	1	—	
2	2	1	-2	Gravity
	$\frac{3}{2}$	4	$-\frac{3}{2}$	
	1	6	-1	
	$\frac{1}{2}$	4	$-\frac{1}{2}$	
	0	1	0	

in the *same* supermultiplet. For this reason it is sometimes called a 'hypermultiplet'. Table 1.2 summarizes these examples of $N = 2$ supermultiplets. Tables 1.3 and 1.4 give $N = 4$ and $N = 8$ examples.

From a phenomenological perspective the hypermultiplet looks rather unattractive. The $N = 2$ supersymmetry requires that the helicity $\pm\frac{1}{2}$ states

Table 1.4 $N = 8$ gravity supermultiplet.

λ	Helicity	Degeneracy
2	2	1
	$\frac{3}{2}$	8
	1	28
	$\frac{1}{2}$	56
	0	70
	$-\frac{1}{2}$	56
	-1	28
	$-\frac{3}{2}$	8
	-2	1

transform in the same way with respect to any gauge symmetry. This would be perfectly acceptable if the only such symmetries were SU(3) of colour and U(1) of electromagnetism, since both chiral components of a quark field do transform in the same way with respect to both groups. However, it is a cast-iron experimental fact that the weak interactions do *not* treat left and right chiral components in the same way. The SU(2) of the electroweak group SU(2) × U(1) is realized non-trivially by the left chiral components: the left-handed component (e_L) of the electron's field belongs to a doublet, whereas the right-handed component (e_R) transforms trivially. Thus if $N = 2$ (or $N > 2$) supersymmetry has any connection with reality we shall need (at some time) to discover the 'mirror' partner E_R of e_L which also belongs to an SU(2) doublet. Since we know that (even $N = 1$) supersymmetry is broken, it is always possible that these hitherto unobserved states have a mass (just) beyond the current experimental bound. However, the fact that the chiral anomaly cancels within each generation[9] is generally taken to be circumstantial evidence that no such mirror states actually exist: if they did the anomaly would cancel separately for each state. It is for this reason that the simple ($N = 1$) supersymmetry is the only one that is thought to have *any* physical relevance, at least at low energies. The 'chiral' supermultiplet has just $\lambda = \frac{1}{2}$ and $\lambda = 0$ states, and we are free to place the $\lambda = -\frac{1}{2}$ state into a *different* representation of the gauge group. The objection to $N \geqslant 2$ supersymmetries is that they are automatically 'non-chiral'.

Exercises

1.1 Verify that the matrices $\frac{1}{2}\Sigma^{\mu\nu}$, defined in (1.11), satisfy the Lorentz algebra (1.8*c*).

1.2 Show that the matrix C, defined in (1.24), satisfies

$$[C^{\mathrm{T}}C^{-1}, \gamma_\mu] = 0$$

and hence deduce that

$$C = -C^{\mathrm{T}}$$

in all representations of the Clifford algebra.

1.3 Show that Poincaré invariance requires the relationship (1.51) for the Poincaré transformed Dirac wave function, with $S(\Lambda)$ obeying (1.52).

1.4 Check that the commutators (1.101) and (1.106) are consistent with the Jacobi identities.

1.5 Show that with the Lagrangian $\mathscr{L}$ given in (1.198) the variation defined in (1.174) *is* a total derivative.

References

General references

Raby S 1986 *Proceedings of the Theoretical Advanced Study Institute (Santa Cruz, 1986)* (Singapore: World Scientific)

Sohnius M F 1985 *Phys. Rep.* **128** 39

Srivastava P P 1986 *Supersymmetry, Superfields, and Supergravity: an Introduction* (Bristol: Institute of Physics Publishing)

Wess J and Bagger J 1983 *Supersymmetry and Supergravity* (Princeton, NJ: Princeton University Press)

References in the text

1 Howe P S, Stelle K S and West P C 1983 *Phys. Lett.* **124B** 55
See also
Jones D R T 1986 *Nucl. Phys.* B **277** 153

2 Dirac P A M 1958 *Quantum Mechanics* 4th edn (Oxford: Oxford University Press) p 309

3 Berezin F 1966 *Method of Second Quantisation* (New York: Academic)

4 Gol'fand Yu A and Likhtman E P 1971 *JETP Lett.* **13** 323
Volkov D V and Akulov V P 1973 *Phys. Lett.* **46B** 109; 1974 *Teor. Mat. Fiz. (USSR)* **18** 39
Wess J and Zumino B 1974 *Nucl. Phys.* B **70** 39

5 Wess J and Zumino B 1974 *Phys. Lett.* **49B** 52

6 Salam A and Strathdee J 1974 *Nucl. Phys.* B **80** 317
Wess J 1974 *Springer Lecture Notes in Physics* vol 37 (Berlin: Springer) p 352
Zumino B 1974 *Proc. Int. Conf. on High Energy Physics (London, 1974)* ed J Smith (Didcot: Rutherford Laboratory)

7 See, for example,
Wess J and Bagger J 1983 *Supersymmetry and Supergravity* (Princeton, NJ: Princeton University Press)
Sohnius M F 1985 *Phys. Rep.* **128** 39

8 Grosser D 1975 *Nucl. Phys.* B **92** 120; 1975 *Nucl. Phys.* B **94** 513
Haag R, Łopusanski J T and Sohnius M 1975 *Nucl. Phys.* B **88** 257

9 See, for example,
Bailin D and Love A 1993 *Introduction to Gauge Field Theory* (Bristol: Institute of Physics Publishing) Chapter 15
for a discussion of the chiral anomaly

2

LAGRANGIANS FOR CHIRAL SUPERFIELDS

2.1 Introduction

Although it is possible to construct supersymmetric Lagrangians directly from the component fields belonging to a supermultiplet as in §1.5, the procedure is greatly facilitated by the introduction of superfields.[(1),(2)] Whereas an ordinary field is a function of the space-time coordinates x only, a superfield $S(x, \theta, \bar{\theta})$ is also a function of anticommuting Grassmann variables θ_α and $\bar{\theta}_{\dot{\alpha}}$ transforming as two-component Weyl spinors:

$$\{\theta_\alpha, \theta_\beta\} = \{\theta_\alpha, \bar{\theta}_{\dot{\beta}}\} = \{\bar{\theta}_{\dot{\alpha}}, \bar{\theta}_{\dot{\beta}}\} = 0. \tag{2.1}$$

(A discussion of Grassmann variables is given following (1.71).) The fields of the supermultiplet then arise as the coefficients in an expansion of $S(x, \theta, \bar{\theta})$ in powers of θ and $\bar{\theta}$ (which necessarily terminates after a finite number of terms for anticommuting Grassmann variables).

The general superfield $S(x, \theta, \bar{\theta})$ turns out to contain more than one supermultiplet, and the chiral and vector supermultiplets of §1.4 are obtained by reducing the number of component fields in S by imposing appropriate constraints. In this chapter it will be shown that we obtain the chiral supermultiplet of §§1.4, 5 by requiring an appropriate covariant derivative of S to vanish. In the case of the vector supermultiplet, as we shall see in Chapter 3, the appropriate constraint is just $S^\dagger = S$.

When potentially realistic supersymmetric theories are constructed, the chiral superfields discussed in this chapter provide the matter fields, i.e. the quarks, leptons and Higgs scalars and associated with them in the same multiplet their supersymmetric partners, which in the case of the quarks and leptons are the scalar squarks and sleptons and in the case of the Higgses are the fermionic Higgsinos. The gauge fields and their fermionic supersymmetric partners, the gauginos, are to be assigned to the vector superfields discussed in the next chapter. The presence of supersymmetric partners for particles is the source of the remarkable non-renormalization theorems, discussed later in this chapter, which are at the root of the supersymmetry solution to the hierarchy problem.

Since the supersymmetric partners of particles are not observed at low energy it is necessary for any potentially realistic theory to contain a mechanism that will spontaneously break supersymmetry to provide mass splittings within supermultiplets. One such mechanism, F-term supersymmetry breaking, will be discussed in this chapter, and another such, D-term

DOI: 10.1201/9780367805807-2

supersymmetry breaking, in the next, while supersymmetry breaking by gaugino condensates will be deferred to chapter 5.

2.2 Superfield representations of the supersymmetry algebra

The supersymmetry algebra of §1.3 is generated by the momentum operators P_μ and the Weyl spinor supersymmetry generators Q_α and $\bar{Q}_{\dot{\alpha}}$:

$$[Q_\alpha, P_\mu] = [\bar{Q}_{\dot{\alpha}}, P_\mu] = [P_\mu, P_\nu] = 0 \tag{2.2}$$

$$\{Q_\alpha, Q_\beta\} = \{\bar{Q}_{\dot{\alpha}}, \bar{Q}_{\dot{\beta}}\} = 0 \tag{2.3}$$

and

$$\{Q_\alpha, \bar{Q}_{\dot{\beta}}\} = 2\sigma^\mu_{\alpha\dot{\beta}} P_\mu . \tag{2.4}$$

A finite element of the corresponding group is

$$G(x^\mu, \theta, \bar{\theta}) = \exp(\mathrm{i}(\theta Q + \bar{\theta}\bar{Q} - x^\mu P_\mu)) \tag{2.5}$$

where θ_α and $\bar{\theta}_{\dot{\alpha}}$ are Grassmann variable parameters. We wish to construct linear representations of this group (of the supersymmetry algebra). This can be done by considering the action induced in $(x^\mu, \theta, \bar{\theta})$ parameter space by the group elements as follows. It is not difficult to show that (Exercise 2.1)

$$G(x^\mu, \theta, \bar{\theta})G(a^\mu, \xi, \bar{\xi})$$
$$= G(x^\mu + a^\mu - \mathrm{i}\,\xi\sigma^\mu\bar{\theta} + \mathrm{i}\,\theta\sigma^\mu\bar{\xi}, \theta + \xi, \bar{\theta} + \bar{\xi}) \tag{2.6}$$

because the Hausdorff formula

$$\mathrm{e}^A\mathrm{e}^B = \exp(A + B + \tfrac{1}{2}[A, B] + \cdots) \tag{2.7}$$

terminates at the first commutator for the group elements considered here. Thus (acting on the right) the supersymmetry generators induce the motion in group parameter space:

$$\exp(\mathrm{i}(\xi Q + \bar{\xi}\bar{Q} - a^\mu P_\mu)):$$
$$(x^\mu, \theta, \bar{\theta}) \to (x^\mu + a^\mu - \mathrm{i}\,\xi\sigma^\mu\bar{\theta} + \mathrm{i}\,\theta\sigma^\mu\bar{\xi}, \theta + \xi, \bar{\theta} + \bar{\xi}). \tag{2.8}$$

For a function $S(x^\mu, \theta, \bar{\theta})$ (referred to as a superfield) we have

$$S(x^\mu + a^\mu - \mathrm{i}\,\xi\sigma^\mu\bar{\theta} + \mathrm{i}\,\theta\sigma^\mu\bar{\xi}, \theta + \xi, \bar{\theta} + \bar{\xi})$$
$$= S(x^\mu, \theta, \bar{\theta}) + (a^\mu - \mathrm{i}\,\xi\sigma^\mu\bar{\theta} + \mathrm{i}\,\theta\sigma^\mu\bar{\xi})\frac{\partial S}{\partial x^\mu}$$
$$+\xi^\alpha\frac{\partial S}{\partial\theta^\alpha} + \bar{\xi}_{\dot{\alpha}}\frac{\partial S}{\partial\bar{\theta}_{\dot{\alpha}}} + \cdots \tag{2.9}$$

from which it follows that the action of the supersymmetry algebra on superfields

$$S(x^\mu, \theta, \bar{\theta}) \rightarrow \exp(\mathrm{i}(\xi Q + \bar{\xi}\bar{Q} - a^\mu P_\mu))S(x^\mu, \theta, \bar{\theta}) \tag{2.10}$$

is generated by

$$P_\mu = \mathrm{i}\, \partial_\mu \tag{2.11}$$

$$\mathrm{i}\, Q_\alpha = \frac{\partial}{\partial \theta^\alpha} - \mathrm{i}\, \sigma^\mu_{\alpha\dot{\alpha}} \bar{\theta}^{\dot{\alpha}}\, \partial_\mu \tag{2.12}$$

$$\mathrm{i}\, \bar{Q}_{\dot{\alpha}} = -\frac{\partial}{\partial \bar{\theta}^{\dot{\alpha}}} + \mathrm{i}\, \theta^\alpha \sigma^\mu_{\alpha\dot{\alpha}}\, \partial_\mu . \tag{2.13}$$

This is the linear representation of the supersymmetry algebra we were seeking. It is easy to check (Exercise 2.2) that P_μ, Q_α, and $\bar{Q}_{\dot{\alpha}}$ of (2.11), (2.12) and (2.13) realize the algebra (2.2)–(2.4), as consistency requires.

The general superfield $S(x^\mu, \theta, \bar{\theta})$ may be expanded as a power series in θ and $\bar{\theta}$ involving not more than two powers of θ and $\bar{\theta}$, because θ and $\bar{\theta}$ are two-component Grassmann variables. The coefficients of the various powers of θ and $\bar{\theta}$ in this expansion are ordinary fields (functions of x^μ). Such a superfield provides a representation of the supersymmetry algebra that is in the first instance reducible. Irreducible representations of the supersymmetry algebra are obtained by imposing on the superfields constraints that are covariant under the supersymmetry algebra. The simplest such constraint is $S = S^\dagger$, which we use in the next chapter to construct the vector superfield, whose component fields form the vector supermultiplet of §1.4. In this section, we use supersymmetric covariant derivatives to construct the chiral superfield whose component fields form the chiral supermultiplet of §1.4.

Fermionic derivatives D_α, $\bar{D}_{\dot{\alpha}}$ which anticommute with the generators (acting on superfields) of the supersymmetry algebra may be defined by

$$D_\alpha = \frac{\partial}{\partial \theta^\alpha} + \mathrm{i}\, \sigma^\mu_{\alpha\dot{\alpha}} \bar{\theta}^{\dot{\alpha}}\, \partial_\mu \tag{2.14}$$

$$\bar{D}_{\dot{\alpha}} = -\frac{\partial}{\partial \bar{\theta}^{\dot{\alpha}}} - \mathrm{i}\, \theta^\alpha \sigma^\mu_{\alpha\dot{\alpha}}\, \partial_\mu . \tag{2.15}$$

The proof that

$$\{D_\alpha, Q_\beta\} = \{D_\alpha, \bar{Q}_{\dot{\beta}}\} = \{\bar{D}_{\dot{\alpha}}, Q_\beta\} = \{\bar{D}_{\dot{\alpha}}, \bar{Q}_{\dot{\beta}}\} = 0 \tag{2.16}$$

follows directly (Exercise 2.3) using the explicit expressions for Q_α and $\bar{Q}_{\dot{\alpha}}$ of (2.12) and (2.13). These covariant derivatives also have the algebra

$$\{D_\alpha, \bar{D}_{\dot{\alpha}}\} = 2\,\mathrm{i}\, \sigma^\mu_{\alpha\dot{\alpha}}\, \partial_\mu \tag{2.17}$$

$$\{D_\alpha, D_\beta\} = \{\bar{D}_{\dot{\alpha}}, \bar{D}_{\dot{\beta}}\} = 0. \tag{2.18}$$

The operators D_α and $\bar{D}_{\dot{\alpha}}$ may be used to impose covariant constraints on superfields because they anticommute with the generators (acting on superfields) of the supersymmetry algebra Q and $\bar{Q}$ and so commute with $\xi Q + \bar{\xi}\bar{Q}$ which occurs in supersymmetry transformations. It is thus possible to apply the covariant condition

$$\bar{D}_{\dot{\alpha}} S = 0. \tag{2.19}$$

A superfield on which the constraint (2.19) has been imposed we shall denote by Φ and refer to as a chiral superfield:

$$\bar{D}_{\dot{\alpha}} \Phi = 0. \tag{2.20}$$

2.3 Expansion of the chiral superfield in component fields

With $\bar{D}_{\dot{\alpha}}$ given by (2.15), we see that *any* function of θ and

$$y^\mu = x^\mu + \mathrm{i}\,\theta\sigma^\mu\bar{\theta} \tag{2.21}$$

satisfies the constraint (2.20), because

$$\bar{D}_{\dot{\alpha}} \theta = 0 \tag{2.22}$$

and

$$\bar{D}_{\dot{\alpha}} y^\mu = 0. \tag{2.23}$$

Indeed $\Phi(y^\mu, \theta)$ is the most general solution of (2.20), because, after changing variables from $(x^\mu, \theta, \bar{\theta})$ to $(y^\mu, \theta, \bar{\theta})$,

$$\bar{D}_{\dot{\alpha}} = -\frac{\partial}{\partial \bar{\theta}^{\dot{\alpha}}}\bigg|_{y,\theta}. \tag{2.24}$$

Expanding $\Phi(y^\mu, \theta)$ in powers of the two-component Grassmann variable θ gives

$$\Phi(y^\mu, \theta) = \varphi(y) + \sqrt{2}\,\theta\psi(y) + \theta\theta F(y) \tag{2.25}$$

where φ and F are complex scalar fields and ψ is a left-handed Weyl spinor field. Thus, the general expansion of a chiral superfield in component fields is

$$\Phi(x^\mu, \theta, \bar{\theta}) = \varphi(x) + \sqrt{2}\,\theta\psi(x) + \theta\theta F(x) + \mathrm{i}\,\partial_\mu\varphi\,\theta\sigma^\mu\bar{\theta}$$
$$+ \mathrm{i}\sqrt{2}\,\theta\,\partial_\mu\psi\,\theta\,\sigma^\mu\bar{\theta} - \tfrac{1}{2}\partial_\mu\,\partial_\nu\varphi\,\theta\sigma^\mu\bar{\theta}\theta\sigma^\nu\bar{\theta}. \tag{2.26}$$

With the aid of the identities of §1.2 and Appendix A, this may be rearranged as

$$\Phi(x^\mu, \theta, \bar{\theta}) = \varphi + \sqrt{2}\,\theta\psi + \theta\theta F + \mathrm{i}\,\partial_\mu\varphi\,\theta\sigma^\mu\bar{\theta} \\ - \frac{\mathrm{i}}{\sqrt{2}}\theta\theta\,\partial_\mu\psi\,\sigma^\mu\bar{\theta} - \tfrac{1}{4}\partial_\mu\,\partial^\mu\varphi\,\theta\theta\bar{\theta}\bar{\theta}. \tag{2.27}$$

It follows immediately that the conjugate superfield $\Phi^\dagger$ has the component field expansion

$$\Phi^\dagger = \varphi^\dagger + \sqrt{2}\,\bar{\theta}\bar{\psi} + \bar{\theta}\bar{\theta}F^\dagger - \mathrm{i}\,\partial_\mu\varphi^\dagger\,\theta\sigma^\mu\bar{\theta} \\ + \frac{\mathrm{i}}{\sqrt{2}}\bar{\theta}\bar{\theta}\theta\sigma^\mu\,\partial_\mu\bar{\psi} - \tfrac{1}{4}\partial_\mu\,\partial^\mu\varphi^\dagger\,\theta\theta\bar{\theta}\bar{\theta} \tag{2.28}$$

and satisfies the constraint

$$D_\alpha\Phi^\dagger = 0. \tag{2.29}$$

We shall sometimes refer to Φ as a left chiral superfield and $\Phi^\dagger$ as a right chiral superfield (because they involve the left- and right-handed Weyl spinors ψ and $\bar{\psi}$, respectively).

From (2.10)–(2.13) the behaviour of the superfield Φ under an infinitesimal supersymmetry transformation is

$$\Phi \rightarrow \Phi + \delta\Phi \tag{2.30}$$

where

$$\delta\Phi = \mathrm{i}(\xi Q + \bar{\xi}\bar{Q})\Phi \tag{2.31}$$

that is

$$\delta\Phi = \xi^\alpha\left(\frac{\partial}{\partial\theta^\alpha} - \mathrm{i}\,\sigma^\mu_{\alpha\dot{\alpha}}\bar{\theta}^{\dot{\alpha}}\,\partial_\mu\right)\Phi + \bar{\xi}^{\dot{\alpha}}\left(\frac{\partial}{\partial\bar{\theta}^{\dot{\alpha}}} - \mathrm{i}\,\theta^\alpha\sigma^\mu_{\alpha\dot{\alpha}}\,\partial_\mu\right)\Phi. \tag{2.32}$$

From (2.32) may be derived the supersymmetry transformations of the component fields in the expansion (2.27) by comparing

$$\delta\Phi = \delta\varphi + \sqrt{2}\,\theta\,\delta\psi + \theta\theta\,\delta F + \cdots \tag{2.33}$$

with

$$\delta\Phi = \sqrt{2}\,\xi\psi + 2\xi\theta F + 2\,\mathrm{i}\,\partial_\mu\varphi\,\theta\sigma^\mu\bar{\xi} \\ + \frac{\mathrm{i}}{\sqrt{2}}\theta\theta\,\partial_\mu\psi\,\sigma^\mu\bar{\xi} + \cdots. \tag{2.34}$$

Thus,

$$\delta\varphi = \sqrt{2}\,\xi\psi \tag{2.35}$$

$$\delta\psi = \sqrt{2}\,\xi F - \sqrt{2}\,\partial_\mu\varphi\,\sigma^\mu\bar{\xi} \tag{2.36}$$

and

$$\delta F = \mathrm{i}\sqrt{2}\,\partial_\mu \psi\,\sigma^\mu \bar{\xi} \tag{2.37}$$

in agreement with equations (1.193*a*), (1.196) and (1.197). It will be important for the purpose of constructing supersymmetric Lagrangians in §2.5 to notice that the change in F under a supersymmetry transformation is a total divergence.

2.4 Products of chiral superfields

For the construction of renormalizable supersymmetric Lagrangians in §2.5 it will prove necessary to study the products of chiral superfields $\Phi_i\Phi_j$, $\Phi_i\Phi_j\Phi_k$ and $\Phi_i^\dagger\Phi_j$, where the index i distinguishes the various left chiral superfields Φ_i in the theory. Because Q, $\bar{Q}$ and $\bar{D}$ are linear differential operators on superspace it is immediate that any product of two left chiral superfields is again a left chiral superfield, i.e. transforms as (2.10) and satisfies the constraint (2.20). (Consequently any product of left chiral superfields is a left chiral superfield.) By direct calculation $\Phi_i(y,\theta)\Phi_j(y,\theta)$ and $\Phi_i(y,\theta)\Phi_j(y,\theta)\Phi_k(y,\theta)$ have the expansions in the form (2.25),

$$\begin{aligned}\Phi_i(y,\theta)\Phi_j(y,\theta) &= \varphi_i(y)\varphi_j(y) + \sqrt{2}\,\theta(\psi_i(y)\varphi_j(y) + \varphi_i(y)\psi_j(y))\\ &\quad + \theta\theta(\varphi_i(y)F_j(y) + \varphi_j(y)F_i(y) - \psi_i(y)\psi_j(y))\end{aligned} \tag{2.38}$$

and

$$\begin{aligned}\Phi_i(y,\theta)\Phi_j(y,\theta)\Phi_k(y,\theta) &= \varphi_i(y)\varphi_j(y)\varphi_k(y)\\ &\quad + \sqrt{2}\,\theta(\psi_i\varphi_j\varphi_k + \varphi_i\psi_j\varphi_k + \varphi_i\varphi_j\psi_k)\\ &\quad + \theta\theta(\varphi_i\varphi_jF_k + \varphi_iF_j\varphi_k + F_i\varphi_j\varphi_k\\ &\quad - \psi_i\psi_j\varphi_k - \psi_i\psi_k\varphi_j - \psi_j\psi_k\varphi_i).\end{aligned} \tag{2.39}$$

These expansions may then be written in terms of x, θ and $\bar{\theta}$ by substituting for y from (2.21). Also

$$\begin{aligned}\Phi_i^\dagger(y,\theta)\Phi_j(y,\theta) &= \varphi_i^\dagger(y)\varphi_j(y) + \sqrt{2}\,\theta\psi_j(y)\varphi_i^\dagger(y)\\ &\quad + \sqrt{2}\,\bar{\theta}\bar{\psi}_i(y)\varphi_j(y) + 2\bar{\theta}\bar{\psi}_i(y)\theta\psi_j(y) + F_j(y)\varphi_i^\dagger(y)\theta\theta\\ &\quad + F_i^\dagger(y)\varphi_j(y)\bar{\theta}\bar{\theta} + \sqrt{2}\,\theta\theta\bar{\theta}\bar{\psi}_i(y)F_j(y)\\ &\quad + \sqrt{2}\,\bar{\theta}\bar{\theta}\theta\psi_j(y)F_i^\dagger(y) + \bar{\theta}\bar{\theta}\theta\theta F_i^\dagger(y)F_j(y)\end{aligned} \tag{2.40}$$

which is *not* a left chiral superfield. (It is a vector superfield as can be seen by comparing with §3.2.) Again, the final expansion in x, θ and $\bar{\theta}$ is obtained by substituting for y from (2.21).

Of particular importance for the construction of supersymmetric Lagrangians in the next section is the coefficient of $\theta\theta$ in the expansion of $\Phi_i(x, \theta, \bar{\theta})\Phi_j(x, \theta, \bar{\theta})$ and $\Phi_i(x, \theta, \bar{\theta})\Phi_j(x, \theta, \bar{\theta})\Phi_k(x, \theta, \bar{\theta})$, referred to as the F-term (by analogy with (2.26)) and denoted by $[\Phi_i\Phi_j]_F$ or $[\Phi_i\Phi_j\Phi_k]_F$. Also important is the coefficient of $\bar{\theta}\bar{\theta}\theta\theta$ in the expansion of $\Phi_i^\dagger(x, \theta, \bar{\theta})\Phi_j(x, \theta, \bar{\theta})$, referred to as the D-term (by analogy with (3.4) or (3.27) for the vector superfield, apart from a factor of $\frac{1}{2}$). After substituting for y from (2.21) we find that

$$[\Phi_i\Phi_j]_F \equiv [\Phi_i(x, \theta, \bar{\theta})\Phi_j(x, \theta, \bar{\theta})]_{\text{coeff. of }\theta\theta}$$

$$= \varphi_i(x)F_j(x) + \varphi_j(x)F_i(x) - \psi_i(x)\psi_j(x) \tag{2.41}$$

$$[\Phi_i\Phi_j\Phi_k]_F \equiv [\Phi_i(x, \theta, \bar{\theta}), \Phi_j(x, \theta, \bar{\theta}), \Phi_k(x, \theta, \bar{\theta})_{\text{coeff. of }\theta\theta}$$

$$= \varphi_i\varphi_jF_k + \varphi_iF_j\varphi_k + F_i\varphi_j\varphi_k - \psi_i\psi_j\varphi_k - \psi_i\psi_k\varphi_j - \psi_j\psi_k\varphi_i \tag{2.42}$$

and

$$[\Phi_i^\dagger\Phi_j]_D \equiv [\Phi_i^\dagger(x, \theta, \bar{\theta})\Phi_j(x, \theta, \bar{\theta})]_{\text{coeff. of }\bar{\theta}\bar{\theta}\theta\theta}$$

$$= F_i^\dagger F_j + \tfrac{1}{2}\partial_\mu\varphi_i^\dagger\,\partial^\mu\varphi_j - \tfrac{1}{4}\varphi_i^\dagger\,\partial_\mu\,\partial^\mu\varphi_j - \tfrac{1}{4}\partial_\mu\,\partial^\mu\varphi_i^\dagger\varphi_j$$
$$+\frac{\mathrm{i}}{2}\bar{\psi}_i\bar{\sigma}^\mu\,\partial_\mu\psi_j - \frac{\mathrm{i}}{2}\partial_\mu\bar{\psi}_i\,\bar{\sigma}^\mu\psi_j\,. \tag{2.43}$$

This last equation requires extensive use of the identities of §1.2 and Appendix A for its proof (Exercise 2.4).

2.5 Renormalizable supersymmetric Lagrangians for chiral superfields

When constructing a supersymmetric Lagrangian using products of chiral superfields Φ_i it is necessary to use terms that are invariant under a supersymmetry transformation up to a total divergence. As observed after (2.37), the variation under a supersymmetry transformation of the F-term in a chiral superfield is a total divergence. Also, as will be seen in §3.2, the variation under a supersymmetry transformation of the D-term in a vector superfield is a total divergence and this applies in particular to the D-term in $\Phi_i^\dagger\Phi_i$. Thus we may construct supersymmetric Lagrangians by using the D-term of $\Phi_i^\dagger\Phi_i$ and the F-term of products of left chiral superfields (together with their Hermitian conjugates). In general, the supersymmetric Lagrangian $\mathscr{L}$ has the form

$$\mathscr{L} = \sum_i [\Phi_i^\dagger\Phi_i]_D + ([W(\Phi)]_F + \text{HC}) \tag{2.44}$$

where $W(\Phi)$, which is referred to as the superpotential, must involve only up to the third power of the superfields Φ_i to obtain a renormalizable Lagrangian. ($[\Phi_i\Phi_j\Phi_k]_F$ is a product of ordinary fields of dimension four, and F-terms involving more factors of the Φ_i will have dimension greater than four.) Apart from a possible tadpole term linear in the Φ_i, we may write (with sums over i, j and k understood)

$$W(\Phi) = \tfrac{1}{2} m_{ij}\Phi_i\Phi_j + \tfrac{1}{3}\lambda_{ijk}\Phi_i\Phi_j\Phi_k \tag{2.45}$$

with m_{ij} and λ_{ijk} real and symmetric in their indices. Then, apart from surface terms,

$$\begin{aligned}\mathscr{L} &= \partial_\mu\varphi_i^\dagger\, \partial^\mu\varphi_i + \mathrm{i}\,\bar{\psi}_i\bar{\sigma}^\mu\, \partial_\mu\psi_i + F_i^\dagger F_i \\ &\quad + (m_{ij}\varphi_i F_j - \tfrac{1}{2} m_{ij}\psi_i\psi_j + \lambda_{ijk}\varphi_i\varphi_j F_k - \lambda_{ijk}\psi_i\psi_j\varphi_k + \text{HC}).\end{aligned} \tag{2.46}$$

The field equations arising from this Lagrangian are

$$\partial_\mu\, \partial^\mu\varphi_i = m_{ij}F_j^\dagger + 2\lambda_{ijk}\varphi_j^\dagger F_k^\dagger \tag{2.47}$$

$$\mathrm{i}\,\bar{\sigma}^\mu\, \partial_\mu\psi_i = -\, m_{ij}\bar{\psi}_j - 2\lambda_{ijk}\bar{\psi}_j\varphi_k^\dagger \tag{2.48}$$

and

$$F_i^\dagger = -\, m_{ij}\varphi_j - \lambda_{ijk}\varphi_j\varphi_k = -\,\frac{\partial W(\varphi)}{\partial\varphi_i} \tag{2.49}$$

where $W(\varphi)$ is the superpotential with each Φ_k replaced by φ_k. The last of these equations shows that the fields F_i are merely auxiliary fields which may now be eliminated. (This results from there being *no* derivatives of the F_i in the Lagrangian.) Using (2.49) the Lagrangian becomes

$$\begin{aligned}\mathscr{L} &= \partial_\mu\varphi_i^\dagger\, \partial^\mu\varphi_i + \mathrm{i}\,\bar{\psi}_i\bar{\sigma}^\mu\, \partial_\mu\psi_i - F_i^\dagger F_i \\ &\quad - (\tfrac{1}{2} m_{ij}\psi_i\psi_j + \lambda_{ijk}\psi_i\psi_j\varphi_k + \text{HC}) \\ &= \partial_\mu\varphi_i^\dagger\, \partial^\mu\varphi_i + \mathrm{i}\,\bar{\psi}_i\bar{\sigma}^\mu\, \partial_\mu\psi_i - |m_{ij}\varphi_j + \lambda_{ijk}\varphi_j\varphi_k|^2 \\ &\quad - (\tfrac{1}{2} m_{ij}\psi_i\psi_j + \lambda_{ijk}\psi_i\psi_j\varphi_k + \text{HC}).\end{aligned} \tag{2.50}$$

In particular, the tree level effective potential V is given by

$$V = F_i^\dagger F_i \equiv |F_i|^2 \tag{2.51}$$

with

$$F_i^\dagger = -\,\frac{\partial W(\varphi)}{\partial\varphi_i} \tag{2.52}$$

as in (2.49). It is not too difficult to show (Exercise 2.5) that (2.51) and (2.52) hold for *any* superpotential $W(\Phi)$ (*not* just the renormalizable form of (2.45)) *provided* that the 'kinetic term' in the Lagrangian is of the minimal form $[\Phi_i^{\dagger}\Phi_i]_D$ of (2.45), and there are no interactions coupling the chiral superfields to other fields. The tree level effective potential for supersymmetric theories of chiral superfields thus has the remarkable property of being positive semi-definite. (We shall see in Chapter 3 that V remains positive semi-definite when the chiral superfields are coupled to vector superfields.)

2.6 Feynman rules for chiral supermultiplets

For the purpose of performing Feynman diagram calculations it is usually more convenient to use Majorana spinor fields rather than Weyl spinor fields (partly because most particle physicists have developed a facility for using Dirac γ-matrices rather than Pauli spin matrices.) The Lagrangian (2.46) may be cast in terms of Majorana spinors by using the identities (see (1.86))

$$\bar{\Psi}_i\Psi_j = \psi_i\psi_j + \bar{\psi}_i\bar{\psi}_j \tag{2.53}$$

$$\bar{\Psi}_i\gamma_5\Psi_j = -\psi_i\psi_j + \bar{\psi}_i\bar{\psi}_j \tag{2.54}$$

and

$$\bar{\Psi}\gamma^\mu\partial_\mu\Psi = \psi\sigma^\mu\partial_\mu\bar{\psi} + \bar{\psi}\bar{\sigma}^\mu\partial_\mu\psi = 2\bar{\psi}\bar{\sigma}^\mu\partial_\mu\psi \tag{2.55}$$

dropping a total divergence, where Ψ denotes Majorana spinors and ψ denotes Weyl spinors. Thus,

$$\begin{aligned}\mathscr{L} = {} & \partial_\mu\varphi_i^{\dagger}\partial^\mu\varphi_i - |m_{ij}\varphi_j + \lambda_{ijk}\varphi_j\varphi_k|^2 \\ & + \frac{\mathrm{i}}{2}\bar{\Psi}_i\gamma^\mu\partial_\mu\Psi_i - \tfrac{1}{2}m_{ij}\bar{\Psi}_i\Psi_j - \frac{\lambda_{ijk}}{2}(\bar{\Psi}_i\Psi_j - \bar{\Psi}_i\gamma_5\Psi_j)\varphi_k \\ & - \frac{\lambda_{ijk}}{2}(\bar{\Psi}_i\Psi_j + \bar{\Psi}_i\gamma_5\Psi_j)\varphi_k^{\dagger}\end{aligned} \tag{2.56}$$

and the Feynman rules for vertices are

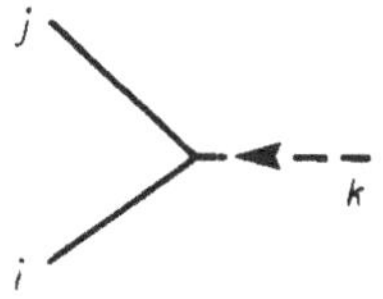

$$: \quad -\mathrm{i}\lambda_{ijk}(I - \gamma_5) \tag{2.57}$$

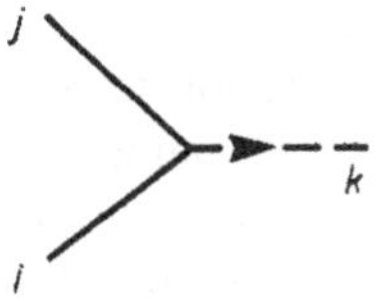

$$: -\mathrm{i}\,\lambda_{ijk}(I+\gamma_5)\,. \tag{2.58}$$

When

$$m_{ij} = m\delta_{ij} \tag{2.59}$$

the Majorana spinor propagator is

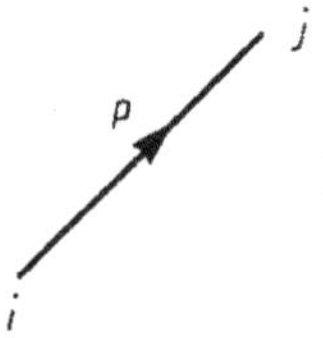

$$: \frac{\mathrm{i}\delta_{ij}}{(\not{p} - m + \mathrm{i}\,\varepsilon)}. \tag{2.60}$$

2.7 Mass and coupling constant renormalization

A very important property of supersymmetric theories is the lack of infinite-mass renormalization other than by renormalization of the wave function. As we shall see in Chapter 6, this means that once the hierarchy of mass scales between the electroweak scale (10^2–10^3 GeV) and the grand unification scale (10^{15}–10^{19} GeV) has been established at tree level, the hierarchy is *not* destroyed by radiative corrections. This phenomenon may be illustrated by considering a simple model with three superfields Φ_x, Φ_y and Φ_z, with the first two of these superfields having *no* mass term, and the third having a large mass. In a non-supersymmetric model we would normally expect mass renormalization to induce a large mass, proportional to m_z, for Φ_x and Φ_y. Here we shall find that it does *not*.

The superpotential for the model is

$$W(\Phi_x, \Phi_y, \Phi_z) = \lambda\Phi_x\Phi_y\Phi_z + m_z\Phi_z\Phi_z \tag{2.61}$$

with corresponding F-terms

$$F_x^{\dagger} = -\frac{\partial W}{\partial\varphi_x} = -\lambda\varphi_y\varphi_z \tag{2.62}$$

$$F_y^{\dagger} = -\frac{\partial W}{\partial\varphi_y} = -\lambda\varphi_x\varphi_z \tag{2.63}$$

$$F_z^{\dagger} = -\frac{\partial W}{\partial\varphi_z} = -\lambda\varphi_x\varphi_y - m_z\varphi_z \tag{2.64}$$

and Lagrangian

$$\begin{aligned}
\mathscr{L} = {} & \partial_\mu \varphi_x^{\dagger} \partial^\mu \varphi_x + \partial_\mu \varphi_y^{\dagger} \partial^\mu \varphi_y + \partial_\mu \varphi_z^{\dagger} \partial^\mu \varphi_z \\
& + \frac{\mathrm{i}}{2} \bar{\Psi}_x \gamma^\mu \partial_\mu \Psi_x + \frac{\mathrm{i}}{2} \bar{\Psi}_y \gamma^\mu \partial_\mu \Psi_y + \frac{\mathrm{i}}{2} \bar{\Psi}_z \gamma^\mu \partial_\mu \Psi_z \\
& - \lambda^2 (\varphi_y^{\dagger} \varphi_y \varphi_z^{\dagger} \varphi_z + \varphi_x^{\dagger} \varphi_x \varphi_z^{\dagger} \varphi_z + \varphi_x^{\dagger} \varphi_x \varphi_y^{\dagger} \varphi_y) \\
& - m_z^2 \varphi_z^{\dagger} \varphi_z - \lambda m_z (\varphi_x^{\dagger} \varphi_y^{\dagger} \varphi_z + \varphi_x \varphi_y \varphi_z^{\dagger}) - \frac{m_z}{2} \bar{\Psi}_z \Psi_z \\
& - \frac{\lambda}{2} (\bar{\Psi}_x \Psi_y - \bar{\Psi}_x \gamma_5 \Psi_y) \varphi_z - \frac{\lambda}{2} (\bar{\Psi}_x \Psi_y + \bar{\Psi}_x \gamma_5 \Psi_y) \varphi_z^{\dagger} \\
& - \frac{\lambda}{2} (\bar{\Psi}_x \Psi_z - \bar{\Psi}_x \gamma_5 \Psi_z) \varphi_y - \frac{\lambda}{2} (\bar{\Psi}_x \Psi_z + \bar{\Psi}_x \gamma_5 \Psi_z) \varphi_y^{\dagger} \\
& - \frac{\lambda}{2} (\bar{\Psi}_y \Psi_z - \bar{\Psi}_y \gamma_5 \Psi_z) \varphi_x - \frac{\lambda}{2} (\bar{\Psi}_y \Psi_z + \bar{\Psi}_y \gamma_5 \Psi_z) \varphi_x^{\dagger} .
\end{aligned} \tag{2.65}$$

The corresponding Feynman rules for vertices are then

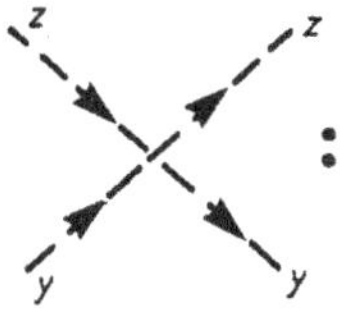

$: -\mathrm{i}\lambda^2$ etc (2.66)

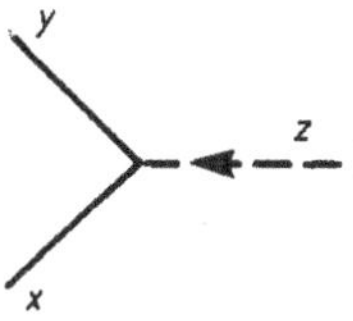

$: -\frac{\mathrm{i}\lambda}{2}(I - \gamma_5)$ etc (2.67)

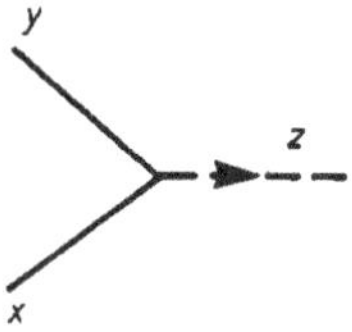

$: -\frac{\mathrm{i}\lambda}{2}(I + \gamma_5)$ etc (2.68)

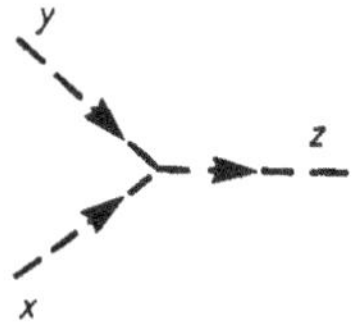

$: -\mathrm{i}\lambda m_z$ (2.69)

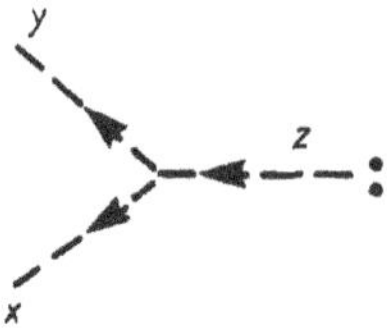

$$: \; -\mathrm{i}\,\lambda m_z. \tag{2.70}$$

In supersymmetric theories it is necessary to use a regularization procedure for Feynman diagrams that manifestly preserves supersymmetry (at least at low loop orders). A suitable procedure is dimensional *reduction* regularization[(3), (4)], in which the Dirac gamma matrix algebra is worked out first in four dimensions, and the momentum integrations are then performed in $d = 2\omega$ dimensions. Using dimensional reduction regularization, the one-loop diagrams involving m_z contributing to the renormalization of the φ_x mass are

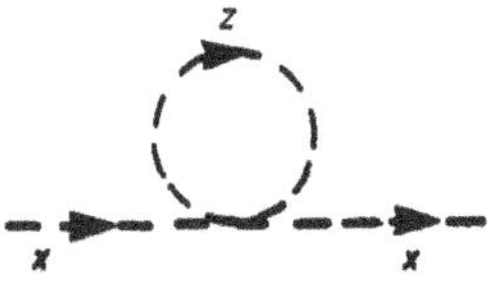

$$= \lambda^2 \int \frac{\mathrm{d}^{2\omega}q}{(2\pi)^{2\omega}} \frac{1}{(q^2 - m_z^2)} \tag{2.71}$$

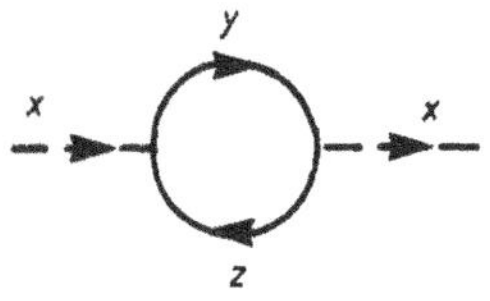

$$= -\frac{\lambda^2}{4} \int \frac{\mathrm{d}^{2\omega}q}{(2\pi)^{2\omega}} \mathrm{Tr}\left((I + \gamma_5)\frac{1}{\not{q}}(I - \gamma_5)\frac{1}{(\not{q} - m_z)}\right)$$

$$= -2\lambda^2 \int \frac{\mathrm{d}^{2\omega}q}{(2\pi)^{2\omega}} \frac{1}{(q^2 - m_z^2)} \tag{2.72}$$

and

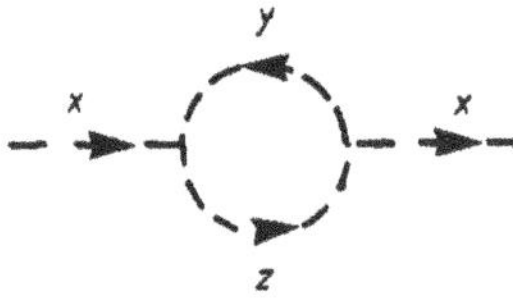

$$= \lambda^2 m_z^2 \int \frac{\mathrm{d}^{2\omega}q}{(2\pi)^{2\omega}} \frac{1}{q^2(q^2 - m_z^2)}. \tag{2.73}$$

Thus,

$$\text{sum of diagrams} = -\lambda^2 \int \frac{\mathrm{d}^{2\omega}q}{(2\pi)^d} \frac{1}{q^2} = 0 \tag{2.74}$$

using the usual prescription for this quadratically divergent integral. We therefore conclude that there is *no* contribution to the mass renormalization of m_x^2 proportional to m_z^2. Consequently, the presence of the large mass m_z in the theory does *not* induce large masses for m_x and m_y through radiative corrections, and a mass hierarchy can be preserved. This is an essentially supersymmetric effect because cancellations are occurring between the diagrams involving fermion loops and those involving boson loops, and the cancellation depends on a particular relationship between the Yukawa coupling and the φ^3 and φ^4 couplings.

In supersymmetric theories of chiral superfields there is also *no* infinite renormalization of coupling constants other than by wave function renormalizations. This can also be illustrated by considering the simple model (2.61). The one-loop diagram contributing to the renormalization of the vertex $\bar{\Psi}_x(I - \gamma_5)\Psi_y\varphi_z$ is

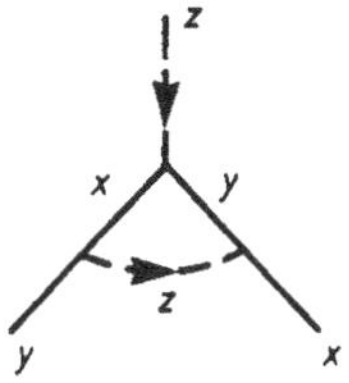

$$
\begin{aligned}
&= \left(\frac{\lambda}{2}\right)^3 \int \frac{d^{2\omega}q}{(2\pi)^{2\omega}} \frac{1}{(q^2 - m_z^2)} \\
&\quad \times (I - \gamma_5) \frac{1}{\not{q}} (I - \gamma_5) \frac{1}{\not{q}} (I + \gamma_5) \\
&= 0
\end{aligned}
\tag{2.75}
$$

because $(I - \gamma_5)(I + \gamma_5)$ is zero. In the next section we shall indicate the origin of these remarkable results.

2.8 Non-renormalization theorems

The results of §2.7 are a special case of a general non-renormalization theorem which may be stated as follows.

The superpotential (for an $N = 1$ supersymmetric theory) is *not* renormalized, except by finite amounts, in any order of perturbation theory, other than by wave function renormalizations.

The proof of this theorem depends on supergraph techniques that allow several Feynman diagrams involving different component fields belonging to the same supermultiplets to be calculated simultaneously. Although we do not discuss these techniques here (a detailed discussion may be found in, e.g., Srivastava[(5)], Wess and Bagger[(6)] or West[(7)]), we sketch the ideas involved in deriving the above theorem. In the absence of massless fields even renormalization of the superpotential by finite amounts is disallowed. However, subtleties arise in the presence of massless fields[(8)] and finite renormalizations can occur beyond one-loop level even in the massless Wess–Zumino model.

It is necessary first to write the Lagrangian as an integral over the

Grassmann variables θ and $\bar{\theta}$ (over superspace). As discussed in references 9, integration over Grassmann variables is defined such that

$$\int d\theta^\alpha = \int d\bar{\theta}^{\dot{\alpha}} = 0 \tag{2.76}$$

and

$$\int d\theta^\alpha \theta^\alpha = \int d\bar{\theta}^{\dot{\alpha}} \bar{\theta}^{\dot{\alpha}} = 1 \tag{2.77}$$

with *no* summation over α or $\dot{\alpha}$ implied. This allows an arbitrary function of θ and $\bar{\theta}$ to be integrated, because for Grassmann variables we need never consider powers higher than the first power of any component of θ or $\bar{\theta}$. Multiple integrals are interpreted as iterated integrals. Volume elements in superspace are defined by

$$d^2\theta = -\tfrac{1}{4} d\theta\, d\theta = -\tfrac{1}{4} d\theta^\alpha\, d\theta_\alpha \tag{2.78}$$

$$d^2\bar{\theta} = -\tfrac{1}{4} d\bar{\theta}\, d\bar{\theta} = -\tfrac{1}{4} d\bar{\theta}_{\dot{\alpha}}\, d\bar{\theta}^{\dot{\alpha}} \tag{2.79}$$

and

$$d^4\theta \equiv d^2\theta\, d^2\bar{\theta}. \tag{2.80}$$

It then follows from (2.76) and (2.77) that the non-zero integrals over superspace are (Exercise 2.5)

$$\int d^2\theta\, \theta^2 = \int d^2\bar{\theta}\, \bar{\theta}^2 = 1. \tag{2.81}$$

The Lagrangian (2.44) may now be written as

$$\mathcal{L} = \int d^4\theta \sum_i \Phi_i^\dagger \Phi_i + \left(\int d^2\theta\, W(\Phi) + \text{HC} \right) \tag{2.82}$$

because the superspace integrations project out D- and F-terms as a result of (2.76) and (2.81).

The non-renormalization theorem derives from the observation that in supergraph perturbation theory any radiative correction to the effective action can be written as a single superspace integration $\int d^4\theta$ over a product of quantities that are local in θ and $\bar{\theta}$ with *no* factors of superspace δ-functions. The superpotential term in (2.82) is *not* of this form because it involves only $\int d^2\theta$ and can only be written as an integration over $d^4\theta$ by introducing a superspace δ-function $\delta(\bar{\theta})$. Consequently, the superpotential undergoes *no* (direct) renormalization. On the other hand, the $\Phi_i^\dagger\Phi_i$ terms in (2.82) are renormalized, in general, resulting in wave function renormalizations. Thus, any renormalization of masses and coupling constants is due entirely to wave function renormalization.

2.9 Spontaneous supersymmetry breaking

There is clearly a need for supersymmetry to be broken in realistic models since we do not see scalar particles accompanied by fermions degenerate in mass with them, nor vice versa. In this section, we discuss how supersymmetry breaking may arise by spontaneous breakdown of symmetry in a theory with a supersymmetric Lagrangian. Let us consider first ways of recognizing when supersymmetry is spontaneously broken. The criterion for spontaneous supersymmetry breaking is that the physical vacuum state $|0\rangle$ should *not* be invariant under a general supersymmetry transformation. Equivalently, $|0\rangle$ should *not* be annihilated by all the supersymmetry generators, i.e.

$$Q_\alpha|0\rangle \neq 0 \text{ or } \bar{Q}_{\dot{\alpha}}|0\rangle \neq 0 \tag{2.83}$$

for some α, for spontaneous supersymmetry breaking. This has implications for the energy of the ground state because (2.4) can be used to relate the Hamiltonian to the supersymmetry generators. With the aid of the identity

$$\mathrm{Tr}(\sigma^\mu\bar{\sigma}^\nu) = \sigma^\mu_{\alpha\dot{\beta}}(\bar{\sigma}^\nu)^{\dot{\beta}\alpha} = 2\eta^{\mu\nu} \tag{2.84}$$

equation (2.4) can be inverted to obtain

$$P^\nu = \tfrac{1}{4}(\bar{\sigma}^\nu)^{\dot{\beta}\alpha}\{Q_\alpha, \bar{Q}_{\dot{\beta}}\} \tag{2.85}$$

as discussed in §1.3. In particular, the Hamiltonian is

$$H = P^0 = \tfrac{1}{4}(Q_1\bar{Q}_{\dot{1}} + \bar{Q}_{\dot{1}}Q_1 + Q_2\bar{Q}_{\dot{2}} + \bar{Q}_{\dot{2}}Q_2) \tag{2.86}$$

where $\bar{Q}_{\dot{\alpha}}$ is the Hermitian adjoint of Q_α. Thus, H is positive semi-definite. When supersymmetry is unbroken in the vacuum state this state has zero energy, and when supersymmetry is spontaneously broken in the vacuum state it has positive energy. As a result, whenever a supersymmetric vacuum state exists as a local minimum of the effective potential it is the global minimum. If there is more than one supersymmetric vacuum, they are all degenerate in energy with zero energy. For the global minimum of the effective potential (the physical vacuum) to be non-supersymmetric it is therefore necessary for the effective potential V to possess *no* supersymmetric minimum. Generic forms of V for unbroken supersymmetry and spontaneously broken supersymmetry are shown in figures 2.1, 2.2 and 2.3.

Another perspective on spontaneous breaking of supersymmetry is obtained by observing that the supersymmetry breaking must arise from some fields in the theory having vacuum expectation values (VEVs) that are *not* invariant under supersymmetry transformations. In a theory of chiral

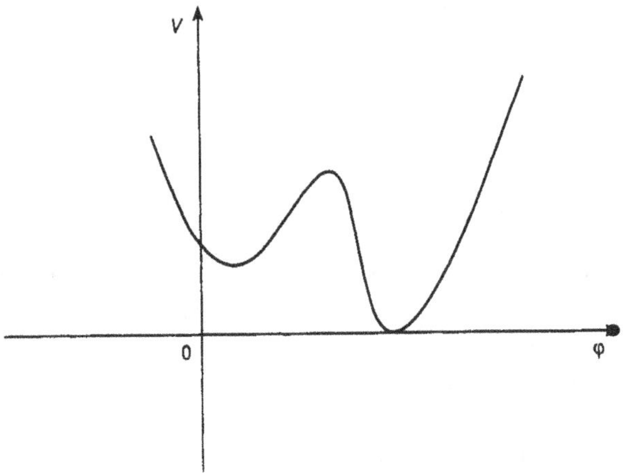

Figure 2.1 The generic effective potential V for unbroken supersymmetry, where φ is the VEV of some scalar field.

superfields, the only one of the supersymmetry transformation laws (2.35)–(2.37) whose expectation value can have a non-zero right-hand side, without breaking Lorentz invariance, is (2.36). Thus, the mechanism for spontaneous supersymmetry breaking is for

$$\langle 0|\delta\psi_i|0\rangle = \sqrt{2}\,\xi\langle 0|F_i|0\rangle \tag{2.87}$$

to be non-zero for some chiral supermultiplet Φ_i, i.e. for one of the auxiliary

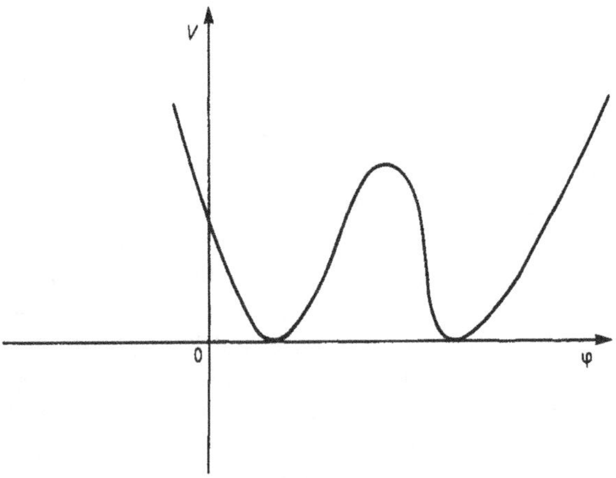

Figure 2.2 The generic effective potential V for unbroken supersymmetry with two supersymmetric vacua where φ is the VEV of some scalar field.

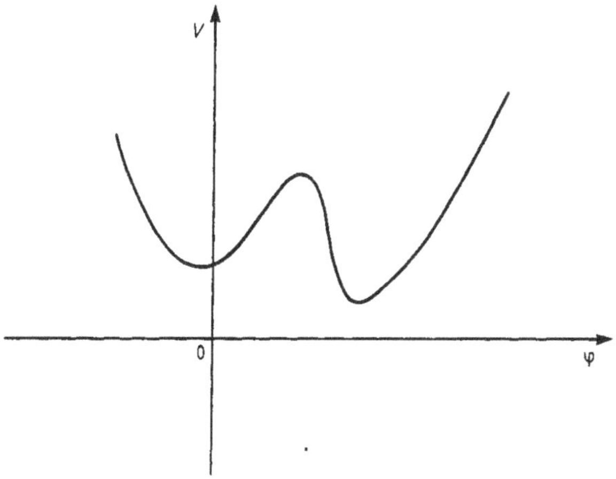

Figure 2.3 The generic effective potential V for spontaneously broken supersymmetry where φ is the VEV of some scalar field.

fields F_i to have a non-zero vacuum expectation value, where in terms of physical fields F_i is given by (2.52). Thus in theories of chiral superfields F_i is the order parameter for spontaneous supersymmetry breaking.

$$\langle 0|F_i|0\rangle \neq 0 \qquad \text{for spontaneous supersymmetry breaking.} \tag{2.88}$$

It may be seen from (2.51) that when such F-term supersymmetry breaking occurs the value of the effective potential in the vacuum state is positive. This is consistent with the general conclusion derived from (2.86). (We shall see in Chapter 3 that another mechanism for spontaneous supersymmetry breaking exists for supersymmetric gauge theories with a U(1) factor in the gauge group. This involves a vacuum expectation value for the auxiliary field D of a vector supermultiplet instead of the auxiliary field F of a chiral supermultiplet. We shall also see in Chapters 4 and 5 that other mechanisms for supersymmetry breaking are possible in supergravity theories.)

Once spontaneous supersymmetry breaking has occurred a massless Goldstone fermion is expected to appear because the supersymmetry generator is fermionic, much as a Goldstone boson appears when ordinary global symmetries are spontaneously broken. When a single auxiliary field F_i acquires a VEV the Goldstone fermion will be the spinor ψ_i in the supermultiplet Φ_i to which F_i belongs. This is in exact analogy to the Goldstone boson in a theory of spontaneously broken ordinary global symmetry being in the same multiplet as the scalar that develops a VEV. The Goldstone fermion is potentially a problem for globally supersymmetric theories. However, we shall see in Chapters 4 and 5 that when global supersymmetry becomes local supersymmetry in supergravity theories the Goldstone fermion is 'eaten' by the gravitino to give the gravitino a mass,

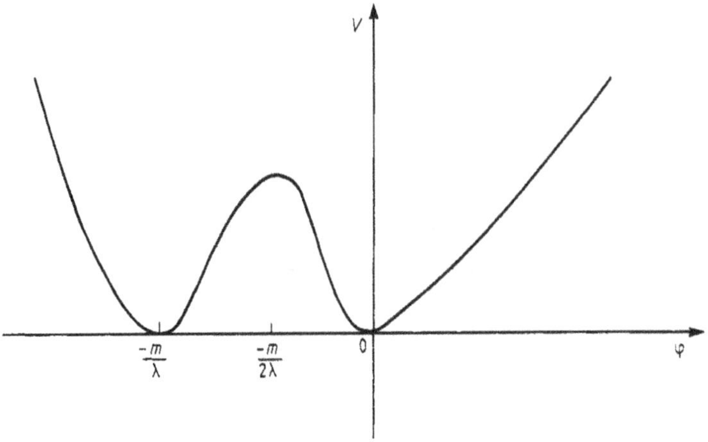

Figure 2.4 The effective potential V for the Wess–Zumino model (for $m > 0$).

just as Goldstone bosons are 'eaten' by gauge fields when an ordinary global symmetry becomes a (local) gauge symmetry.

2.10 *F*-term supersymmetry breaking

It has been observed in §2.9 that in globally supersymmetric theories of chiral superfields supersymmetry breaking occurs via the auxiliary field F_i in some chiral supermultiplet Φ_i acquiring a VEV. It has also been observed that, in a globally supersymmetric theory, whenever a supersymmetric vacuum state exists as a local minimum of the effective potential it will be the global minimum, since it has zero energy whereas all vacua in which supersymmetry is spontaneously broken have positive energy. Thus, to obtain a theory with spontaneously broken supersymmetry it is necessary to ensure that the effective potential has *no* supersymmetric minimum. This considerably restricts the possible forms of the superpotential. In particular, any superpotential of the form (2.45) that does not contain terms linear in the superfields can never produce spontaneous supersymmetry breaking because there is always at least one solution of

$$F_i^\dagger = -\frac{\partial W}{\partial \varphi_i} = 0 \qquad \text{for all } i \tag{2.89}$$

obtained by taking all expectation values φ_i equal to zero, which is a supersymmetric minimum. For instance, for the Wess–Zumino model

$$W(\Phi) = \tfrac{1}{2}m\Phi^2 + \tfrac{1}{3}\lambda\Phi^3 \tag{2.90}$$

there are two supersymmetric minima, one at $\varphi = 0$ and the other at $\varphi = -m/\lambda$. (See figure 2.4.)

The renormalizable superpotential (2.45) may be generalized to include terms linear in the superfields by writing

$$W(\Phi) = f_i\Phi_i + \tfrac{1}{2} m_{ij}\Phi_i\Phi_j + \tfrac{1}{3}\lambda_{ijk}\Phi_i\Phi_j\Phi_k . \tag{2.91}$$

Even then supersymmetric minima of the effective potential generally arise except for carefully chosen numbers of superfields and values of coefficients in (2.91). The simplest example of a model without any supersymmetric minima is the O'Raifeartaigh model which has three chiral superfields Φ_1, Φ_2 and Φ_3 with superpotential

$$W(\Phi_1, \Phi_2, \Phi_3) = \lambda_1\Phi_1(\Phi_3^2 - M^2) + \mu\Phi_2\Phi_3 . \tag{2.92}$$

For this model,

$$-F_1^{\dagger} = \frac{\partial W}{\partial \varphi_1} = \lambda_1({\varphi_3}^2 - M^2) \tag{2.93}$$

$$-F_2^{\dagger} = \frac{\partial W}{\partial \varphi_2} = \mu\varphi_3 \tag{2.94}$$

$$-F_3^{\dagger} = \frac{\partial W}{\partial \varphi_3} = 2\lambda_1\varphi_1\varphi_3 + \mu\varphi_2 \tag{2.95}$$

and there is *no* solution with F_1, F_2 and F_3 all zero. Thus, supersymmetry is spontaneously broken.

For $M^2 < \mu^2/2{\lambda_1}^2$, the absolute minimum of the effective potential

$$V = \sum_{i=1}^{3} |F_i|^2 = \lambda_1^2|\varphi_3^2 - M^2|^2 + \mu^2|\varphi_3|^2 + |\mu\varphi_2 + 2\lambda_1\varphi_1\varphi_3|^2 \tag{2.96}$$

occurs at

$$\langle\varphi_2\rangle = \langle\varphi_3\rangle = 0 \tag{2.97}$$

and $\langle\varphi_1\rangle$ is undetermined (i.e. the potential has a flat direction). At this absolute minimum,

$$F_1^{\dagger} = \lambda_1 M^2 \qquad F_2^{\dagger} = F_3^{\dagger} = 0 \tag{2.98}$$

and

$$V = \lambda_1^2 M^4 > 0. \tag{2.99}$$

Because F_1 is non-zero we expect that ψ_1 is the Goldstone fermion. This may be verified by looking at the fermion mass matrix given by the Lagrangian terms $\mathscr{L}_m^F$ as in (2.50),

$$\mathscr{L}_m^F = -\tfrac{1}{2}(m_{ij} + 2\lambda_{ijk}\langle\varphi_k\rangle)\psi_i\psi_j + \text{HC} \tag{2.100}$$

which is unaffected by the terms linear in the Φ_i in the superpotential. For the present model,

$$m_{23} = m_{32} = \mu \tag{2.101}$$

and

$$\lambda_{133} = \lambda_{313} = \lambda_{331} = \lambda_1 \tag{2.102}$$

with all other m_{ij} and λ_{ijk} equal to zero. Then,

$$\mathscr{L}_m^F = -\tfrac{1}{2}(2\mu\psi_2\psi_3 + 2\lambda_1\langle\varphi_1\rangle\psi_3^2 + \text{HC}). \tag{2.103}$$

Thus, ψ_1 is massless as expected, and there are two massive combinations of ψ_2 and ψ_3. If we assume for convenience that

$$\langle\varphi_1\rangle = 0 \tag{2.104}$$

these are two Weyl spinors of mass μ. Equivalently, defining

$$\Psi = \begin{pmatrix}\psi_2\\ \bar{\psi}_3\end{pmatrix} \tag{2.105}$$

we have

$$\mathscr{L}_m^F = -\mu\bar{\Psi}\Psi \tag{2.106}$$

corresponding to a single Dirac spinor of mass μ.

The bosonic masses may be obtained from the $|F_i|^2$ terms in the superpotential after shifting the complex scalar fields by the VEVs. Then, with the expectation values (2.97) and (2.104), the bosonic mass terms $\mathscr{L}_m^B$ are

$$\mathscr{L}_m^B = \lambda_1^2 M^2(\varphi_3^2 + (\varphi_3{}^\dagger)^2) - \mu^2(\varphi_2{}^\dagger\varphi_2 + \varphi_3{}^\dagger\varphi_3). \tag{2.107}$$

If we define real scalar fields a_3 and b_3 by

$$\varphi_3 = \frac{1}{\sqrt{2}}(a_3 + \mathrm{i}\, b_3) \tag{2.108}$$

this becomes

$$\mathscr{L}_m^B = -\tfrac{1}{2}(\mu^2 - 2\lambda_1^2M^2)a_3^2 - \tfrac{1}{2}(\mu^2 + 2\lambda_1^2M^2)b_3^2 - \mu^2\varphi_2{}^\dagger\varphi_2. \tag{2.109}$$

Thus, the complex scalar field φ_1 is massless, the complex scalar field φ_2 has mass μ, and the real scalar fields a_3 and b_3 have masses

$$m_{a_3}^2 = \mu^2 - 2\lambda_1^2M^2 \tag{2.110}$$

and

$$m_{b_3}^2 = \mu^2 + 2\lambda_1^2M^2. \tag{2.111}$$

This means that φ_1 and φ_2 are still degenerate in mass with their superpartner fermions ψ_1 and ψ_2 even in the presence of supersymmetry breaking. On the other hand, supersymmetry breaking manifests itself by a_3 and b_3 having masses that differ from the mass μ of their superpartner ψ_3. The reason for this outcome is that only the superfield Φ_3 couples to the superfield Φ_1 containing the Goldstone fermion in (2.92).

It will be noticed that although m_{a_3} and m_{b_3} differ from m_{ψ_3}, it is nonetheless the case that

$$m_{a_3}^2 + m_{b_3}^2 = 2\mu^2 = 2m_{\psi_3}^2 \tag{2.112}$$

just as in the absence of supersymmetry breaking. This is a special case of the general tree level result[10] for theories of chiral superfields

$$\mathrm{STr}\, M^2 \equiv \sum_J (-1)^{2J}(2J+1)m_J^2 = 0 \tag{2.113}$$

where STr M^2 (the supertrace) denotes the trace of the mass-squared matrix over the real fields, of spin J. In the presence of supersymmetry breaking equation (2.113) is modified by radiative corrections. It is also modified in theories containing vector superfields if D-terms develop VEVS.

The supersymmetry-breaking scale M_s^2 may be defined as the expectation value of the F-term responsible for supersymmetry breaking. Thus, in the present case,

$$M_s^2 = \lambda_1 M^2. \tag{2.114}$$

Casting (2.110) and (2.111) in terms of M_s^2,

$$m_{a_3}^2 = \mu^2 - 2\lambda_1 M_s^2 \tag{2.115}$$

and

$$m_{b_3}^2 = \mu^2 + 2\lambda_1 M_s^2. \tag{2.116}$$

Thus, the squared-mass splittings within supermultiplets resulting from supersymmetry breaking are on the scale of $\lambda_1 M_s^2$. It is possible for this scale to be much smaller than the supersymmetry-breaking scale M_s if the coupling λ_1 of the other chiral superfields to the Goldstone fermion superfield is small.

Exercises

2.1 Derive (2.6) for the product of two group elements associated with the supersymmetry algebra.

2.2 Show that P_μ, Q_α and $\bar{Q}_{\dot\alpha}$ defined by (2.11)–(2.13) realize the supersymmetry algebra.

2.3 Derive the anti-commutator (2.16) amongst fermionic covariant derivatives and generators of the supersymmetry algebra.

2.4 Show that the tree level effective potential is

$$V = |F_i|^2$$

where

$$F_i^\dagger = -\frac{\partial W}{\partial \varphi_i}$$

for *any* superpotential $W(\Phi_i)$ when the 'kinetic term' in the Lagrangian is of the minimal form $[\Phi_i^\dagger \Phi_i]_D$.

2.5 Show that

$$\int \mathrm{d}^2\theta\, \theta^2 = \int \mathrm{d}^2\bar{\theta}\, \bar{\theta}^2 = 1.$$

References

General references

Srivastava P P 1986 *Supersymmetry, Superfields and Supergravity: An Introduction* (Bristol: Institute of Physics Publishing)

Wess J and Bagger J 1983 *Supersymmetry and Supergravity* (Princeton, NJ: Princeton University Press)

West P 1986 *Introduction to Supersymmetry and Supergravity* (Singapore: World Scientific)

References in the text

1 Salam A and Strathdee J 1974 *Nucl. Phys.* B **76** 477

2 Ferrara S, Wess J and Zvaino B 1974 *Phys. Lett.* **51B** 239

3 Siegel W 1979 *Phys. Lett.* **84B** 193; 1980 *Phys. Lett.* **94B** 37

4 Capper D A, Jones D R T and Nieuwenhuizen P 1980 *Nucl. Phys.* B **167** 479

5 Srivastava P P 1986 *Supersymmetry, Superfields and Supergravity: An Introduction* (Bristol: Institute of Physics Publishing) Chapter 10

6 Wess J and Bagger J 1983 *Supersymmetry and Supergravity* (Princeton, NJ: Princeton University Press) Chapter 10

7 West P 1986 *Introduction to Supersymmetry and Supergravity* (Singapore: World Scientific) Chapter 17

8 Howe P S and West P 1989 *Phys. Lett.* **227B** 379
Jack I and Jones D R T 1991 *Phys. Lett.* **258B** 382
West P 1991 *Phys. Lett.* **258B** 375

9 Bailin D and Love A 1986 *Introduction to Gauge Field Theory* (Bristol: Institute of Physics Publishing) §3.4
Berezin F A 1966 *Method of the Second Quantization* (London: Academic)

10 Ferrara S, Girardello L and Palumbo F 1979 *Phys. Rev.* D **20** 403

3

LAGRANGIANS FOR VECTOR SUPERFIELDS

3.1 Introduction

We saw in Chapter 1 that the general (on-shell) representation of supersymmetry has (particle) states $|p, \lambda\rangle$, of momentum p and helicity λ, together with the ('sparticle') states $Q_{\dot{2}}|p, \lambda\rangle \sim |p, \lambda - \frac{1}{2}\rangle$, having the same momentum but half a unit less helicity. Thus the simplest realization involves a (Weyl) spinor field $\psi_\alpha(x)$ and a scalar field $\varphi(x)$, so the known fermion fields (quarks and leptons) have (yet to be discovered) scalar superpartners (the squarks and sleptons). We have seen also that to extend this to an off-shell realization it was necessary to introduce an auxiliary field $F(x)$, and that the three fields may be elegantly viewed as the component fields of a 'chiral' superfield $\Phi(y, \theta)$. Any renormalizable supersymmetric field theory involving just scalar and spinor fields can then be formulated most succinctly in terms of products of chiral superfields, as was shown in Chapter 2. However, the quarks and leptons, and therefore the squarks and sleptons, all participate in gauge theories, quantum chromodynamics and electroweak theory, which of course involve gauge vector bosons. Thus to formulate a supersymmetric version of quantum electrodynamics, for example, we shall certainly have to involve the vector supermultiplet, which involves the photon and its spin-$\frac{1}{2}$ partner the photino, and we might reasonably expect that a superfield description of these should also exist.

In the following section we shall construct the vector superfield; in general it includes four auxiliary scalar fields as well as an auxiliary fermion field. However, we can utilize the gauge invariance of the theories in which this superfield is deployed to eliminate all except one of these fields, the so-called D-field. This is shown in §3.3. In fact this D-term provides another way to break supersymmetry, besides the F-term method described in §2.9. This method is described in §3.5. In order to construct gauge-invariant kinetic terms for the vector potential $A_\mu(x)$ in quantum electrodynamics, it is of course essential to introduce the electromagnetic field strength tensor $F_{\mu\nu}(x)$. To make this theory supersymmetric, therefore, we must also construct a field strength superfield, as well as the U(1) gauge-invariant and supersymmetric Lagrangian. This too is done in §3.3. In §3.4 we investigate spontaneous symmetry breaking of a global and local U(1) gauge invariance in the context of a supersymmetric theory. Then besides the Goldstone boson which arises when the global symmetry is broken, the (unbroken)

DOI: 10.1201/9780367805807-3

supersymmetry ensures that there is a massless Goldstone fermion. Similarly, when the local symmetry is broken, besides the massive gauge boson the supersymmetry ensures the existence of a massive gaugino. The generalization of these ideas to non-abelian gauge theories is discussed in §3.7 and illustrated in §3.8, in which we construct a supersymmetric version of electroweak theory. Finally, in §3.8, we derive the renormalization group equations for the gauge coupling constants in a general supersymmetry gauge theory. Applying this to the standard $SU(3) \times SU(2) \times U(1)$ model we find that the gauge coupling constants achieve a common value at a unification scale of under 10^{16} GeV, thereby supplying the best (circumstantial) evidence to date for supersymmetry (and grand unification).

3.2 The vector superfield

The vector supermultiplet includes a (massless) vector particle (with helicity eigenstates ± 1) together with the fermionic gaugino (having helicity eigenstates $\pm\frac{1}{2}$). We therefore seek a 'vector' superfield involving a real gauge field $V_\mu(x)$ and its fermionic partner $\lambda_\alpha(x)$. It suffices to start with a Lorentz *invariant* superfield $F(x, \theta, \bar{\theta})$. This may be expanded in powers of θ, $\bar{\theta}$ up to and including quadratic terms in both θ and $\bar{\theta}$. (Any cubic or higher powers necessarily vanish because of the anti-commutation properties.) Using just these fermionic coordinates, we saw in §1.2 that we can construct the Lorentz scalar quantities $\theta\theta$ and $\bar{\theta}\bar{\theta}$, and the vector $\theta\sigma^\mu\bar{\theta} = -\bar{\theta}\sigma^\mu\theta$. The only candidate tensor quantities $\theta\sigma^{\mu\nu}\theta$ and $\bar{\theta}\bar{\sigma}^{\mu\nu}\bar{\theta}$ vanish identically. Thus without loss of generality we may write any Lorentz invariant superfield in the form

$$F(x, \theta, \bar{\theta}) = f(x) + \theta\varphi(x) + \bar{\theta}\bar{\chi}(x) + \theta\theta m(x) + \bar{\theta}\bar{\theta}n(x)$$
$$+ \theta\sigma^\mu\bar{\theta}V_\mu(x) + \theta\theta\bar{\theta}\bar{\lambda}(x) + \bar{\theta}\bar{\theta}\theta\psi(x) + \theta\theta\bar{\theta}\bar{\theta}d(x) \qquad (3.1)$$

where f, m, n, d are scalar fields, V_μ is a vector field, and $\varphi, \psi, \bar{\chi}, \bar{\lambda}$ are Weyl spinor fields. If we require that $F(x, \theta, \bar{\theta})$ is real:

$$F(x, \theta, \bar{\theta}) = F(x, \theta, \bar{\theta})^\dagger \qquad (3.2)$$

then the properties derived in (1.78), (1.80) give

$$f = f^* \qquad V_\mu = V_\mu{}^* \qquad d = d^* \qquad (3.3a)$$

$$m^* = n \qquad \varphi = \chi \qquad \lambda = \psi. \qquad (3.3b)$$

Then the superfield F has two (Weyl) spinors each with two complex components, giving a total of eight real fermionic degrees of freedom. These are matched by the total of eight (real) bosonic degrees of freedom comprised by $f(1)$, $d(1)$, $m(2)$, $V_\mu(4)$. It is convenient to rewrite this superfield using special field combinations for the coefficients of the $\theta\theta\bar{\theta}$, $\bar{\theta}\theta\theta$ and $\theta\theta\bar{\theta}\bar{\theta}$ components of F. Instead of (3.1) we write

$$\begin{aligned} V(x, \theta, \bar{\theta}) = {} & C(x) + \mathrm{i}\,\theta\chi(x) - \mathrm{i}\,\bar{\theta}\bar{\chi}(x) + \tfrac{1}{2}\mathrm{i}\,\theta\theta[M(x) + \mathrm{i}\,N(x)] \\ & - \tfrac{1}{2}\mathrm{i}\,\bar{\theta}\bar{\theta}[M(x) - \mathrm{i}\,N(x)] + \theta\sigma^{\mu}\bar{\theta}V_{\mu}(x) \\ & + \mathrm{i}\,\theta\theta\bar{\theta}\left[\bar{\lambda}(x) + \frac{\mathrm{i}}{2}\bar{\sigma}^{\mu}\,\partial_{\mu}\chi(x)\right] - \mathrm{i}\,\bar{\theta}\bar{\theta}\theta\left[\lambda(x) + \frac{\mathrm{i}}{2}\sigma^{\mu}\,\partial_{\mu}\bar{\chi}(x)\right] \\ & + \tfrac{1}{2}\theta\theta\bar{\theta}\bar{\theta}[D - \tfrac{1}{2}\partial_{\mu}\,\partial^{\mu}C] \end{aligned} \tag{3.4}$$

where C, M, N, D are real scalar fields. As before, χ, λ are Weyl spinor fields and V_{μ} is a (real) vector field. There is, of course, no loss of generality in using the form (3.4), rather than (3.1), since the extra terms in the coefficients merely use (correctly Lorentz transforming) fields constructed from those used elsewhere. However, there is also no immediately apparent advantage. The advantage becomes clear when we derive the transformation properties of the component fields under a supersymmetry transformation. (We shall find that the fields $V_{\mu\nu} \equiv \partial_{\mu}V_{\nu} - \partial_{\nu}V_{\mu}$, λ, $\bar{\lambda}$, D form a representation of the supersymmetry algebra by themselves.) The reason for this is that with the form (3.4) the components can all be written as covariant derivatives of the superfield V, evaluated at $\theta = \bar{\theta} = 0$. For example, it is obvious using the definitions (2.14) and (2.15) that

$$V| = C \tag{3.5}$$

$$D_{\alpha}V| = \mathrm{i}\,\chi_{\alpha} \qquad \bar{D}_{\dot{\alpha}}V| = -\,\mathrm{i}\,\bar{\chi}_{\dot{\alpha}} \tag{3.6}$$

where the vertical line signifies evaluation at $\theta = \bar{\theta} = 0$. With a little more work, and the use of the results of §1.3, we can also see that

$$D^2\bar{D}_{\dot{\alpha}}V| = -\,4\mathrm{i}\,\bar{\lambda}_{\dot{\alpha}} \qquad \bar{D}^2 = D_{\alpha}V| = 4\mathrm{i}\,\lambda_{\alpha} \tag{3.7}$$

where we use the notation

$$D^2 \equiv D^{\alpha}D_{\alpha} \qquad \bar{D}^2 = \bar{D}_{\dot{\alpha}}\bar{D}^{\dot{\alpha}} \tag{3.8}$$

introduced in (1.75), and

$$D^{\alpha} \equiv \epsilon^{\alpha\beta}D_{\beta} = -\frac{\partial}{\partial\theta_{\alpha}} + \mathrm{i}\,\epsilon^{\alpha\beta}\sigma^{\mu}_{\beta\dot{\beta}}\bar{\theta}^{\dot{\beta}}\,\partial_{\mu} \tag{3.9a}$$

$$\bar{D}^{\dot{\alpha}} \equiv \epsilon^{\dot{\alpha}\dot{\beta}}\bar{D}_{\dot{\beta}} = \frac{\partial}{\partial\bar{\theta}_{\dot{\alpha}}} - \mathrm{i}\,\epsilon^{\dot{\alpha}\dot{\beta}}\theta^{\beta}\sigma^{\mu}_{\beta\dot{\beta}}\,\partial_{\mu}. \tag{3.9b}$$

Similarly we find

$$[D_{\alpha}, \bar{D}_{\dot{\alpha}}]V| = 2\sigma^{\mu}_{\alpha\dot{\alpha}}V_{\mu} \tag{3.10}$$

$$D^\beta \bar{D}^2 D_\alpha V| = 4\delta^\beta_\alpha D + 2\mathrm{i}(\sigma^\mu\bar{\sigma}^\nu)_\alpha{}^\beta V_{\mu\nu} \tag{3.11}$$

where

$$V_{\mu\nu} \equiv \partial_\mu V_\nu - \partial_\nu V_\mu \tag{3.12}$$

is the (U(1) gauge-invariant) field strength. It follows immediately from (3.11) that

$$D^\alpha \bar{D}^2 D_\alpha V| = 8D. \tag{3.13}$$

It is now straightforward to determine the transformation properties of the components under an (infinitesimal) supersymmetry transformation. We use

$$\delta_\xi V = \mathrm{i}(\xi Q + \bar{\xi}\bar{Q})V \tag{3.14}$$

with Q, $\bar{Q}$ given by (2.12), (2.13). For example

$$\delta C = \mathrm{i}(\xi Q + \bar{\xi}\bar{Q})V| = (\xi D + \bar{\xi}\bar{D})V| = \mathrm{i}(\xi\chi - \bar{\xi}\bar{\chi}) \tag{3.15}$$

using (3.6). In the same way we find

$$\delta\lambda_\alpha = -\frac{1}{4\mathrm{i}}(\xi D + \bar{\xi}\bar{D})\bar{D}^2 D_\alpha V| = -\mathrm{i}\, D\xi_\alpha - \tfrac{1}{2}(\sigma^\mu\bar{\sigma}^\nu)_\alpha{}^\beta \xi_\beta V_{\mu\nu} \tag{3.16}$$

$$\delta V^\mu = \mathrm{i}(\xi\sigma^\mu\bar{\lambda} - \lambda\sigma^\mu\bar{\xi}) - \partial^\mu(\xi\chi + \bar{\xi}\bar{\chi}) \tag{3.17}$$

$$\delta D = \partial_\mu(-\xi\sigma^\mu\bar{\lambda} + \lambda\sigma^\mu\bar{\xi}). \tag{3.18}$$

Then it follows from (3.17) that the transformation property of the field strength $V_{\mu\nu}$ is given by

$$\delta V^{\mu\nu} = \mathrm{i}\,\partial^\mu(\xi\sigma^\nu\bar{\lambda} - \lambda\sigma^\nu\bar{\xi}) - \mathrm{i}\,\partial^\nu(\xi\sigma^\mu\bar{\lambda} - \lambda\sigma^\mu\bar{\xi}). \tag{3.19}$$

Thus, as claimed, λ, $\bar{\lambda}$, $V_{\mu\nu}$ and D form an (irreducible) representation of the supersymmetry algebra, by themselves. Note also that the variation of the D-field is a total divergence. The dimensions of the fields in the 'vector supermultiplet' are fixed by requiring that the vector field V_μ has its canonical dimension

$$[V] = 1. \tag{3.20}$$

Then since D_α, like Q_α, has dimension $\frac{1}{2}$ it follows that

$$[\theta] = [\bar{\theta}] = -\tfrac{1}{2} \tag{3.21}$$

and then

$$[C] = 0 \tag{3.22a}$$

$$[\chi] = \tfrac{1}{2} = [\bar{\chi}] \tag{3.22b}$$

$$[M] = 1 = [N] \tag{3.22c}$$

$$[\lambda] = \tfrac{3}{2} = [\bar{\lambda}] \tag{3.22d}$$

$$[D] = 2. \tag{3.22e}$$

So, as noted in Chapter 1, the highest-dimension field (D) in the supermultiplet must transform as a total divergence. Further, since $\partial_\mu \partial^\mu C$ also transforms as a total divergence, we see that $D - \frac{1}{2}\partial_\mu \partial^\mu C$ does too, and so therefore does the entire coefficient of $\theta\theta\bar{\theta}\bar{\theta}$ in $V(x, \theta, \bar{\theta})$. This is the justification of the claim in §2.5 that the variation of the D-term in a vector superfield is a total divergence.

Since the only requirement (3.2) for a vector superfield is that it be real, it is easy to construct a particular example of one using the chiral superfield Φ and the anti-chiral superfield $\Phi^\dagger$ given in (2.27), (2.28). For instance

$$\begin{aligned} \mathrm{i}(\Phi - \Phi^\dagger) = {} & \mathrm{i}(\varphi - \varphi^\dagger) + \mathrm{i}\sqrt{2}(\theta\psi - \bar{\theta}\bar{\psi}) + \mathrm{i}\,\theta\theta F - \mathrm{i}\,\bar{\theta}\bar{\theta}F^\dagger \\ & - \theta\sigma^\mu\bar{\theta}\,\partial_\mu(\varphi + \varphi^\dagger) - \frac{1}{\sqrt{2}}\theta\theta\bar{\theta}\bar{\sigma}^\mu\,\partial_\mu\psi + \frac{1}{\sqrt{2}}\bar{\theta}\bar{\theta}\theta\sigma^\mu\,\partial_\mu\bar{\psi} \\ & - \tfrac{1}{4}\mathrm{i}\,\theta\theta\bar{\theta}\bar{\theta}\,\partial_\mu\partial^\mu(\varphi - \varphi^\dagger) \end{aligned} \tag{3.23}$$

has the form (3.4) with

$$C = \mathrm{i}(\varphi - \varphi^\dagger) \tag{3.24a}$$

$$\chi = \sqrt{2}\,\psi \tag{3.24b}$$

$$\tfrac{1}{2}(M + \mathrm{i}N) = F \tag{3.24c}$$

$$V_\mu = -\,\partial_\mu(\varphi + \varphi^\dagger) \tag{3.24d}$$

$$\lambda = 0 \tag{3.24e}$$

$$D = 0. \tag{3.24f}$$

Of course, for this identification to work the dimension of the fields φ, ψ, F must be shifted by one unit from the canonical dimensions (1.175), (1.179), (1.194) which they are assigned in order to make the usual identification with quarks, leptons etc. Nevertheless the force of the observation (3.24) becomes clear when we note that the vector potential V_μ for the superfield $\mathrm{i}(\Phi - \Phi^\dagger)$ is a pure U(1) gauge transformation, and this suggests how to make a supersymmetric generalization of gauge invariance.

3.3 Supersymmetric gauge invariance

We start with the familiar local U(1) gauge invariance (of QED). Under such a gauge transformation the vector potential transforms as

$$V_\mu(x) \to V_\mu{}'(x) = V_\mu(x) + \partial_\mu\Lambda(x) \tag{3.25}$$

where $\Lambda(x)$ is a 'gauge function'. The discussion at the end of §3.2 suggests an immediate way to supersymmetrize the transformation (3.25). Since V_μ is a component of the vector superfield (3.4), and $\partial_\mu\Lambda \equiv \partial_\mu(\varphi + \varphi^\dagger)$ is in $\mathrm{i}(\Phi - \Phi^\dagger)$, Wess and Zumino[1] suggested that the superfield transforms as

$$V(x, \theta, \bar{\theta}) \to V'(x, \theta, \bar{\theta}) = V(x, \theta, \bar{\theta}) + \mathrm{i}[\Phi(x, \theta, \bar{\theta}) - \Phi^\dagger(x, \theta, \bar{\theta})] \quad (3.26)$$

under a U(1) gauge transformation. In fact, it is clear from (3.24) that in a gauge theory the fields C, χ, M, N are not physical degrees of freedom, since they can be 'gauged away' by a suitable choice of $\varphi - \varphi^\dagger, \psi, F$ while still leaving $\Lambda = \varphi + \varphi^\dagger$ arbitrary. Then in the 'Wess–Zumino gauge' the vector superfield is

$$V_{\mathrm{WZ}}(x, \theta, \bar{\theta}) = \theta\sigma^\mu\bar{\theta}V_\mu(x) + \mathrm{i}\,\theta\theta\bar{\theta}\bar{\lambda}(x) - \mathrm{i}\,\bar{\theta}\bar{\theta}\theta\lambda(x) + \tfrac{1}{2}\theta\theta\bar{\theta}\bar{\theta}D(x) \quad (3.27)$$

and from (3.24) the fields λ, $\bar{\lambda}$, D are gauge invariant while V_μ transforms as in (3.25). Note that in the Wess–Zumino gauge the field D, which from (3.18) transforms as a total divergence, *is* the coefficient of $\theta\theta\bar{\theta}\bar{\theta}$. Also all powers V^n_{WZ} with $n > 2$ vanish, since they will involve at least θ^3.

The only non-zero power is

$$V^2_{\mathrm{WZ}}(x, \theta, \bar{\theta}) = -(\theta\sigma^\mu\bar{\theta})(\bar{\theta}\bar{\sigma}^\nu\theta)V_\mu V_\nu = \tfrac{1}{2}\theta\theta\bar{\theta}\bar{\theta}V^\mu V_\mu \quad (3.28)$$

using (1.74*a*) and (A7). Such a term supplies a mass for the vector field, and thereby breaks the gauge invariance. Since the massive vector theory is not gauge invariant, the degrees of freedom C, χ, M, N *are* physical and cannot be gauged away. In fact, as is clear from their dimensionality (3.22), the field C supplies the longitudinal mode of the vector field, while $\chi, \bar{\chi}$ supply the extra degrees of freedom for the massive gaugino field.

To construct a supersymmetric gauge field theory we need first to construct the field strength superfield, and secondly to couple the vector superfield to the charged (chiral) matter superfield in a gauge-invariant way. We have already observed that the fields $\lambda, \bar{\lambda}, V_{\mu\nu}$ and D form an irreducible representation of the supersymmetry algebra, and that all of these fields are gauge invariant. This suggests that the field strength superfield is a spinor (chiral) superfield, since the lowest-dimension field is λ_α with $[\lambda_\alpha] = \frac{3}{2} = [\bar{\lambda}]$ while $[V_{\mu\nu}] = 2 = [D]$. It is easy to construct the required superfield W_α using covariant derivatives. Let

$$W_\alpha \equiv \bar{D}^2 D_\alpha V. \quad (3.29)$$

Then from (3.7)

$$W_\alpha| = 4\mathrm{i}\,\lambda_\alpha \quad (3.30)$$

and we see that the lowest-dimension field is λ, as required. Also, it follows from (3.29) that

$$\bar{D}_{\dot{\beta}}W_\alpha = 0 \quad (3.31)$$

since

$$\bar{D}_{\dot{\beta}}\bar{D}_{\dot{\gamma}}\bar{D}_{\dot{\delta}} = 0 \tag{3.32}$$

automatically. Thus W_α *is* a chiral superfield satisfying the constraint (2.20), which means that it has the general form

$$W_\alpha(y, \theta) = 4\mathrm{i}\,\lambda_\alpha(y) + \theta^\beta \varphi_{\alpha\beta}(y) + \theta\theta F_\alpha(y) \tag{3.33a}$$

as in (2.25), with

$$y^\mu = x^\mu + \mathrm{i}\,\theta\sigma^\mu\bar{\theta} \tag{3.33b}$$

but now $\varphi_{\alpha\beta}$ is a bosonic field and F_α a spinor field. It follows from (3.32), (3.28) and (3.11) that

$$D_\beta W_\alpha| = \varphi_{\alpha\beta} = D_\beta \bar{D}^2 D_\alpha V| = \epsilon_{\beta\gamma}[4\delta^\alpha_\alpha D + 2\mathrm{i}(\sigma^\mu\bar{\sigma}^\nu)_\alpha{}^\gamma V_{\mu\nu}]. \tag{3.34}$$

Also, using (2.17) and (3.7)

$$\begin{aligned} D^2 W_\alpha| &= -4F_\alpha = D^2\bar{D}^2 D_\alpha V| = D^2[\bar{D}^2, D_\alpha]V| \\ &= D^2(\bar{D}_{\dot{\beta}}\{D_\alpha, \bar{D}^{\dot{\beta}}\} - \{D_\alpha, \bar{D}_{\dot{\beta}}\}\bar{D}^{\dot{\beta}})V| \\ &= -\,4\mathrm{i}\,\sigma^\mu_{\alpha\dot{\beta}}\,\partial_\mu D^2\bar{D}^{\dot{\beta}} V| = -\,16\sigma^\mu_{\alpha\dot{\alpha}}\,\partial_\mu \bar{\lambda}^{\dot{\alpha}}. \end{aligned} \tag{3.35}$$

So substituting into (3.33*a*) gives the field strength superfield

$$\begin{aligned} W_\alpha(y, \theta) = {} & 4\mathrm{i}\,\lambda_\alpha(y) - [4\delta_\alpha{}^\beta D(y) + 2\mathrm{i}(\sigma^\mu\bar{\sigma}^\nu)_\alpha{}^\beta V_{\mu\nu}(y)]\theta_\beta \\ & + 4\theta^2\sigma^\mu_{\alpha\dot{\alpha}}\,\partial_\mu\bar{\lambda}^{\dot{\alpha}} \end{aligned} \tag{3.36}$$

with y given by (3.33*b*). To construct the (gauge-invariant) supersymmetric pure gauge theory we want the F-component of $W^\alpha W_\alpha$, since, as shown in Chapter 2, this transforms as a total divergence under supersymmetry transformations and therefore yields an invariant action. A simple calculation yields

$$\tfrac{1}{32}(W^\alpha W_\alpha)_F = -\,\tfrac{1}{4}V^{\mu\nu}V_{\mu\nu} + \mathrm{i}\,\lambda\sigma^\mu\,\partial_\mu\bar{\lambda} - \tfrac{1}{4}V^{\mu\nu}({}^*V_{\mu\nu}) + \tfrac{1}{2}D^2 \tag{3.37}$$

where

$${}^*V_{\mu\nu} \equiv \frac{\mathrm{i}}{2}\,\epsilon_{\mu\nu\rho\sigma}V^{\rho\sigma} \tag{3.38}$$

is the dual field strength tensor. We can use (3.37) as the supersymmetric generalization of the familiar kinetic terms $-\tfrac{1}{4}V_{\mu\nu}V^{\mu\nu}$ of the U(1) gauge field, since the term involving ${}^*V_{\mu\nu}$ is a total divergence and so does not affect the equations of motion. The D-field is an auxiliary field which can be eliminated using the equations of motion. The gaugino contribution can be rewritten in terms of the (four-component) Majorana spinor

$$\Lambda_{\mathrm{M}} = \begin{pmatrix} \lambda_\alpha \\ \bar{\lambda}^{\dot{\alpha}} \end{pmatrix} \tag{3.39a}$$

$$\mathrm{i}\,\lambda\sigma^\mu\,\partial_\mu\bar{\lambda} = \tfrac{1}{2}\bar{\Lambda}_{\mathrm{M}}\gamma^\mu\,\partial_\mu\Lambda_{\mathrm{M}} \tag{3.39b}$$

(dropping a total divergence again).

To go beyond a pure gauge theory we also require a supersymmetric version of the interaction of the gauge field with (charged) matter. The conventional wisdom is that the known matter fields (quarks, leptons, Higgs (?)) are all described by chiral superfields, whose properties were discussed in Chapter 2.

To describe a charged massive field, such as an electron, we must include both its left and right chiral components (and of course the left and right chiral components of the anti-particle). Thus for a massive charged superfield we need to employ two (left) chiral superfields, Φ_1 and Φ_2. Then the complex chiral superfield

$$S = \frac{1}{\sqrt{2}}(\Phi_1 + \mathrm{i}\,\Phi_2) \tag{3.40}$$

transforms under a U(1) gauge transformation as

$$S \rightarrow S' = \exp(-2\mathrm{i}\,q\Lambda)S \tag{3.41}$$

where to avoid confusion now (and henceforth) we denote the scalar chiral superfield associated with the gauge transformation by Λ (rather than Φ). The reason for the factor 2 in the exponent will become apparent later. Then

$$S^\dagger = \frac{1}{\sqrt{2}}(\Phi_1{}^\dagger - \mathrm{i}\,\Phi_2{}^\dagger) \tag{3.42}$$

transforms as

$$S^\dagger \rightarrow S^{\dagger\prime} = S^\dagger \exp(2\mathrm{i}\,q\Lambda^\dagger). \tag{3.43}$$

Now consider the combination $S^\dagger\exp(2qV)S$ with V the vector superfield (3.4) which transforms as

$$V \rightarrow V' = V + \mathrm{i}(\Lambda - \Lambda^\dagger) \tag{3.44}$$

under a U(1) gauge transformation, as in (3.26). Then it is easy to demonstrate that the quoted combination is U(1) gauge invariant, since

$$\begin{aligned} S^{\dagger\prime}\exp(2qV')S' &= S^\dagger\exp(2\mathrm{i}\,q\Lambda^\dagger)\exp[2qV + 2\mathrm{i}\,q(\Lambda - \Lambda^\dagger)] \\ &\quad\times\exp(-2\mathrm{i}\,q\Lambda)S = S^\dagger\exp(2qV)S. \end{aligned} \tag{3.45}$$

In the same way we can show that if we define

$$T \equiv \frac{1}{\sqrt{2}}(\Phi_1 - \mathrm{i}\,\Phi_2) \tag{3.46}$$

then

$$T \to T' = \exp(2\mathrm{i}\,q\Lambda)T \tag{3.47}$$

and the combination

$$T^\dagger \exp(-2qV)T = T^{\dagger\prime} \exp(-2qV')T' \tag{3.48}$$

is also gauge invariant. Both (3.45) and (3.48) are real superfields, since V is, so the D-terms of each yield a supersymmetry-invariant action. As in Chapter 2, we construct mass terms using just left (or just right) chiral superfields. From (3.41), (3.47) ST, and therefore also $S^\dagger T^\dagger$, is U(1) gauge invariant, and the required mass term is given by the F-term of $m(ST + S^\dagger T^\dagger)$. Putting all this together yields the Lagrangian for the supersymmetric U(1) gauge-invariant theory

$$\mathscr{L} = \tfrac{1}{32}(W^\alpha W_\alpha)_F + (S^\dagger \mathrm{e}^{2qV} S + T^\dagger \mathrm{e}^{-2qV} T)_D + m(ST + S^\dagger T^\dagger)_F. \tag{3.49}$$

In the Wess–Zumino gauge (3.27), the exponential

$$\mathrm{e}^{2qV_{\mathrm{WZ}}} = 1 + 2qV_{\mathrm{WZ}} + 2q^2 V_{\mathrm{WZ}}^2 \tag{3.50}$$

since $V_{\mathrm{WZ}}^n = 0, n > 2$, as already noted. The leading term of the exponential contributes

$$(S^\dagger S + T^\dagger T)_D = (\Phi_1{}^\dagger \Phi_1 + \Phi_2{}^\dagger \Phi_2)_D \tag{3.51}$$

and the mass terms

$$m(ST + S^\dagger T^\dagger)_F = \tfrac{1}{2} m(\Phi_1{}^2 + \Phi_2{}^2)_F + \mathrm{HC} \tag{3.52}$$

just as in (2.44), (2.45). The appearance of interaction terms proportional to q and q^2 is also to be expected since in a supersymmetric theory there must also appear interactions of the gauge field with the (charged) scalar particles (squarks, sleptons) which are the supersymmetric partners of the known matter fields.

It is straightforward in principle, although tedious in practice, to express (3.49) in terms of the component fields of superfields S, T, V. We write S in the form (2.27) involving (φ_S, ψ_S, F_S), T in the same way but involving (φ_T, ψ_T, F_T), and V in the Wess–Zumino gauge (3.27). Then

$$\begin{aligned}(S^\dagger \mathrm{e}^{2qV} S)_D = {} & (D_\mu \varphi_S)^\dagger (D^\mu \varphi_S) + \mathrm{i}\,\psi_S \sigma^\mu D_\mu{}^\dagger \bar{\psi}_S \\ & + F_S{}^\dagger F_S + \mathrm{i}\sqrt{2} q(\varphi_S{}^\dagger \psi_S \lambda - \varphi_S \bar{\psi}_S \bar{\lambda}) + q\varphi_S{}^\dagger \varphi_S D \end{aligned} \tag{3.53a}$$

where

$$D^\mu \equiv \partial^\mu + \mathrm{i}\,qV^\mu. \tag{3.53b}$$

So using the earlier results (2.46), (3.37) we find

$$\begin{aligned}\mathcal{L} = (D_\mu\varphi_S)^\dagger(D^\mu\varphi_S) &+ (D_\mu\varphi_T^\dagger)(D^\mu\varphi_T^\dagger)^\dagger + \mathrm{i}\,\psi_S\sigma^\mu\, D_\mu^{\ \dagger}\bar{\psi}_S \\ &+ \mathrm{i}\,\psi_T\sigma^\mu\, D_\mu\bar{\psi}_T + F_S^\dagger F_S + F_T^\dagger F_T + \mathrm{i}\sqrt{2}q(\varphi_S^\dagger\psi_S - \varphi_T^\dagger\psi_T)\lambda \\ &+ \mathrm{i}\sqrt{2}q(\varphi_T\bar{\psi}_T - \varphi_S\bar{\psi}_S)\bar{\lambda} + q(\varphi_S^\dagger\varphi_S - \varphi_T^\dagger\varphi_T)D \\ &+ m(\varphi_S F_T + \varphi_T F_S + \varphi_S^\dagger F_T^\dagger + \varphi_T^\dagger F_S^\dagger - \psi_S\psi_T - \bar{\psi}_S\bar{\psi}_T) \\ &- \tfrac{1}{4}V_{\mu\nu}V^{\mu\nu} + \mathrm{i}\,\lambda\sigma^\mu\,\partial_\mu\bar{\lambda} + \tfrac{1}{2}D^2. \end{aligned} \tag{3.54}$$

The fields F_S, F_T, D are auxiliary, since their derivatives do not occur. Using their field equations

$$F_S + m\varphi_T^\dagger = 0 = F_T + m\varphi_S^\dagger \tag{3.55a}$$

$$D + q(\varphi_S^\dagger\varphi_S - \varphi_T^\dagger\varphi_T) = 0 \tag{3.55b}$$

we may eliminate them and obtain

$$\begin{aligned}\mathcal{L} = (D_\mu\varphi_S)^\dagger(D^\mu\varphi_S) &- m^2\varphi_S^\dagger\varphi_S + (D_\mu^{\ \dagger}\varphi_T)(D^\mu\varphi_T^\dagger) - m^2\varphi_T^\dagger\varphi_T \\ &+ \mathrm{i}\,\psi_S\sigma^\mu\, D_\mu^{\ \dagger}\bar{\psi}_S + \mathrm{i}\,\psi_T\sigma^\mu\, D_\mu\bar{\psi}_T - m\psi_S\psi_T - m\bar{\psi}_S\bar{\psi}_T \\ &+ \mathrm{i}\sqrt{2}q(\varphi_S^\dagger\psi_S - \varphi_T^\dagger\psi_T)\lambda + \mathrm{i}\sqrt{2}q(\varphi_T\bar{\psi}_T - \varphi_S\bar{\psi}_S)\bar{\lambda} \\ &- \tfrac{1}{2}q^2(\varphi_S^\dagger\varphi_S - \varphi_T^\dagger\varphi_T)^2 - \tfrac{1}{4}V_{\mu\nu}V^{\mu\nu} + \mathrm{i}\,\lambda\sigma^\mu\,\partial_\mu\bar{\lambda}. \end{aligned} \tag{3.56}$$

Finally we combine the Weyl spinors ψ_S, $\bar{\psi}_T$ to construct a Dirac spinor, as in (1.83). We take

$$\Psi \equiv \begin{pmatrix}\psi_{S\alpha} \\ \bar{\psi}_T^{\ \dot{\alpha}}\end{pmatrix} \tag{3.57}$$

and use the Majorana spinor Λ_M defined in (3.39). Then

$$\begin{aligned}\mathcal{L} = \mathrm{i}\,\bar{\Psi}\gamma^\mu\, D_\mu\Psi &- m\bar{\Psi}\Psi + (D_\mu\varphi_S)^\dagger(D^\mu\varphi_S) - m^2\varphi_S^\dagger\varphi_S \\ &+ (D_\mu\varphi_T^\dagger)^\dagger(D^\mu\varphi_T^\dagger) - m^2\varphi_T^\dagger\varphi_T - \tfrac{1}{2}q^2(\varphi_S^\dagger\varphi_S - \varphi_T^\dagger\varphi_T)^2 \\ &+ \frac{q}{\sqrt{2}}[\bar{\Lambda}_\mathrm{M}\Psi\mathrm{i}(\varphi_S^\dagger + \varphi_T) + \bar{\Lambda}_\mathrm{M}\,\mathrm{i}\,\gamma_5\Psi(\varphi_T - \varphi_S^\dagger) \\ &- \bar{\Psi}\Lambda_\mathrm{M}\mathrm{i}(\varphi_S + \varphi_T^\dagger) + \bar{\Psi}\,\mathrm{i}\,\gamma_5\Lambda_\mathrm{M}(\varphi_T^\dagger - \varphi_S)] - \tfrac{1}{4}V_{\mu\nu}V^{\mu\nu} \\ &+ \frac{\mathrm{i}}{2}\bar{\Lambda}_\mathrm{M}\gamma^\mu\,\partial_\mu\Lambda_\mathrm{M}. \end{aligned} \tag{3.58}$$

Comparing with the gauge-invariant U(1) field theory described in Chapter 9 of Bailin and Love[2], for example, we see that the Dirac field Ψ has charge q, as do φ_S, $\varphi_T^\dagger$. Besides the usual minimal couplings of the charged fields

Ψ, φ_S, φ_T we note that the supersymmetry has forced the fields to have the same mass, as expected, but has also completely determined the self-couplings of the scalars as well as coupling the scalars to the (massless) gaugino field Λ_M. The Feynman rules for the vertices are

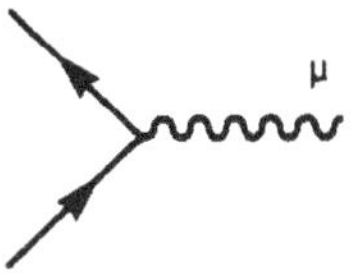

$$: \quad -\mathrm{i}\, q\gamma_\mu \tag{3.59}$$

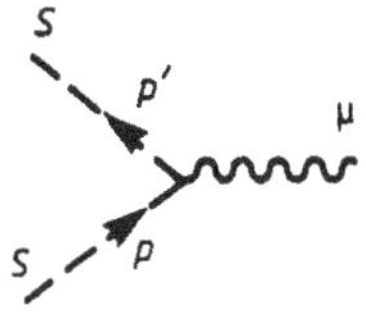

$$: \quad -\mathrm{i}\, q(p+p')_\mu \tag{3.60}$$

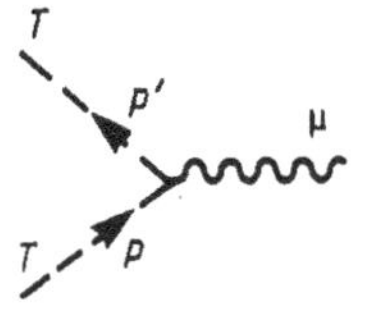

$$: \quad \mathrm{i}\, q(p+p')_\mu \tag{3.61}$$

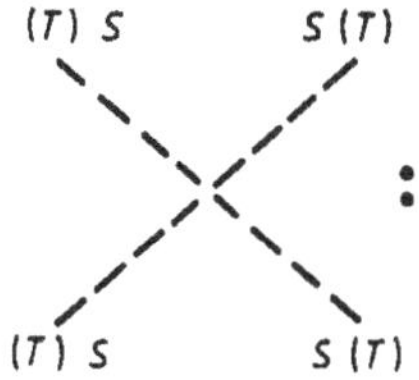

$$: \quad -2\mathrm{i}\, q^2 \tag{3.62}$$

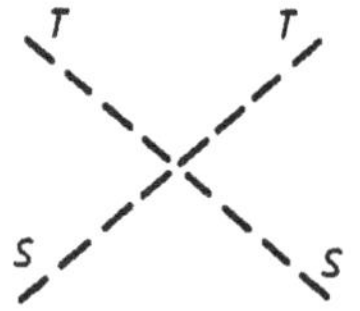

$$: \quad \mathrm{i}\, q^2 \tag{3.63}$$

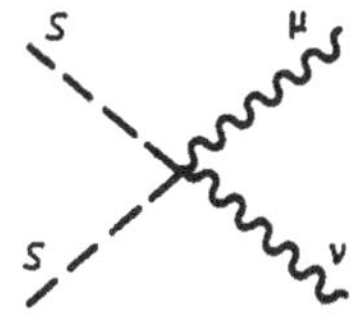

$$: \quad \mathrm{i}\, q^2 g_{\mu\nu} \tag{3.64}$$

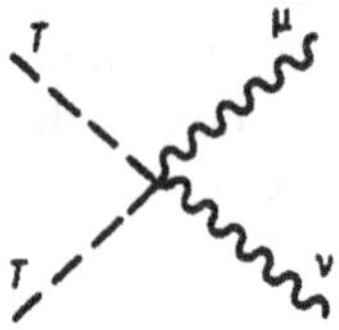

$$: \; \mathrm{i}\, q^2 g_{\mu\nu} \tag{3.65}$$

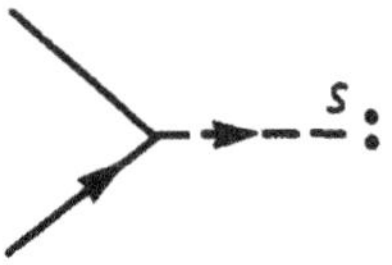

$$: \; \frac{-q}{\sqrt{2}}(1-\gamma_5) \tag{3.66}$$

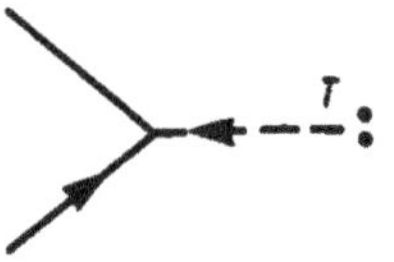

$$: \; \frac{-q}{\sqrt{2}}(1+\gamma_5). \tag{3.67}$$

3.4 Spontaneously broken gauge invariance

We saw in Bailin and Love[(2)] that when a theory with a global internal symmetry is spontaneously broken, by a scalar field developing a non-zero vacuum expectation value, then massless Goldstone bosons appear in the theory (and there are relationships between the trilinear and quadrilinear couplings of the scalars). When the internal symmetry is promoted to a local (gauge) symmetry, the spontaneously broken theory has no Goldstone bosons. Instead erstwhile massless gauge fields 'eat' the putative Goldstone bosons and we have a theory with massive vector bosons. It is this mechanism that is used in the standard electroweak theory. It is clearly of interest to study this phenomenon in the case where we have a supersymmetric gauge theory.

The test bed for the investigation of spontaneous symmetry breaking is a theory with a single complex scalar field, so the simplest supersymmetric extension is a theory with chiral superfields Φ_i, as in §2.5. The potential is

$$V(\varphi_i) = F_i^{\dagger} F_i \tag{3.68a}$$

where

$$F_i^{\dagger} = -(m_{ij}\varphi_j + \lambda_{ijk}\varphi_j\varphi_k + f_i) \tag{3.68b}$$

when we allow for a linear (tadpole) term $f_i\Phi_i$ in the superpotential $W(\Phi_i)$ given in (2.45). If the equations

$$F_i = 0 \tag{3.69}$$

have solutions in which the scalar fields φ_i have non-zero values, then, as discussed in §2.9, the supersymmetry will be unbroken, but the internal (global) symmetry may be broken.

For example, we may take a theory with *three* chiral superfields S, T, N, with S and T transforming as in (3.41), (3.47) under a U(1) global transformation, and N invariant:

$$S \to S' = e^{-2i q \Lambda_0} S \tag{3.70}$$

$$T \to T' = e^{2i q \Lambda_0} T \tag{3.71}$$

$$N \to N' = N \tag{3.72}$$

where Λ_0 is the constant chiral superfield. Then

$$W(S, T, N) = fN + \lambda STN \tag{3.73}$$

is invariant under the U(1) transformation. The requirement (3.69) that the F-terms all vanish gives the simultaneous equations

$$\lambda tn = 0 \tag{3.74a}$$

$$\lambda sn = 0 \tag{3.74b}$$

$$\lambda st + f = 0 \tag{3.74c}$$

for the vacuum expectation values s, t, n of the scalar components

$$\langle \varphi_s \rangle_0 = s \tag{3.75a}$$

$$\langle \varphi_T \rangle_0 = t \tag{3.75b}$$

$$\langle \varphi_N \rangle_0 = n \tag{3.75c}$$

of the three superfields. The solution

$$n = 0 \qquad st = f/\lambda \tag{3.76}$$

requires s and t to be non-zero and hence (spontaneously) breaks the U(1) invariance. We therefore shift the fields φ_S, φ_T by

$$\varphi_S = s + \hat{\varphi}_S \tag{3.77a}$$

$$\varphi_T = t + \hat{\varphi}_T. \tag{3.77b}$$

The potential is then

$$\begin{aligned} V(\hat{\varphi}_S, \hat{\varphi}_T, \varphi_N) &= \lambda^2 |(t + \hat{\varphi}_T)\varphi_N|^2 + \lambda^2 |(s + \hat{\varphi}_S)\varphi_N|^2 \\ &\quad + |\lambda(s + \hat{\varphi}_S)(t + \hat{\varphi}_T) + f|^2 \\ &= \lambda^2(s^2 + t^2)\varphi_N^{\dagger}\varphi_N \\ &\quad + \lambda^2(s\hat{\varphi}_T + t\hat{\varphi}_S)^{\dagger}(s\hat{\varphi}_T + t\hat{\varphi}_S) + \cdots \end{aligned} \tag{3.78}$$

where we have used (3.74), and the unspecified terms . . . are the cubic and quartic couplings of the scalar fields. Evidently there are two massive scalar fields with squared mass $\lambda^2(s^2+t^2)$, as well as the massless Goldstone boson mode (proportional to) $-t\hat{\varphi}_T + s\hat{\varphi}_S$. (It is easy to see that this is in accord with the general treatment given in Chapter 13 of Bailin and Love[(2)] in which we showed that (in a U(1) theory) the mode $\Sigma q_i v_i \hat{\varphi}_i$ is massless, where v_i are the VEVs of the scalar fields φ_i.)

Since supersymmetry is unbroken, there must also be a massless Goldstone fermionic mode. This can be verified using (2.56), which exhibits the Yukawa couplings in a theory involving chiral superfields. The scalar couplings generate fermion mass terms when we make the shifts (3.77)

$$-\tfrac{1}{2}\lambda_{ijk}\bar{\psi}_i\psi_j(\varphi_k+\varphi_k{}^{\dagger}) = -\lambda\bar{\psi}_N(s\psi_T+t\psi_S)+\cdots \tag{3.79}$$

where ... refers to the Yukawa couplings of the shifted fields $\hat{\varphi}_S, \hat{\varphi}_T, \varphi_N$. As anticipated the fermionic mode proportional to $-t\Psi_T + s\Psi_S$ is the massless Goldstone fermion.

Now let us consider the spontaneous breaking when the U(1) symmetry is local. We know, of course, that the massless gauge boson will become massive, and because supersymmetry is unbroken there must also be a massive gaugino: the erstwhile massless Majorana field Λ_M combines with the massless Goldstone fermion mode to generate a massive Dirac fermion. This is the supersymmetrized Higgs mechanism.

The Lagrangian for the U(1) local gauge-invariant version of the simple model we are studying is

$$\mathscr{L} = \tfrac{1}{32}(W^\alpha W_\alpha)_F + (S^\dagger e^{2qV}S + T^\dagger e^{-2qV}T + N^\dagger N)_D + ([W(S,T,N)]_F + \text{HC}). \tag{3.80}$$

The potential is now

$$V(\varphi_S,\varphi_T,\varphi_N) = F_S{}^\dagger F_S + F_T{}^\dagger F_T + F_N{}^\dagger F_N + \tfrac{1}{2}D^2 \tag{3.81a}$$

where

$$F_S{}^\dagger = -\frac{\partial W}{\partial \varphi_S} = -\lambda\varphi_T\varphi_N \tag{3.81b}$$

$$F_T{}^\dagger = -\lambda\varphi_S\varphi_N \tag{3.81c}$$

$$F_N{}^\dagger = -(\lambda\varphi_S\varphi_T + f) \tag{3.81d}$$

$$D = -q(\varphi_S{}^\dagger\varphi_S - \varphi_T{}^\dagger\varphi_T). \tag{3.81e}$$

Thus the (supersymmetry-preserving) minimum is when

$$s = t = -\sqrt{f/\lambda} \qquad n = 0. \tag{3.82}$$

As usual the covariant derivatives of the scalar fields in (3.58) generate the vector boson mass terms

$$m_V{}^2 = 2q^2(s^2 + t^2) = 4q^2 f/\lambda. \tag{3.83}$$

However, the same scalars' VEVs also generate a bilinear coupling of the gaugino field Λ_M to the fermion matter field Ψ in (3.58). This gives

$$m_\Lambda = 2q\sqrt{f/\lambda} \tag{3.84}$$

as anticipated because of the preserved supersymmetry.

3.5 *D*-term supersymmetry breaking

So far, we have insisted that the non-zero VEVs of the scalar fields preserve the supersymmetry, and certainly if this is possible it will be the global minimum of the potential, as discussed in §2.10. However, for the gauge theories that we are considering there is an additional method of breaking supersymmetry, besides the F-term method of Chapter 2. The new method utilizes the field $D(x)$ of the vector superfield.

We noted in §2.9 that spontaneous breaking of supersymmetry requires that some field in the theory, which is *not* invariant under the supersymmetry transformation, acquires a non-zero VEV. For the chiral superfield the only possibility is the field $F(x)$. When there is a vector superfield present there is the possibility that the field $D(x)$ has a VEV

$$\langle 0|D(x)|0\rangle = d \neq 0. \tag{3.85}$$

In other words the D-field is an order parameter for spontaneous supersymmetry breaking via the vector superfield, just as the fields F_i are for breaking with chiral superfields. Notice that (3.85) is the only way to achieve a non-zero variation under the supersymmetry transformations (3.15), (3.16), (3.17), (3.18) *without* breaking Lorentz invariance. This means that the gaugino field $\lambda_\alpha(x)$ has a variation

$$\langle 0|\delta_\xi \lambda_\alpha(x)|0\rangle = -\mathrm{i}\,\xi_\alpha d \neq 0. \tag{3.86}$$

This non-vanishing of d is consistent with our earlier observation that to break supersymmetry it is necessary for the effective potential V to possess no supersymmetric minimum. The field $D(x)$ contributes $\frac{1}{2}D^2$ to V, as is apparent in (3.81*a*), so with $d \neq 0$ we have

$$V \geqslant \tfrac{1}{2}d^2 > 0. \tag{3.87}$$

At first sight, the D-term method of supersymmetry breaking looks unpromising since for a U(1) gauge theory interacting with chiral superfields Φ_i having charges e_i, the generalization of the Lagrangian (3.49) yields

$$D = -\sum_i e_i \varphi_i{}^\dagger \varphi_i \tag{3.88}$$

as a generalization of (3.55*b*).

Thus d is zero if the scalars φ_i have zero VEVs, and supersymmetry is unbroken. Further, since the supersymmetric state is always stable, it will be the preferred state. (Of course, depending on the superpotential $W(\Phi_i)$ it may be that zero VEVs for all φ_i is not allowed, and then supersymmetry *is* broken.) However, the D-term method that we wish to discuss does not rely on the superpotential. Indeed it is most simply realized when W, and therefore the F-terms, are absent. It utilizes the fact that (for a U(1) gauge theory only) there is an additional gauge-invariant supersymmetric term that may be added to the Lagrangian. This is (proportional to) the D-term of the vector superfield

$$\mathscr{L}_1 = \xi D(x). \tag{3.89}$$

We have already noted that the D-term of any vector supermultiplet yields a supersymmetric action, and for the U(1) gauge theory, we noted also that the D-term is gauge invariant. The addition of such a Fayet–Iliopoulos[3] D-term changes the equation (3.88) for the auxiliary D-field to

$$D = -\left(\xi + \sum_i e_i \varphi_i^\dagger \varphi_i\right). \tag{3.90}$$

It is then possible to break supersymmetry in a gauge theory with just a *single* chiral field Φ having charge e. The Lagrangian is

$$\mathscr{L} = \tfrac{1}{32}(W^\alpha W_\alpha)_F + (\Phi^\dagger \, \mathrm{e}^{2eV}\Phi)_D + \xi(V)_D.$$

The superpotential, and therefore the F-term, is forced to vanish because of U(1) invariance and

$$D = -(\xi + e\varphi^\dagger\varphi). \tag{3.91}$$

If $\xi e < 0$ we get a vanishing D with a non-zero VEV for φ. This provides another example of the supersymmetric Higgs mechanism discussed in the previous section. However, if $\xi e > 0$, we minimize $V(\varphi) = \frac{1}{2}D^2$ by choosing a zero VEV for φ. Thus the U(1) gauge invariance is unbroken, but since $d = -\xi \neq 0$ the minimum of V is $\frac{1}{2}\xi^2$ and supersymmetry is (spontaneously) broken. In fact

$$V(\varphi) = \tfrac{1}{2}(\xi + e\varphi^\dagger\varphi)^2 \tag{3.92}$$

so the scalar field φ acquires a non-zero mass

$$m_\varphi{}^2 = e\xi \tag{3.93}$$

while its fermionic partner ψ remains massless, thereby verifying that supersymmetry is broken. The unbroken gauge invariance ensures that the gauge field V_μ remains massless, as does its superpartner the gaugino field λ. However, the masslessness of λ is due to it being the Goldstone fermion associated with the spontaneous breaking of global supersymmetry; it is called the 'Goldstino'.

A similar phenomenon arises in the (more realistic?) model (3.49) which represents a supersymmetric extension of quantum electrodynamics. When we add the Fayet–Iliopoulos term (3.89) the only effect is upon the scalar sector: the fermion sector does not feel the broken supersymmetry. The D-term in (3.58) is modified to

$$\tfrac{1}{2}[\xi + e(\varphi_S{}^\dagger\varphi_S - \varphi_T{}^\dagger\varphi_T)]^2 \tag{3.94}$$

so

$$m_S^2 = m^2 + e\xi \tag{3.95a}$$

$$m_T^2 = m^2 - e\xi \tag{3.95b}$$

but

$$m_\psi = m. \tag{3.96}$$

3.6 Supersymmetric non-abelian gauge theories

If supersymmetry is realized in nature, it is certainly at an energy scale that is higher than that of the electroweak scale. It is therefore essential to have a supersymmetric extension not only of the U(1) abelian gauge invariance, as discussed in the preceding sections, but also of the non-abelian gauge invariance that occurs in electroweak theory, quantum chromodynamics and grand unified theories.

Suppose that we have a chiral superfield Φ transforming as an (irreducible) representation of a non-abelian group G. Under a gauge transformation

$$\Phi \to \Phi' = \exp(-2\mathrm{i}\, gt^a\Lambda^a)\Phi \tag{3.97}$$

where the Λ^a are chiral superfields and the Hermitian matrices t^a constitute the representation of G to which Φ belongs. That is to say,

$$[t^a, t^b] = \mathrm{i} f^{abc} t^c \tag{3.98}$$

where the f^{abc} are the totally antisymmetric structure constants of G. Then

$$\Phi^\dagger \to \Phi^{\dagger\prime} = \Phi^\dagger \exp(2\mathrm{i}\, gt^a\Lambda^{a\dagger}). \tag{3.99}$$

The (non-abelian) gauge-invariant combination of superfields that is analogous to the abelian case (3.45) is

$$\Phi^\dagger \exp(2gt^aV^a)\Phi \tag{3.100}$$

where V^a are the vector superfields containing the non-abelian vector bosons belonging to the adjoint representation of G. The gauge invariance of the above combination follows provided that the gauge-transformed vector superfields $V^{a\prime}$ satisfy

$$\exp(V') = \exp(-\mathrm{i}\,\Lambda^\dagger)\exp(V)\exp(\mathrm{i}\,\Lambda) \tag{3.101a}$$

where

$$\Lambda \equiv 2g\Lambda^a t^a \qquad V = 2gV^a t^a. \tag{3.101b}$$

It should be noted that the above equation *does* have a solution for $V' = 2gV'^a t^a$ precisely because the matrices t^a represent the generators of a group. As in the non-supersymmetric case, it suffices to consider an infinitesimal gauge transformation in which we neglect terms of order Λ^2. Even so, the general solution of (3.101) is non-trivial and requires the use of the Baker–Hausdorff formula. However, we shall content ourselves with the first few terms. We write

$$\exp(V') - \exp(V) = \delta V + \tfrac{1}{2}(\delta V\,V + V\,\delta V) + \tfrac{1}{6}[\delta V\,V^2 + V\,\delta V\,V + V^2\,\delta V] + \cdots \tag{3.102a}$$

where

$$\delta V \equiv V' - V. \tag{3.102b}$$

For infinitesimal Λ

$$\exp(-\mathrm{i}\,\Lambda^\dagger)\exp(V)\exp(\mathrm{i}\,\Lambda) - \exp V = \mathrm{i}(\Lambda - \Lambda^\dagger) + \mathrm{i}(V\Lambda - \Lambda^\dagger V) + \frac{\mathrm{i}}{2}(V^2\Lambda - \Lambda^\dagger V^2) + \cdots. \tag{3.103}$$

Then substituting into (3.101) we can solve (perturbatively) for δV giving

$$\delta V = \mathrm{i}(\Lambda - \Lambda^\dagger) + \frac{\mathrm{i}}{2}[V, \Lambda + \Lambda^\dagger] + \frac{\mathrm{i}}{12}[V,[V,\Lambda - \Lambda^\dagger]] + \cdots. \tag{3.104}$$

In terms of the superfields we get

$$V^{a\prime} - V^a = \mathrm{i}(\Lambda^a - \Lambda^{a\dagger}) - gf^{abc}V^b(\Lambda^c + \Lambda^{c\dagger}) - \frac{\mathrm{i}}{3}g^2 f^{abc}f^{cde}V^bV^d(\Lambda^e - \Lambda^{e\dagger}) + \cdots. \tag{3.105}$$

Then it is easy to see that the first *two* terms generate the familiar gauge transformation of the non-abelian vector potential

$$V'^a_\mu = V^a_\mu + \partial_\mu(\varphi^a + \varphi^{a\dagger}) + gf^{abc}(\varphi^b + \varphi^{b\dagger})V^c_\mu \tag{3.106}$$

where φ^a is the (leading) scalar term of Λ^a. As before, we shall work in the Wess–Zumino gauge of the superfields V^a. However, a general choice of the gauge superfield Λ^a will yield a $V^{a\prime}$ that is *not* in the Wess–Zumino gauge, although it can be transformed to it by a supersymmetry transformation. If we require that both V'^a and V^a are in the Wess–Zumino gauge, so that

$$V'^a V'^b V'^c = 0 \tag{3.107}$$

then we must choose Λ^a such that

$$V^a V^b \Lambda^c = 0 \tag{3.108}$$

in which case the third and succeeding terms of (3.104) vanish and we get

$$\delta V^{\mathrm{WZ}} = \mathrm{i}(\Lambda - \Lambda^\dagger) + \frac{\mathrm{i}}{2}[V^{\mathrm{WZ}}, \Lambda + \Lambda^\dagger]. \tag{3.109}$$

To construct the non-abelian field strength superfield, analogous to (3.29), we first need to generalize the supersymmetric covariant derivatives to be gauge *and* supersymmetric covariant[4]. We define these as

$$\nabla_A \qquad (A = \mu, \alpha, \dot{\alpha}) \tag{3.110}$$

and having the property that if a (matter) superfield Φ transforms as in (3.97) under a gauge transformation, then the (supersymmetry- and) gauge-covariant derivative Φ transforms as

$$(\nabla_A \Phi)' = \exp(-\mathrm{i}\,\Lambda)(\nabla_A \Phi) \tag{3.111}$$

using the notation (3.101). Thus

$$\nabla_A{}' = \mathrm{e}^{-\mathrm{i}\,\Lambda} \nabla_A\, \mathrm{e}^{\mathrm{i}\,\Lambda} \tag{3.112}$$

Since Λ is a chiral superfield,

$$\bar{D}_{\dot{\alpha}} \Lambda = 0 = D_\alpha \Lambda^\dagger \tag{3.113}$$

we may choose

$$\nabla_{\dot{\alpha}} = \bar{D}_{\dot{\alpha}} \tag{3.114}$$

and then

$$\nabla_{\dot{\alpha}}' = \nabla_{\dot{\alpha}}. \tag{3.115}$$

We also define

$$\nabla_\alpha \equiv \mathrm{e}^{-V} D_\alpha\, \mathrm{e}^{V} \tag{3.116}$$

in the notation of (3.101) so that using (3.113)

$$\begin{aligned} \nabla_\alpha{}' &= \mathrm{e}^{-V'} D_\alpha\, \mathrm{e}^{V'} = \mathrm{e}^{-\mathrm{i}\,\Lambda}\, \mathrm{e}^{-V}\, \mathrm{e}^{\mathrm{i}\,\Lambda^\dagger} D_\alpha\, \mathrm{e}^{-\mathrm{i}\,\Lambda^\dagger}\, \mathrm{e}^{V}\, \mathrm{e}^{\mathrm{i}\,\Lambda} \\ &= \mathrm{e}^{-\mathrm{i}\,\Lambda}\, \mathrm{e}^{-V} D_\alpha\, \mathrm{e}^{V}\, \mathrm{e}^{\mathrm{i}\,\Lambda} = \mathrm{e}^{-\mathrm{i}\,\Lambda}\, \nabla_\alpha\, \mathrm{e}^{\mathrm{i}\,\Lambda} \end{aligned} \tag{3.117}$$

as required. Finally we can define ∇_μ by a gauge-covariant generalization of the supersymmetry algebra:

$$\{\nabla_\alpha, \nabla_{\dot{\alpha}}\} \equiv 2\mathrm{i}\, \sigma^\mu_{\alpha\dot{\alpha}}\, \nabla_\mu. \tag{3.118}$$

We leave the construction of ∇_μ as an exercise. The form (3.116) of ∇_α can be expanded as

$$\mathrm{i}\,\nabla_\alpha = \mathrm{i}\,D_\alpha + \mathrm{i}\,\mathrm{e}^{-V}(D_\alpha\,\mathrm{e}^V) \tag{3.119}$$

which invites the interpretation of the second term as a (supersymmetric) gauge connection:

$$\Gamma_\alpha \equiv \mathrm{i}\,\mathrm{e}^{-V}(D_\alpha\,\mathrm{e}^V). \tag{3.120}$$

Under a gauge transformation we get, as in (3.117),

$$\Gamma_\alpha' = \mathrm{i}\,\mathrm{e}^{-V'}(D_\alpha\,\mathrm{e}^{V'}) = \mathrm{e}^{-\mathrm{i}\Lambda}\Gamma_\alpha\,\mathrm{e}^{\mathrm{i}\Lambda} + \mathrm{i}\,\mathrm{e}^{-\mathrm{i}\Lambda}(D_\alpha\,\mathrm{e}^{\mathrm{i}\Lambda}) \tag{3.121}$$

characteristic of a gauge connection. In the abelian case Γ_α reduces to

$$\Gamma_\alpha = 2\mathrm{i}\,g\,D_\alpha V \tag{3.122}$$

and consulting (3.29) this suggests that the non-abelian generalization of the field strength (spinor) superfield is given by

$$W_\alpha \equiv (2\mathrm{i}\,g)^{-1}\bar{D}^2\Gamma_\alpha = (2g)^{-1}\bar{D}^2\,\mathrm{e}^{-V}(D_\alpha\,\mathrm{e}^V). \tag{3.123}$$

It is easy to see that W_α transforms covariantly, since the (inhomogeneous) last term of (3.121) drops out because Λ is chiral.

Expanding in powers of V gives

$$-\mathrm{i}\,\Gamma_\alpha = D_\alpha V + \tfrac{1}{2}[D_\alpha V, V] + \cdots \tag{3.124}$$

and in the Wess–Zumino gauge only the first two terms survive. In terms of the component superfields we get

$$W^a_\alpha = \bar{D}^2\,D_\alpha V^a + \mathrm{i}\,g f^{abc}\bar{D}^2(D_\alpha V^b)V^c \tag{3.125}$$

and the effect of the second term is to convert the ordinary derivatives in (3.36) into gauge-covariant derivatives. Thus we get

$$W^a_\alpha = 4\mathrm{i}\,\lambda^a_\alpha + [4\delta^\beta_\alpha\,D^a(y) + 2\mathrm{i}(\sigma^\mu\bar{\sigma}^\nu)_\alpha{}^\beta\,V^a_{\mu\nu}(y)]\theta_\beta$$
$$+\,4\theta^2\sigma^\mu_{\alpha\dot{\alpha}}\,\mathscr{D}_\mu\bar{\lambda}^{a\dot{\alpha}}(y) \tag{3.126a}$$

where

$$V^a_{\mu\nu} \equiv \partial_\mu V_\nu{}^a - \partial_\nu V_\mu{}^a - g f^{abc} V^b_\mu V^c_\nu \tag{3.126b}$$

$$\mathscr{D}_\mu\bar{\lambda}^{a\dot{\alpha}} = \partial_\mu\bar{\lambda}^{a\dot{\alpha}} - g f^{abc} V_\mu{}^b\bar{\lambda}^{c\dot{\alpha}}. \tag{3.126c}$$

Then, as in (3.37), the (pure) gauge-invariant and supersymmetric contribution to the Lagrangian is given by

$$\mathscr{L}_V = \tfrac{1}{64}[(W^{a\alpha}W^a{}_\alpha) + (W^a{}_\alpha{}^\dagger)(W^{a\alpha\dagger})]_F$$
$$= -\tfrac{1}{4}V^a_{\mu\nu}V^{a\mu\nu} + \mathrm{i}\,\lambda^a\sigma^\mu\,\mathscr{D}_\mu\bar{\lambda}^a + \tfrac{1}{2}D^aD^a \tag{3.127}$$

in the Wess–Zumino gauge. Similarly when we calculate the gauge interaction (3.100) with the chiral matter supermultiplet Φ the only difference from (3.53) is that the gauge-covariant derivatives are now those of the non-abelian theory:

$$\mathcal{L}_\Phi = [\Phi^\dagger \mathrm{e}^V \Phi]_D = (\mathcal{D}_\mu \varphi)^\dagger (\mathcal{D}^\mu \varphi) + \mathrm{i}\,\psi \sigma^\mu \mathcal{D}_\mu{}^\dagger \bar{\psi} + F^\dagger F$$
$$+ \mathrm{i}\sqrt{2} g(\varphi^\dagger t^a \lambda^a \psi - \bar{\psi} t^a \bar{\lambda}^a \varphi) + g\varphi^\dagger t^a D^a \varphi \tag{3.128a}$$

where

$$\mathcal{D}_\mu \varphi = \partial_\mu \varphi + \mathrm{i}\, g t^a V^a_\mu \varphi . \tag{3.128b}$$

In general, in physical applications, there are several chiral supermultiplets $\Phi_{(i)}$ transforming as possibly different representations of the gauge group G. Then, of course, for each $\mathcal{L}_{\Phi_{(i)}}$ we use for V the matrix defined in (3.101b) but constructed with the representation $t^a_{(i)}$ appropriate to $\Phi_{(i)}$. Then the final part of the Lagrangian is given, as before, by the F-part of the superpotential

$$\mathcal{L}_{int} = [W(\Phi_{(i)}) + \text{HC}]_F \tag{3.129}$$

which is required to be invariant under the action of G (as well as no more than cubic in the superfields $\Phi_{(i)}$). Then the auxiliary fields are given by

$$F^\dagger_{(i)} = -\frac{\partial W}{\partial \varphi_{(i)}} \tag{3.130a}$$

$$D^a = -\sum_i g\varphi^\dagger_{(i)} t^a_{(i)} \varphi_{(i)} \tag{3.130b}$$

and the tree level approximation to the effective potential is

$$V(\varphi_{(i)}) = \sum_i \left|\frac{\partial W}{\partial \varphi_{(i)}}\right|^2 + \tfrac{1}{2} g^2 \sum_a \left(\sum_i \varphi^\dagger_{(i)} t^a_{(i)} \varphi_{(i)}\right)^2 . \tag{3.131}$$

The simplest illustration of this is supersymmetric QCD with a single quark flavour. As for the U(1) case, we need two chiral supermultiplets which we denote by Φ_S and Φ_T where Φ_S includes the left chiral component of the quark field and Φ_T the left chiral component of the anti-quark field; $\Phi_T{}^\dagger$ then includes the right chiral component of the quark field. Under a gauge transformation Φ_S transforms as in (3.97)

$$\Phi_S \rightarrow \Phi_S' = \exp(-2\mathrm{i}\, g_3 t^a \Lambda^a)\Phi_S \tag{3.132}$$

where t^a are the eight 3×3 matrices constituting the **3** representation of SU(3). Since Φ_T constitutes a $\overline{\mathbf{3}}$ representation

$$\Phi_T \rightarrow \Phi_T' = \exp(2\mathrm{i}\, g_3 t^{a*} \Lambda^a)\Phi_T . \tag{3.133}$$

It follows that the transpose

$$\Phi_T{}^{\mathrm{T}} \to \Phi_T{}^{\mathrm{T}} \exp(2\mathrm{i}\, g_3 t^a \Lambda^a) \tag{3.134}$$

and so

$$W(\Phi_S, \Phi_T) = -\, m \Phi_T{}^{\mathrm{T}} \Phi_S \tag{3.135}$$

is SU(3) invariant, and

$$V(\varphi_S, \varphi_T) = m^2(\varphi_S{}^{\dagger}\varphi_S + \varphi_T{}^{\dagger}\varphi_T) + \tfrac{1}{2}g_3^2[\varphi_S{}^{\dagger}t^a\varphi_S - \varphi_T{}^{\dagger}t^{a*}\varphi_T]^2 \tag{3.136}$$

is the (tree level) effective potential. (In the above we are denoting by g_3 the (SU(3)) QCD coupling constant, to avoid confusion with the electroweak coupling constants (g, g') that arise in §3.7.)

As in (3.58) we may express the Lagrangian in terms of the quark's Dirac spinor field Ψ and the gaugino Majorana spinor fields Λ^a_{M}

$$\begin{aligned}\mathcal{L} = {} & \mathrm{i}\,\bar{\Psi}\gamma^\mu \mathcal{D}_\mu \Psi - m\bar{\Psi}\Psi + (\mathcal{D}^\mu\varphi_S)^\dagger(\mathcal{D}_\mu\varphi_S) - m^2\varphi_S{}^\dagger\varphi_S \\ & + (\mathcal{D}^\mu\varphi_T{}^*)^\dagger(\mathcal{D}_\mu\varphi_T{}^*) - m^2\varphi_T{}^\dagger\varphi_T - \tfrac{1}{2}g_3^2(\varphi_S{}^\dagger t^a\varphi_S - \varphi_T{}^{\mathrm{T}}t^a\varphi_T{}^*)^2 \\ & + \frac{g_3}{\sqrt{2}}[\bar{\Lambda}^a_{\mathrm{M}}(\varphi_S{}^\dagger + \varphi_T{}^{\mathrm{T}})t^a\Psi + \bar{\Lambda}^a_{\mathrm{M}}(\varphi_T{}^{\mathrm{T}} - \varphi_S{}^\dagger)\,\mathrm{i}\,\gamma_5 t^a\Psi \\ & - \bar{\Psi}t^a(\varphi_S + \varphi_T{}^*)\Lambda^a_{\mathrm{M}} + \bar{\Psi}t^a(\varphi_T{}^* - \varphi_S)\,\mathrm{i}\,\gamma_5\Lambda^a_{\mathrm{M}}] \\ & - \tfrac{1}{4}V^a_{\mu\nu}V^{a\mu\nu} + \frac{\mathrm{i}}{2}\bar{\Lambda}^a_{\mathrm{M}}\gamma^\mu \mathcal{D}_\mu\Lambda^a_{\mathrm{M}}\end{aligned} \tag{3.137}$$

where

$$\mathcal{D}_\mu\varphi_S = \partial_\mu\varphi_S + \mathrm{i}\, g_3 t^a V^a_\mu \varphi_S \tag{3.138a}$$

$$\mathcal{D}_\mu\Lambda^a_{\mathrm{M}} = \partial_\mu\Lambda^a_{\mathrm{M}} - g_3 f^{abc}V^b_\mu\Lambda^c_{\mathrm{M}}. \tag{3.138b}$$

3.7 Supersymmetric electroweak theory

It is also instructive to write down a supersymmetric version of the $\mathrm{SU}(2)_{\mathrm{L}} \times \mathrm{U}(1)_Y$ gauge theory which has been so spectacularly verified by experiments during the past twenty or so years. In particular we must construct a supersymmetric version of the (non-abelian) Higgs mechanism that is required to break the gauge symmetry and generate masses. Thus besides introducing chiral superfields for each of the chiral components of the known fermions, we must also assign the fields of the electroweak Higgs scalars to chiral superfields. It would be nice if we could economize by placing these in the same superfields as some of the known fermions. For example, the chiral supermultiplet $\Phi(E_{\mathrm{L}})$ which contains the electron doublet $(\nu_{\mathrm{eL}}, e_{\mathrm{L}})$ will also have a scalar doublet component with the same electric charges $(\tilde{\nu}_{\mathrm{eL}}, \tilde{e}_{\mathrm{L}})$, which *a priori* could be used to generate masses for the down-like quarks and charged leptons. In the non-supersymmetric theory the up-like quarks acquire their masses using the charge-conjugate doublet $(-\tilde{e}^{c}_{L}, \tilde{\nu}^{c}_{eL})$, but in the supersymmetric theory this will be associated with the *right* chiral antiparticle doublet $(-e^{c}_{R}, \nu^{c}_{eR})$ which appears in $\Phi^{\dagger}(E_L)$. It therefore cannot be used in the superpotential $W(\Phi_{(i)})$ which, as was shown in Chapter 2, must be constructed entirely from left chiral superfields. Thus, in order to generate masses for the up quarks, we are forced to introduce a new left chiral supermultiplet $\Phi(H_1)$ having weak isospin $\frac{1}{2}$ but hypercharge $\frac{1}{2}$, which includes a scalar doublet (H_1^+, H_1^0). Further since there is also a new charged chiral fermion in this supermultiplet, we must ensure that it can be given a Dirac mass, since a Majorana mass will break charge conservation; in any case we require a mass term for the scalar doublet in order to drive the spontaneous symmetry breaking. The cheapest solution is therefore to introduce a further chiral supermultiplet $\Phi(H_2)$ having weak isospin $\frac{1}{2}$ and hypercharge $-\frac{1}{2}$, which contains a scalar doublet (H_2^0, H_2^-). Then $\Phi(H_2)$ is used to give masses to the down-like quarks and charged leptons and $\Phi(H_1)$ to give masses to the up-like quarks. Thus in the minimal supersymmetric electroweak theory there are *two* chiral supermultiplets besides those needed for the known matter fields.

We denote the three doublet superfields that contain the (left chiral) lepton doublets by $L^{(l)}$ $(l = \mathrm{e}, \mu, \tau)$. Similarly the quark doublets are in the superfields $Q^{(f)}$ $(f = 1, 2, 3)$ with an undisplayed colour index running over the three labels which constitute the **3** representation of SU(3). The singlet superfields are denoted by l^{c}, $U^{c(f)}$, $D^{c(f)}$ where the three family labels indicate the flavours u, c, t in $U^{(f)}$ and d, s, b in $D^{(f)}$; the superscript c indicates charge conjugate, and for the quark fields there is an undisplayed colour index running over the three components of the $\mathbf{\bar{3}}$ representation. We also abbreviate our earlier notation and denote the two Higgs doublet superfields by H_1, H_2. Then the Yukawa couplings necessary to generate masses for the charged leptons and quarks arise from the F-part of a superpotential of the form

$$W = \sum_{l} m^{(l)}(L^{(l)\mathrm{T}}\mathrm{i}\,\tau_2 H_2)l^{c} + \sum_{f,g} m^{(d)}_{fg}(Q^{(f)\mathrm{T}}\,\mathrm{i}\,\tau_2 H_2)D^{c(g)}$$
$$+ \sum_{f,g} m^{(u)}_{fg}(Q^{(f)\mathrm{T}}\,\mathrm{i}\,\tau_2 H_1)U^{c(g)} \tag{3.139}$$

where $m^{(u)}$ and $m^{(d)}$ are (proportional to) the up-like and down-like quark mass matrices. (The factor $\mathrm{i}\,\tau^2$ is just the matrix $\epsilon_{\alpha\beta}$ used to construct an SU(2) singlet from two doublets of the internal symmetry group, just as we did in (1.73) for the space-time spinors.) In the two terms involving quark fields an implicit sum over the (three undisplayed) colour labels is assumed.

The remaining parts of the Lagrangian are simply written down using the techniques developed in the earlier sections of this chapter. We denote by

$W_\alpha^{\ i}$ ($i = 1, 2, 3$) and B_α the field strength superfields of the SU(2) and U(1) gauge theories, so as in (3.125) and (3.29)

$$W^i_\alpha = \bar{D}^2 D_\alpha W^i + \mathrm{i}\, g\epsilon^{ijk}\bar{D}^2(D_\alpha W^j)W^k \tag{3.140a}$$

$$B_\alpha = \bar{D}^2 D_\alpha B \tag{3.140b}$$

where W^i, B are the vector superfields. Then the supersymmetric pure gauge Lagrangian is

$$\mathscr{L}_V = \tfrac{1}{64}[W^{i\alpha}W^i_\alpha + W^{i\dagger}_{\dot\alpha}W^{i\dot\alpha\dagger} + 2B^\alpha B_\alpha]_F. \tag{3.141}$$

The interaction of the gauge supermultiplets with the chiral matter and Higgs supermultiplets is fixed by their weak isospin and hypercharge quantum numbers:

$$\begin{aligned}\mathscr{L}_\Phi = \Bigg[\sum_l L^{(l)\dagger}\exp(\mathrm{i}\, g\boldsymbol{\tau}\cdot\boldsymbol{W} - \mathrm{i}\, g'B)L^{(l)} \\ + \sum_f Q^{(f)\dagger}\exp(\mathrm{i}\, g\boldsymbol{\tau}\cdot\boldsymbol{W} + \tfrac{1}{3}\mathrm{i}\, g'B)Q^{(f)} \\ + \sum_f U^{c(f)\dagger}\exp(-\tfrac{4}{3}\mathrm{i}\, g'B)U^{c(f)} + \sum_f D^{c(f)\dagger}\exp(\tfrac{2}{3}\mathrm{i}\, g'B)D^{c(f)} \\ + \sum_l l^{c\dagger}\exp(2\mathrm{i}\, g'B)l^c + H_1^{\ \dagger}\exp(\mathrm{i}\, g\boldsymbol{\tau}\cdot\boldsymbol{W} + \mathrm{i}\, g'B)H_1 \\ + H_2^{\ \dagger}\exp(\mathrm{i}\, g\boldsymbol{\tau}\cdot\boldsymbol{W} - \mathrm{i}\, g'B)H_2\Bigg]_D.\end{aligned} \tag{3.142}$$

We leave it as an (extended) exercise to express the Lagrangian in terms of the component fields, and to eliminate the auxiliary fields.

We might also use the technology developed in this chapter to formulate a supersymmetric grand unified theory, which unifies supersymmetric QCD and supersymmetric electroweak theory in a single (supersymmetric) theory. However, we shall postpone that pleasure until Chapter 6. The reason is that the scale at which such a symmetry is apparent is even higher than in the non-supersymmetric case, discussed in Chapter 16 of Bailin and Love[(2)]. In fact the scale (10^{16} GeV) is comparable with the Planck energy (10^{19} GeV) at which we are forced to discuss quantum gravity. At this scale instead of supersymmetry being a (rigid) global symmetry, we must allow it to be a local symmetry, just like the gauge symmetries. Such theories are called 'supergravity' theories and their formulation is the topic that we begin to address in the following chapter. Before doing that, however, we can at least see why the unification scale is pushed to an even larger energy by (global) supersymmetry.

3.8 The renormalization group equations

We recall first that the renormalized coupling constants necessarily depend upon the scale M used in their definition. Since the physics described by the (bare) Lagrangian is independent of M, it must be that coupling constants 'run' with M, and so physical quantities calculated with different values of M have the same values, provided that they are calculated to a sufficiently high order. The renormalization group equations specify precisely how the renormalized coupling constants vary. We saw in Bailin and Love[(2)] (Chapter 12) that for a general gauge group G with coupling constant g the fine-structure constant

$$\alpha \equiv g^2/4\pi \tag{3.143}$$

satisfies

$$M\frac{\mathrm{d}\alpha}{\mathrm{d}M} = -\frac{b}{2\pi}\alpha^2 + \mathrm{O}(\alpha^3) \tag{3.144a}$$

where

$$b = \tfrac{11}{3}C_1(G) - \tfrac{2}{3}\sum_R C_2(R) - \tfrac{1}{3}\sum_S C_2(S) \tag{3.144b}$$

with

$$C_1(G)\delta^{ab} = f^{acd}f^{bcd} \tag{3.144c}$$

where f^{abc} are the structure constants of G, defined in (3.98). The sum over R is for Weyl fermions in representations T_R of G and

$$C_2(R)\delta^{ab} = \mathrm{tr}(T_R{}^a T_R{}^b)\,. \tag{3.145}$$

The sum over S is for scalars in representations T_S of G and $C_2(S)$ is defined analogously to $C_2(R)$.

In a supersymmetric theory the gauge bosons are accompanied by gauginos, in the same (adjoint) representation of G. Thus the vector supermultiplet contributes

$$b(V) = \tfrac{11}{3}C_1(G) - \tfrac{2}{3}C_1(G) = 3C_1(G) \tag{3.146}$$

to b. Similarly in a chiral supermultiplet each Weyl fermion is accompanied by a scalar in the same representation of G, so their contribution is

$$b(\Phi) = -\tfrac{2}{3}C_2(R) - \tfrac{1}{3}C_2(R) = -C_2(R)\,. \tag{3.147}$$

In all we have

$$b = 3C_1(G) - \sum_R C_2(R) \tag{3.148}$$

with the sum being over the representations R of all chiral supermultiplets. It follows that for the supersymmetric standard model, described in the previous section, we have for the QCD group SU(3)

$$b_3 = 9 - 2n_G \tag{3.149}$$

where n_G is the number of (fermion) generations. Similarly, for the SU(2) group we have

$$b_2 = 6 - 2n_G - \tfrac{1}{2}n_H \tag{3.150}$$

where n_H is the number of Higgs doublets, and for the $U(1)_Y$ group

$$b_1 = -\tfrac{10}{3}n_G - \tfrac{1}{2}n_H. \tag{3.151}$$

Integrating (3.144) between m_Z and (the unification scale) m_X gives

$$\alpha_i^{-1}(m_Z) - \alpha_i^{-1}(m_X) = -\frac{b_i}{2\pi}\ln\frac{m_X}{m_Z}. \tag{3.152}$$

The unification scale (m_X) is defined as the scale at which all three (properly normalized) coupling constants have equal value. Thus

$$\alpha_3(m_X) = \alpha_2(m_X) = \tfrac{5}{3}\alpha_1(m_X) \equiv \alpha_{\mathrm{GUT}}(m_X). \tag{3.153}$$

The origin of the factor $\frac{5}{3}$ was explained in Bailin and Love[(2)], and derives from the requirement that the $U(1)_Y$ is associated with a (diagonal) generator of SU(5) whose normalization is determined by (3.145). The same normalization also arises in SO(10) and 'flipped' SU(5) × U(1). Eliminating the *a priori* unknowns $\alpha_{\mathrm{GUT}}(m_X)$ and $\ln m_X/m_Z$ constrains the values of $\alpha_i(m_Z)$ through the relation

$$[8b_3 - 3(b_1 + b_2)]\sin^2\theta_W(m_Z) = 3(b_3 - b_2) + (5b_2 - 3b_1)\frac{\alpha_{\mathrm{em}}(m_Z)}{\alpha_3(m_Z)} \tag{3.154}$$

given in (16.45) of Bailin and Love[(2)], for example. α_{em} is the ordinary electromagnetic fine-structure constant and θ_W is the weak mixing angle defined by

$$\tan^2\theta_W \equiv \alpha_1/\alpha_2. \tag{3.155}$$

Using the supersymmetric values of the b_i given in (3.149)–(3.151) gives

$$(54 + 3n_H)\sin^2\theta_W(m_Z) = 9 + \tfrac{3}{2}n_H + (30 - n_H)\frac{\alpha_{\mathrm{em}}(m_Z)}{\alpha_3(m_Z)} \tag{3.156}$$

independently of n_G. In the minimal supersymmetric model

$$n_H = 2 \tag{3.157}$$

as we have seen in §3.7, and the fine-structure constants have the values[(5)]

$$\alpha_{em}^{-1}(m_Z) = 128.8 \tag{3.158}$$

$$\alpha_3(m_Z) = 0.108 \pm 0.005. \tag{3.159}$$

Then (3.156) gives

$$\sin^2\theta_W(m_Z) = 0.234 \tag{3.160}$$

whereas experimentally

$$\sin^2\theta_W(m_Z) = 0.2336 \pm 0.0018. \tag{3.161}$$

Thus the supersymmetric renormalization group equations 'predict' $\sin^2\theta_W$ with remarkable accuracy. Put another way, starting with the measured values of the three gauge coupling constants the renormalization group equations predict that all three achieve the same value at the unification scale m_X. In contrast, the non-supersymmetric equations exclude such a single unification scale by seven standard deviations. (The value of $\sin^2\theta_W$ predicted is 0.21.) This looks like excellent circumstantial evidence for supersymmetry and grand unification. The unification scale is also independent of n_G and given by

$$(18 + n_H)\ln\frac{m_X}{m_Z} = 2\pi[\alpha_{em}^{-1}(m_Z) - \tfrac{8}{3}\alpha_3^{-1}(m_Z)]. \tag{3.162}$$

Using

$$m_Z = 91.176 \pm 0.023\ \text{GeV} \tag{3.163}$$

this gives

$$m_X = 1.46 \times 10^{16}\ \text{GeV} \tag{3.164}$$

which is only three orders of magnitude from the Planck scale, and comfortably consistent with the measured lower bounds on the lifetime of the proton, as we shall see. In contrast, the non-supersymmetric theory gives $m_X = 5 \times 10^{14}$ GeV, which is excluded by the data.

The common value $\alpha_{GUT}(m_X)$ of the three coupling constants is given by

$$(18 + n_H)\alpha_{GUT}^{-1}(m_X) = \alpha_3^{-1}(m_Z)(-6 + \tfrac{16}{3}n_G + n_H) + (9 - 2n_H)\alpha_{em}^{-1}(m_Z). \tag{3.165}$$

We take[6]

$$n_G = 3 \tag{3.166}$$

as conclusively shown by the LEP experiments, and then

$$\alpha_{GUT}^{-1}(m_X) = 25.8. \tag{3.167}$$

Exercises

3.1 Verify (3.7).

3.2 Verify that the F-part of $W^{\alpha}W_{\alpha}$ gives the expression (3.37).

3.3 Verify (3.109).

3.4 Construct ∇_{μ} using (3.118).

3.5 Verify (3.126).

3.6* Express the Lagrangian (3.142) in terms of the component fields and eliminate the auxiliary fields.

References

General references

The books and articles that we have found most useful in preparing this chapter are listed below.

Fayet P and Ferrara S 1977 *Phys. Rep.* C **32** 249
Nilles H P 1984 *Phys. Rep.* C **110** 2
Srivastava Y 1986 *Supersymmetry, Superfields and Supergravity: an Introduction* (Bristol: Institute of Physics Publishing)

References and footnotes in the text

1 Wess J and Zumino B 1974 *Phys. Lett.* **49B** 52; 1974 *Nucl. Phys.* B **70** 39; 1974 *Nucl. Phys.* B **78** 1
2 Bailin D and Love A 1993 *Introduction to Gauge Field Theory* (Bristol: Institute of Physics Publishing)
3 Fayet P and Iliopoulos J 1974 *Phys. Lett.* **51B** 461
4 Srivastava Y 1986 *Supersymmetry, Superfields and Supergravity: an Introduction* (Bristol: Institute of Physics Publishing)
5 Ellis J, Kelley S and Nanopoulos D V 1991 *Phys. Lett.* **260B** 131
Amaldi V, de Boer W and Furstenau H 1991 *Phys. Lett.* **260B** 447
6 See, for example,
Particle Data Group 1992 *Phys. Rev.* D **45** 51

4

PURE SUPERGRAVITY

4.1 Introduction

Most symmetries in particle physics are realized as local symmetries (gauge symmetries) rather than mere global symmetries. This suggests that supersymmetry should also be realized as a local symmetry. Then the group parameters ξ and a^μ in (2.10) should be allowed to be functions of the point in space-time. In particular, because the supersymmetry algebra contains the generator of translations P_μ, we should consider translations that vary from point to point in space-time. Thus, we expect local supersymmetry to be, among other things, a theory of general coordinate transformations of space-time, and so a theory of gravity. The theory of local supersymmetry is therefore referred to as supergravity.

It might be thought that supergravity effects could be neglected at the electroweak scale, and that global supersymmetry could then be used rather than local supersymmetry. However, there are phenomenological considerations that suggest that supergravity effects are important even at such low energies. For instance, in theories with F-term supersymmetry breaking, equations (2.115) and (2.116) show that at tree level a fermion such as a lepton has a scalar supersymmetric partner of higher mass than itself and another of lower mass than itself. This does not occur in the real world and if the theory is to be phenomenologically acceptable it is necessary for the quantum corrections to STr M^2 to be significant. In realistic models of this type the supersymmetry-breaking scale M_S (where $M_S{}^2$ is the expectation value of the F-term responsible for supersymmetry breaking) is typically found to be 10^{10}–10^{11} GeV (more details of such models may be found in the review of Nilles[(1)]). In these circumstances, effects of supergravity cannot be neglected. Specifically, we shall find in Chapter 5 that, in the presence of supersymmetry breaking, scalar particles in supergravity theory acquire masses of order $M_S{}^2/m_P$ where m_P is the Planck mass ($\sim 10^{19}$ GeV). For M_S of order 10^{10} to 10^{11} GeV, these contributions to the scalar masses are of order 10^2 GeV and may not be neglected. It is therefore prudent to consider supergravity from the outset.

In this chapter, we shall consider pure supergravity without couplings to matter fields. A pure gravity theory would involve only the spin-2 graviton. A pure supergravity theory will have to include in addition the supersymmetric partner of the graviton. From §1.4 it can be seen that the spin-2 graviton could belong either to a supermultiplet containing a spin-$\frac{3}{2}$ particle or to one containing a spin-$\frac{5}{2}$ particle. It is well known that theories

DOI: 10.1201/9780367805807-4

containing particles with spins greater than 2 tend to have undesirable features, and it is therefore natural to assume in formulating supergravity that the supermultiplet of the graviton (the supergravity multiplet) contains a spin-$\frac{3}{2}$ particle, referred to as the gravitino. (It will be seen in §5.3, when the coupling of the gravity supermultiplet to matter is considered, that this is the correct choice, because the gravitino is the 'gauge field' associated with the local supersymmetry transformation.)

The problem of constructing the locally supersymmetric Lagrangian for the supergravity multiplet (the pure supergravity Lagrangian) can be tackled in two stages. First, construct the globally supersymmetric Lagrangian for the supergravity multiplet, and second use the Noether procedure to derive the locally supersymmetric Lagrangian from the globally supersymmetric Lagrangian. In the next section we shall show how the Noether procedure works for the simpler case of an ordinary global symmetry, before proceeding in the subsequent section to the construction of the supergravity Lagrangian.

4.2 The Noether procedure

The Noether procedure is a systematic technique for deriving an action with a local symmetry from an action with a global symmetry. The simplest example of this procedure is obtained by considering the action for a free massless Dirac field ψ,

$$S_0 = \mathrm{i}\int \mathrm{d}^4x\, \bar{\psi}\gamma^\mu\, \partial_\mu\psi . \tag{4.1}$$

This action is invariant under the transformation

$$\psi \to \mathrm{e}^{-\mathrm{i}\,\varepsilon}\psi \tag{4.2}$$

where ε is a constant phase, i.e. S_0 has an abelian global symmetry. To make the transformation local we allow ε to depend on the space-time coordinates and consider

$$\psi \to \mathrm{e}^{-\mathrm{i}\,\varepsilon(x)}\psi . \tag{4.3}$$

However, as it stands S_0 is not invariant under the local transformation but instead changes by an amount

$$\delta S_0 = \int \mathrm{d}^4x\, \bar{\psi}\gamma^\mu\psi\, \partial_\mu\varepsilon = \int \mathrm{d}^4x\, j^\mu\, \partial_\mu\varepsilon \tag{4.4}$$

where

$$j^\mu = \bar{\psi}\gamma^\mu\psi \tag{4.5}$$

is the Noether current associated with the symmetry (4.2) of S_0.

To restore invariance a gauge field A^μ is introduced that transforms under (4.3) as

$$A_\mu \rightarrow A_\mu + \partial_\mu \varepsilon \tag{4.6}$$

and a term coupling A^μ to the Noether current is added to the action to obtain the modified action

$$S = S_0 - \int d^4x\, j^\mu A_\mu = \int d^4x\, i\, \bar{\psi}\gamma^\mu(\partial_\mu + i A_\mu)\psi . \tag{4.7}$$

The action S is then invariant under the local symmetry (4.3) and (4.6). (Of course, S is just the action for the quantum electrodynamics of the Dirac field, apart from the kinetic term for the electromagnetic field A_μ which is separately invariant, and with the electromagnetic coupling constant absorbed in the definition of A_μ.)

In this simple example, the locally invariant action was achieved in a single stage, because the variation of the term $-\int d^4x\, j^\mu A_\mu$ added to S_0 exactly cancelled the variation of S_0 under the local transformation. More generally, for example for an initial non-abelian global symmetry, the cancellation only occurs correct to lowest order in some expansion parameter, and the process has to be iterated. At each stage, a further term is added to the action to cancel its variation correct to the next order in the expansion parameter, and, in general, for this to occur it is also necessary to add further terms to the transformation law (equation (4.6)) of the gauge field (in such a way that the algebra of transformations still closes). After a *finite* number of iterations (with luck) an action is obtained that is exactly invariant under the local symmetry, i.e. under the form of (4.3) and under the *final* form of (4.6) appropriate to the symmetry being studied.

To fill out the above general remarks, we now carry through the Noether procedure to derive the pure gauge field action for non-abelian gauge theory(2) from the globally symmetric action

$$S_0 = -\tfrac{1}{4}\int G_a{}^{\mu\nu} G_{\mu\nu}{}^a\, d^4x \tag{4.8}$$

where

$$G_a{}^{\mu\nu} \equiv \partial^\mu A_a{}^\nu - \partial^\nu A_a{}^\mu \tag{4.9}$$

and $A_a{}^\mu$, $a = 1, \cdots, r$, are vector fields transforming as the adjoint representation (of dimension r) of some Lie group. If T_a are the generators of the Lie group and f_{abc} are the structure constants defined such that

$$[T_a, T_b] = i f_{abc} T_c \tag{4.10}$$

then under an *infinitesimal* global transformation of the group with parameters ε_a,

$$A_a{}^\mu \to A_a{}^\mu + g f_{abc}\varepsilon_b A_c{}^\mu . \tag{4.11}$$

The action (4.8) is invariant under this transformation.

In the derivation of an action with local non-abelian symmetry from the globally symmetric action, it will prove useful to notice that the action (4.8) is also invariant under the *abelian* local transformation

$$A_a{}^\mu \to A_a{}^\mu + \partial^\mu \varepsilon_a \tag{4.12}$$

where the ε_a now depend on the space-time coordinates.

The action S_0 is *not* invariant under the transformation (4.11) when ε_a is space-time dependent, but changes by an amount

$$\delta S_0 = \int \mathrm{d}^4x\, j_a{}^\mu\, \partial_\mu \varepsilon_a \tag{4.13}$$

where

$$j_a{}^\mu = g f_{abc} G_b{}^{\mu\nu} A_\nu{}^c \tag{4.14}$$

is the Noether current associated with the symmetry generator T_a. Invariance is restored, correct to first order in the coupling constant g, by adding a term coupling $A_a{}^\mu$ to the Noether current to the Lagrangian, and simultaneously modifying the transformation law (4.11). In detail, we replace S_0 by

$$S_1 = S_0 - \tfrac{1}{2}\int \mathrm{d}^4x\, j_a{}^\mu A_\mu{}^a \tag{4.15}$$

and modify the transformation law (4.11) by the addition of a term $\partial^\mu \varepsilon_a$, so that it becomes

$$A_a{}^\mu \to A_a{}^\mu + g f_{abc}\varepsilon_b A_c{}^\mu + \partial^\mu \varepsilon_a . \tag{4.16}$$

The required cancellation of δS_0 against the variation of the term coupling $A_\mu{}^a$ to the Noether current in (4.15) is achieved entirely (Exercise 4.1) by the extra term $\partial^\mu \varepsilon_a$ in (4.16). The modified transformation law combines the non-abelian global symmetry (4.11) and the abelian local symmetry (4.12) into a single non-abelian local symmetry.

The action S_1 is invariant under the non-abelian local symmetry (4.16) correct to order g. To achieve invariance correct to order g^2, a further modification of the action is required, but no further modification of the transformation law (4.16). The variation of S_1 under (4.16) is

$$\delta S_1 = -\, g^2 f_{abc} f_{bde} \int A_\mu{}^a A_\nu{}^c A_e{}^\nu\, \partial^\mu \varepsilon_d\, \mathrm{d}^4x . \tag{4.17}$$

If we replace S_1 by

$$S = S_1 + \frac{g^2}{4} f_{abc} f_{bde} \int A_\mu{}^a A_\nu{}^c A_e{}^\nu A_d{}^\mu \, d^4x \tag{4.18}$$

(substituting $A_d{}^\mu$ for $-\partial^\mu \varepsilon_d$ in the expression (4.17) up to a numerical factor) then the variation of the added term cancels δS_1 correct to order g^2, and S is invariant under the local symmetry transformation (4.16) to this order in g. Indeed, a little more effort shows that S is now exactly invariant (to all orders in g).

Assembling the terms in (4.18), the final action is

$$S = -\tfrac{1}{4} \int d^4x \, F_{\mu\nu}{}^a F_a{}^{\mu\nu} \tag{4.19}$$

with

$$F_a{}^{\mu\nu} = G_a{}^{\mu\nu} - g f_{abc} A_b{}^\mu A_c{}^\nu = \partial^\mu A_a{}^\nu - \partial^\nu A_a{}^\mu - g f_{abc} A_b{}^\mu A^\nu{}_c \tag{4.20}$$

which is the usual action for pure non-abelian gauge field theory[(2)].

We shall use the general procedure described here in §4.4 to derive a locally supersymmetric action for the supergravity multiplet from a globally supersymmetric action.

4.3 The globally supersymmetric Lagrangian for the supergravity multiplet

Before constructing the locally supersymmetric pure supergravity Lagrangian, we first obtain the globally supersymmetric (free) Lagrangian for the supergravity multiplet of the graviton and its spin-$\frac{3}{2}$ partner the gravitino. The on-shell free Lagrangian for a chiral supermultiplet discussed in §1.5 was just the sum of quadratic kinetic terms for the Weyl spinor field and the complex scalar field making up this supermultiplet. By analogy, we might expect to have to use here the sum of a quadratic kinetic term for the gravitino, which we may conveniently describe by a massless Majorana vector spinor field Ψ^μ, and a quadratic kinetic term for the spin-2 graviton, described by a traceless symmetric tensor $g_{\mu\nu}$. (As discussed in §1.3, a supermultiplet has the same number of bosonic and fermionic degrees of freedom. Thus, on-shell the supermultiplet of the graviton contains the two bosonic degrees of freedom of the graviton and the two fermionic degrees of freedom of a massless Weyl vector spinor, or equivalently a massless Majorana vector spinor.) The massless Rarita–Schwinger action[(3)] S_{RS} provides a suitable kinetic term for the gravitino.

$$S_{\mathrm{RS}} = -\tfrac{1}{2}\int \mathrm{d}^4x\, \epsilon^{\mu\nu\rho\sigma}\bar{\Psi}_\mu \gamma_5 \gamma_\nu\, \partial_\rho \Psi_\sigma \tag{4.21}$$

where our conventions for γ_μ and γ_5 are as in §1.2, and $\epsilon^{\mu\nu\rho\sigma}$ is the totally antisymmetric Levi–Cevita symbol defined so that

$$\epsilon^{0123} = 1\,. \tag{4.22}$$

The graviton Lagrangian requires a little more thought.

The Einstein action provides the most natural choice of Lagrangian term for the graviton, but it is *not* quadratic in $g_{\mu\nu}$ (indeed not even polynomial). This is not surprising because the Einstein action describes a theory invariant under *local* coordinate transformations and, at this stage, we are merely trying to construct a theory with *global* supersymmetry. To obtain a Lagrangian term that is quadratic in the graviton field we need to use the action for linearized Einstein gravity. (The linearized field equation can be found in textbooks on gravitation in discussions of gravitational waves[(4)].) Let us write

$$g_{\mu\nu} = \eta_{\mu\nu} + \kappa h_{\mu\nu} \tag{4.23}$$

where $\eta_{\mu\nu}$ is the Minkowski metric, and

$$\kappa^2 \equiv 8\pi G_{\mathrm{N}} \tag{4.24}$$

where G_{N} is the Newtonian gravitational constant. (It will sometimes be convenient later to take units where $\kappa^2 = 1$.) The factor κ in (4.23) is introduced so that $h_{\mu\nu}$ has mass dimensions 1, as appropriate to a bosonic field describing the graviton. Expanding the Einstein field equation to linear order in $h_{\mu\nu}$ leads to the action $S^{\mathrm{L}}_{\mathrm{EINSTEIN}}$ for linearized Einstein gravity (Exercise 4.2)

$$S^{\mathrm{L}}_{\mathrm{EINSTEIN}} = -\tfrac{1}{2}\int \mathrm{d}^4x\, (R^{\mathrm{L}}_{\mu\nu} - \tfrac{1}{2}\eta_{\mu\nu}R^{\mathrm{L}})h^{\mu\nu} \tag{4.25}$$

where indices are raised and lowered using $\eta^{\mu\nu}$ and $\eta_{\mu\nu}$, the linearized Ricci tensor $R^{\mathrm{L}}_{\mu\nu}$ is given by

$$R^{\mathrm{L}}_{\mu\nu} = \tfrac{1}{2}\left(-\frac{\partial^2 h_{\mu\nu}}{\partial x^\lambda\, \partial x_\lambda} + \frac{\partial^2 h^\lambda{}_\nu}{\partial x^\mu\, \partial x^\lambda} + \frac{\partial^2 h^\lambda{}_\mu}{\partial x^\nu\, \partial x^\lambda} - \frac{\partial^2 h^\lambda{}_\lambda}{\partial x^\mu\, \partial x^\nu}\right) \tag{4.26}$$

and the linearized curvature scalar R^{L} by

$$R^{\mathrm{L}} \equiv \eta^{\mu\nu}R^{\mathrm{L}}_{\mu\nu}\,. \tag{4.27}$$

It may now be suspected that the on-shell globally supersymmetric action S_{GLOBAL} for the supergravity multiplet is

$$S_{\mathrm{GLOBAL}} = S_{\mathrm{RS}} + S^{\mathrm{L}}_{\mathrm{EINSTEIN}}\,. \tag{4.28}$$

To check that this is the case it is necessary to construct transformations connecting the graviton and gravitino fields ($h_{\mu\nu}$ and Ψ_μ) which realize the $N = 1$ supersymmetry algebra and leave the action invariant. Since we are constructing an on-shell realization of the supersymmetry algebra we only expect the algebra to close correctly provided that we use the field equations and, in general, any gauge invariances of the theory. In the present case, the field equations are

$$R^{\mathrm{L}}_{\mu\nu} = 0 \tag{4.29}$$

and

$$\epsilon^{\mu\nu\rho\sigma}\gamma_5\gamma_\nu\,\partial_\rho\Psi_\sigma = 0. \tag{4.30}$$

The Rarita–Schwinger action (4.21) is invariant under the gauge transformation

$$\Psi_\mu \rightarrow \Psi_\mu + \partial_\mu\eta \equiv \Psi_\mu + \delta_\eta\Psi_\mu \tag{4.31}$$

where η is an arbitrary Majorana spinor parameter and the linearized Einstein action (4.25) is invariant under the gauge transformation

$$h_{\mu\nu} \rightarrow h_{\mu\nu} + \partial_\mu\varepsilon_\nu + \partial_\nu\varepsilon_\mu \equiv h_{\mu\nu} + \delta_\varepsilon h_{\mu\nu} \tag{4.32}$$

where ε_μ is an arbitrary four-vector parameter, as can be checked directly.

Consider the global supersymmetry transformation $\mathrm{e}^{\mathrm{i}\,\bar{\xi}Q}$, where ξ is a Majorana spinor parameter and Q is the Majorana spinor supersymmetry generator of (1.107). The effect of this global supersymmetry transformation on the supergravity multiplet is expected to be of a form linear in ξ and the fields, and with the correct symmetry in the indices,

$$h_{\mu\nu} \rightarrow h_{\mu\nu} + c_1\bar{\xi}(\gamma_\mu\Psi_\nu + \gamma_\nu\Psi_\mu) + c_2\eta_{\mu\nu}\bar{\xi}\gamma^\rho\Psi_\rho \equiv h_{\mu\nu} + \delta_\xi h_{\mu\nu} \tag{4.33}$$

and

$$\Psi_\mu \rightarrow \Psi_\mu + c_3\sigma^{\rho\tau}\,\partial_\rho h_{\tau\mu}\xi + c_4\,\partial_\rho h_\mu{}^\rho\xi \equiv \Psi_\mu + \delta_\xi\Psi_\mu \tag{4.34}$$

where

$$\sigma^{\mu\nu} \equiv \frac{\mathrm{i}}{2}[\gamma^\mu, \gamma^\nu]. \tag{4.35}$$

(This matrix was denoted by $\Sigma^{\mu\nu}$ in Chapter 1 to avoid confusion with a related 2×2 matrix.) The constant coefficients, c_1, c_2, c_3 and c_4, will now be determined by the requirement that the transformations (4.31), (4.32), (4.33) and (4.34) together with space-time translations form a closed algebra that realizes the supersymmetry algebra correctly on-shell up to gauge transformations.

We look first at the action of the commutator of the Ravita–Schwinger gauge transformation (4.31) and the supersymmetry transformation on $h_{\mu\nu}$,

$$[\delta_\eta, \delta_\xi]h_{\mu\nu} = c_1\bar{\xi}(\gamma_\mu \partial_\nu \eta + \gamma_\nu \partial_\mu \eta) + c_2\eta_{\mu\nu}\bar{\xi}\gamma^\rho \partial_\rho \eta. \quad (4.36)$$

This is just an Einstein gauge transformation (4.32) with

$$\varepsilon_\mu = c_1\bar{\xi}\gamma_\mu \eta \quad (4.37)$$

provided that we take

$$c_2 = 0. \quad (4.38)$$

(Otherwise the algebra does not close.)

Second, we study the action of the commutator of the Einstein gauge transformation (4.32) and the supersymmetry transformation on Ψ_μ:

$$[\delta_\varepsilon, \delta_\xi]\Psi_\mu = c_3\sigma^{\rho\tau}(\partial_\rho \partial_\mu \varepsilon_\tau)\xi + c_4(\partial_\rho \partial^\rho \varepsilon_\mu + \partial_\rho \partial_\mu \varepsilon^\rho)\xi. \quad (4.39)$$

This is a Rarita–Schwinger gauge transformation with

$$\eta = c_3\sigma^{\rho\tau} \partial_\rho \varepsilon_\tau \xi \quad (4.40)$$

provided that we take

$$c_4 = 0. \quad (4.41)$$

Turning next to the action of the commutator of two supersymmetry transformations with parameters ξ_1 and ξ_2 on $h_{\mu\nu}$, we find

$$[\delta_{\xi_1}, \delta_{\xi_2}]h_{\mu\nu} = c_1c_3(\bar{\xi}_2\gamma_\mu\sigma^{\rho\tau} \partial_\rho h_{\tau\nu} \xi_1 - \bar{\xi}_1\gamma_\mu\sigma^{\rho\tau} \partial_\rho h_{\tau\nu} \xi_2) + (\mu \leftrightarrow \nu) \quad (4.42)$$

where we have used (4.38) and (4.41). This may be simplified by using the γ-matrix identity

$$\gamma_\lambda\gamma_\mu\gamma_\nu = \eta_{\lambda\mu}\gamma_\nu + \eta_{\mu\nu}\gamma_\lambda - \eta_{\lambda\nu}\gamma_\mu + \mathrm{i}\,\varepsilon_{\lambda\mu\nu\rho}\gamma_5\gamma^\rho \quad (4.43)$$

and the Majorana spinor identities (1.86). Thus

$$[\delta_{\xi_1}, \delta_{\xi_2}]h_{\mu\nu} = 2\mathrm{i}\, c_1c_3(\bar{\xi}_2\gamma^\tau\xi_1 \partial_\mu h_{\tau\nu} + \bar{\xi}_2\gamma^\tau\xi_1 \partial_\nu h_{\tau\mu} - 2\bar{\xi}_2\gamma^\rho\xi_1 \partial_\rho h_{\mu\nu}) \quad (4.44)$$

which is the sum of an Einstein gauge transformation (4.32) with

$$\varepsilon_\mu = 2\mathrm{i}\, c_1c_3\bar{\xi}_2\gamma^\tau\xi_1 h_{\tau\nu} \quad (4.45)$$

and a space-time translation. The coefficient of the space-time translation in (4.44) is consistent with the supersymmetry algebra (as in (1.184)) provided that

$$2c_1c_3 = -1. \quad (4.46)$$

Finally, we consider the action of the commutator of two supersymmetry transformations on Ψ_μ. After some labour this can be written as

$$[\delta_{\xi_1}, \delta_{\xi_2}]\Psi_\mu = 2\mathrm{i}\, \bar{\xi}_2\gamma^\lambda\xi_1 \partial_\lambda\Psi_\mu - \mathrm{i}\, \partial_\mu(\bar{\xi}_2\gamma^\lambda\xi_1\Psi_\lambda + \tfrac{1}{4}\bar{\xi}_2\gamma^\lambda\xi_1\gamma_\lambda\gamma^\rho\Psi_\rho) \quad (4.47)$$

which is the sum of a space-time translation with the correct coefficient to agree with the supersymmetry algebra of (1.187), and a gauge transformation (4.31) on Ψ_μ. The derivation (Exercise 4.3) requires the use of the Fierz identity for Majorana spinors

$$\bar{\xi}\gamma_\mu \partial_\nu \Psi \sigma^{\rho\tau}\eta = -\tfrac{1}{4}\bar{\xi}\gamma^\lambda \eta \sigma^{\rho\tau}\gamma_\lambda \gamma_\mu \partial_\nu \Psi \tag{4.48}$$

and the Rarita–Schwinger field equation (4.30) which can be written in the alternative forms[3]

$$\varepsilon^{\mu\nu\rho\sigma}\gamma_5\gamma_\nu \partial_\rho \Psi_\sigma = 0 \tag{4.49a}$$

$$\gamma^\mu(\partial_\mu \Psi_\nu - \partial_\nu \Psi_\mu) = 0 \tag{4.49b}$$

or

$$\varepsilon_{\mu\nu\sigma\tau}\gamma_5(\partial^\sigma \Psi^\tau - \partial^\tau \Psi^\sigma) = 2\mathrm{i}(\partial_\mu \Psi_\nu - \partial_\nu \Psi_\mu). \tag{4.49c}$$

In summary, we should take the action of the global supersymmetry transformation $e^{i\bar{\xi}Q}$ on the fields of the on-shell supergravity multiplet to be

$$h_{\mu\nu} \to h_{\mu\nu} + \delta_\xi h_{\mu\nu} = h_{\mu\nu} + c_1 \bar{\xi}(\gamma_\mu \Psi_\nu + \gamma_\nu \Psi_\mu) \tag{4.50}$$

and

$$\Psi_\mu \to \Psi_\mu + \delta_\xi \Psi_\mu = \Psi_\mu - \frac{1}{2c_1}\sigma^{\rho\tau}\partial_\rho h_{\tau\mu}\xi. \tag{4.51}$$

Then, the algebra (4.50), (4.51), (4.31) and (4.32) of supersymmetry transformations and gauge transformations taken together with space-time translations closes, and the commutators of supersymmetry transformations are as required by the supersymmetry algebra up to a gauge transformation, provided that we employ the Rarita–Schwinger field equation (4.49). It remains to check that the action (4.28), (4.25) and (4.21) for the on-shell supergravity multiplet is invariant under the supersymmetry transformations (4.50) and (4.51). After some labour (Exercise 4.4) it can be shown that the action is indeed invariant provided that we choose

$$c_1 = -\mathrm{i}/2. \tag{4.52}$$

Thus, the final form of the global supersymmetry transformation $e^{i\bar{\xi}Q}$ for the on-shell supergravity multiplet is

$$h_{\mu\nu} \to h_{\mu\nu} + \delta_\xi h_{\mu\nu} = h_{\mu\nu} - \frac{\mathrm{i}}{2}\bar{\xi}(\gamma_\mu \Psi_\nu + \gamma_\nu \Psi_\mu) \tag{4.53}$$

and

$$\Psi_\mu \to \Psi_\mu + \delta_\xi \Psi_\mu = \Psi_\mu - \mathrm{i}\,\sigma^{\rho\tau}\partial_\rho h_{\tau\mu}\xi. \tag{4.54}$$

4.4 The locally supersymmetric Lagrangian for the supergravity multiplet

In §4.3, we have seen how to construct an action (4.28) for the on-shell supergravity multiplet that is invariant under global supersymmetry transformations (4.53) and (4.54). In this section, we shall sketch the derivation of the corresponding locally supersymmetric action utilizing the Noether procedure described in §4.2.

Consider the local supersymmetry transformation $e^{i\bar{\xi}(x)Q}$ where $\xi(x)$ is a Majorana spinor parameter that depends on the point in space time and Q is a Majorana spinor supersymmetry generator of (1.107). It is tempting to surmise that the action (4.28) can be made locally supersymmetric rather than just globally supersymmetric by replacing the linearized Einstein Lagrangian by the usual Einstein Lagrangian, and replacing the derivatives in the Rarita–Schwinger Lagrangian by covariant derivatives. It turns out that this is almost the case, but that some extra terms quartic in the Rarita–Schwinger field need to be added to the Lagrangian, or equivalently that an extra term needs to be added to the standard covariant derivative[5].

The action (4.28) is invariant under the supersymmetry transformation $e^{i\bar{\xi}Q}$ when ξ is independent of x, but ceases to be invariant when ξ depends on x. To apply the Noether procedure of §4.2, we begin by modifying the globally supersymmetric action (4.28) and the transformation laws of the fields of the supergravity multiplet so as to obtain an action that is locally supersymmetric to lowest order in κ, which is the appropriate expansion parameter here.

Under the transformations (4.53) and (4.54), but with ξ now dependent on x, the variation of the action (4.28) is

$$\delta S_{\text{GLOBAL}} = \int d^4x\, \bar{j}^{\mu}\, \partial_\mu \xi \tag{4.55}$$

where j^μ is the Majorana vector spinor Noether current

$$\bar{j}^{\mu} = \frac{i}{2}\varepsilon^{\mu\nu\rho\sigma}\bar{\Psi}_\rho \gamma_5 \gamma_\nu \sigma^{\lambda\tau}\, \partial_\lambda h_{\tau\sigma}. \tag{4.56}$$

An action S_1 invariant to order κ may be obtained by modifying the transformation law (4.54) of Ψ_μ to

$$\Psi_\mu \to \Psi_\mu + \delta_\xi \Psi_\mu = \Psi_\mu + a\kappa^{-1}\, \partial_\mu \xi - i\,\sigma^{\rho\tau}\, \partial_\rho h_{\tau\mu}\, \xi \tag{4.57}$$

where a is a constant, and adding a term to S_{GLOBAL} coupling Ψ_μ to the Noether current, so that

$$S_1 = S_{\text{GLOBAL}} - \frac{\kappa}{2a}\int d^4x\, \bar{j}^{\mu}\Psi_\mu. \tag{4.58}$$

No modification of the transformation law for $h_{\mu\nu}$ is required at this stage. As we shall see shortly, equations (4.57) and (4.58) shape up neatly in terms of a covariant derivative provided that we take

$$a = 2. \tag{4.59}$$

Proceeding in this way to obtain actions invariant to higher orders in κ leads (with sufficient labour) to the final local supersymmetry transformation laws[5]

$$e_\mu{}^m \to e_\mu{}^m + \delta_\xi e_\mu{}^m = e_\mu{}^m - \mathrm{i}\,\kappa\bar{\xi}\gamma^m\Psi_\mu \tag{4.60}$$

where $e_\mu{}^m$ is the vierbein[4],[6] with μ a world index and m a local Lorentz index, satisfying $h_{\mu\nu} = e_\mu{}^m e_\nu{}^n \eta_{mn}$, and

$$\Psi_\mu \to \Psi_\mu + \delta_\xi \Psi_\mu = \Psi_\mu + 2\kappa^{-1}\,\tilde{D}_\mu \xi \tag{4.61}$$

where $\tilde{D}_\mu$ is the covariant derivative

$$\tilde{D}_\mu = \partial_\mu - \mathrm{i}\,\tilde{\omega}_{\mu mn}\frac{\sigma^{mn}}{4} \tag{4.62}$$

with

$$\tilde{\omega}_{\mu mn} = \omega_{\mu mn} + \frac{\mathrm{i}\,\kappa^2}{4}(\bar{\Psi}_\mu\gamma_m\Psi_n + \bar{\Psi}_m\gamma_\mu\Psi_n - \bar{\Psi}_\mu\gamma_n\Psi_m) \tag{4.63}$$

and

$$\omega_{\mu mn} = \tfrac{1}{2}e_m{}^\nu(\partial_\mu e_{n\nu} - \partial_\nu e_{n\mu}) + \tfrac{1}{2}e_m{}^\rho e_n{}^\sigma\,\partial_\sigma e_{\rho p}e_\mu{}^p - (m \leftrightarrow n) \tag{4.64}$$

which is the standard spin connection[4],[6]. The covariant derivative $\tilde{D}_\mu$ differs from the minimal covariant derivative of general relativity[4] by the term quadratic in the Rarita–Schwinger field, which is necessary to achieve invariance of the action at order κ^2. The final locally supersymmetric action[5] is

$$S = -\frac{1}{2\kappa^2}\int \mathrm{d}^4x\,|\det e|R - \tfrac{1}{2}\int \mathrm{d}^4x\,\varepsilon^{\mu\nu\rho\sigma}\bar{\Psi}_\mu\gamma_5\gamma_\nu\,\hat{D}_\rho\Psi_\sigma \tag{4.65}$$

which is the sum of the standard Einstein action, with R the curvature scalar, and the action for the Rarita–Schwinger field covariantized using the non-minimal covariant derivative (4.62). It is straightforward to check that the transformation laws (4.60) and (4.61) and action (4.65) reduce to (4.53), (4.57) and (4.58) correct to order κ. The invariance of (4.65) under (4.60) and (4.61) to all orders in κ may be checked directly. This is a somewhat laborious process which can be simplified[7],[8] by using the first-order formalism in which the vierbein and spin connection are treated as independent variables with the spin connection ultimately given in terms of the vierbein by solving the field equations.

Exercises

4.1 Check the invariance of the action (4.15) correct to order g under the transformation (4.16), and of the action (4.18) correct to order g^2.

4.2 Derive the action for linearized Einstein gravity by expanding the Einstein field equations to linear order in $h_{\mu\nu}$.

4.3 Show that the commutator of two supersymmetry transformations acts on a Rarita–Schwinger spinor as in (4.47).

4.4 Show that the choice (4.52) of c_1 gives a globally supersymmetric action for the supergravity multiplet.

References

General references

Srivastava P P 1986 *Supersymmetry, Superfields and Supergravity* (Bristol: Institute of Physics Publishing)

van Nieuwenhuizen P 1981 *Phys. Rep.* **68** 189

West P 1986 *Introduction to Supersymmetry and Supergravity* (Singapore: World Scientific)

References in the text

1 Nilles H P 1984 *Phys. Rep.* **110** §5

2 See, for example,
Bailin D and Love A 1993 *Introduction to Gauge Field Theory* (Bristol: Institute of Physics Publishing) Chapter 9

3 Rarita W and Schwinger J 1941 *Phys. Rev.* **60** 61

4 Weinberg S 1972 *Gravitation and Cosmology* (New York: Wiley)

5 Freedman D Z, van Nieuwenhuizen P and Ferrara S 1976 *Phys. Rev.* D **13** 335

6 Srivastava P P 1986 *Supersymmetry, Superfields and Supergravity* (Bristol: Institute of Physics Publishing) §11.2.2

7 van Nieuwenhuizen P 1981 *Phys. Rep.* **68** 189, §1.6

8 West P 1986 *Introduction to Supersymmetry and Supergravity* (Singapore: World Scientific) Chapter 10

5

COUPLING OF SUPERGRAVITY TO MATTER

5.1 Introduction

To make contact with the physics of quarks, leptons, gauge fields and Higgses it is necessary to extend the discussion of the pure supergravity Lagrangian of the graviton and gravitino of Chapter 4 to include couplings to these 'matter fields'. In Chapter 4, the locally supersymmetric Lagrangian for the supergravity multiplet was derived from the globally supersymmetric Lagrangian using the Noether procedure of §4.2. In this chapter, we shall use the same technique to illustrate how couplings of the supergravity multiplet to matter fields may be obtained, before proceeding to a discussion of supersymmetry breaking in supergravity theories. Whereas in globally supersymmetric theories supersymmetry breaking manifested itself in the appearance of a massless Goldstone fermion, in locally supersymmetric theories we shall find that the corresponding effect is the appearance of a mass for the gravitino, which is the gauge particle of local supersymmetry. Supersymmetry breaking in a 'hidden sector' is communicated to the 'observable sector' (of quarks, leptons etc) in the form of supersymmetry-breaking terms in an otherwise globally supersymmetric low-energy Lagrangian. For the supersymmetry-breaking mass splittings within supermultiplets to be small enough ($\leqslant 1$ TeV) for the hierarchy problem of grand unified theories to be solved (as discussed in §5.6) it will be found to be necessary for the gravitino mass $m_{3/2}$ to be small on the Planck scale m_{P}. This is most naturally achieved in the no-scale supergravity theories, with which we conclude the chapter, in which $m_{3/2}$ is undetermined at tree level with non-gravitational radiative corrections leading to a value of $m_{3/2}$ hierarchically suppressed relative to the Planck scale.

5.2 The supergravity Lagrangian for the Wess–Zumino model

The simplest matter field system to couple to supergravity by making the supersymmetry local is the free massless chiral supermultiplet of the Wess–Zumino model. From (2.50), the corresponding globally supersymmetric Lagrangian is

$$\mathscr{L}_0 = \partial_\mu \varphi^* \, \partial^\mu \varphi + \mathrm{i}\, \bar{\psi} \bar{\sigma}^\mu \, \partial_\mu \psi \tag{5.1}$$

DOI: 10.1201/9780367805807-5

where ψ is a Weyl spinor, and the on-shell global supersymmetry transformation laws corresponding to (2.36) and (2.37) are

$$\delta\varphi = \sqrt{2}\xi_{W}\psi \tag{5.2}$$

and

$$\delta\psi = -\mathrm{i}\sqrt{2}\,\partial_\mu\varphi\,\sigma^\mu\bar{\xi}_W \tag{5.3}$$

where ξ_W is a Weyl spinor parameter. Alternatively, the Lagrangian may be written in terms of a Majorana spinor Ψ as in (2.56):

$$\mathscr{L}_0 = \partial_\mu\varphi^*\,\partial^\mu\varphi + \frac{\mathrm{i}}{2}\bar{\Psi}\gamma^\mu\,\partial_\mu\Psi. \tag{5.4}$$

It is not difficult (Exercise 5.1) to show that the global supersymmetry transformation laws (5.2) and (5.3) may be recast in terms of the Majorana spinor Ψ and real fields A and B defined by

$$\varphi = \frac{1}{\sqrt{2}}(A + \mathrm{i}\,B) \tag{5.5}$$

as

$$\delta A = \bar{\xi}\Psi \tag{5.6}$$

$$\delta B = \mathrm{i}\,\bar{\xi}\gamma_5\Psi \tag{5.7}$$

and

$$\delta\Psi = -\,\mathrm{i}\,\gamma^\mu\,\partial_\mu(A + \mathrm{i}\,\gamma_5 B)\,\xi \tag{5.8}$$

where ξ is the Majorana spinor parameter

$$\xi = \begin{pmatrix}\xi_W\\ \bar{\xi}_W\end{pmatrix}. \tag{5.9}$$

To obtain a corresponding locally supersymmetric Lagrangian we need to replace ξ by $\xi(x)$, i.e. to allow ξ to depend on the point of space-time, and, employing the Noether procedure, to add terms to the Lagrangian and to the supersymmetry transformation laws until invariance is restored. In the absence of space-time dependence of ξ, the variation $\delta\mathscr{L}_0$ of $\mathscr{L}_0$ under a supersymmetry transformation is a total derivative and so $\mathscr{L}_0$ is globally supersymmetric. Allowing ξ to have space-time dependence, but retaining for the moment the transformation laws (5.6), (5.7) and (5.8) (where ∂_μ differentiates $A + \mathrm{i}\,\gamma_5 B$ but *not* ξ), we have instead (Exercise 5.2)

$$\delta\mathscr{L}_0 = \partial_\mu\bar{\xi}\,j^\mu \tag{5.10}$$

up to a total divergence, where

$$j^\mu \equiv \partial\!\!\!/(A - \mathrm{i}\,\gamma_5 B)\gamma^\mu\Psi. \tag{5.11}$$

Cancellation of $\delta\mathscr{L}_0$ may be achieved by adding a term to the Lagrangian coupling the (Noether) current j^μ to a gauge field for the local supersymmetry transformation, by analogy with the Yang–Mills case of §4.2. Since the supersymmetry generator is a Majorana spinor, the corresponding gauge field should be a vector spinor Ψ^μ, which we identify with the gravitino of §4.3. The transformation law of Ψ^μ under a local supersymmetry transformation is already known from (4.61). To leading order in κ it is

$$\Psi_\mu \rightarrow \Psi_\mu + 2\kappa^{-1}\,\partial_\mu\xi. \tag{5.12}$$

Thus, if we add the Lagrangian term

$$\mathscr{L}_1 = a\bar{\Psi}_\mu j^\mu \tag{5.13}$$

the supersymmetry transformation variations $\delta\mathscr{L}_0$ and $\delta\mathscr{L}_1$ cancel to zeroth order in κ for

$$a = -\frac{\kappa}{2}. \tag{5.14}$$

To cancel $\delta\mathscr{L}_0 + \delta\mathscr{L}_1$ to next to leading order (order κ) it is necessary to add further terms to the Lagrangian and to the supersymmetry transformation laws. To this order (Exercise 5.3)

$$\delta\mathscr{L}_1 = -\,\partial_\mu\bar{\xi}\,j^\mu + \mathrm{i}\,\kappa\bar{\Psi}^\mu\gamma^\nu\xi T_{\mu\nu} + \frac{\mathrm{i}\,\kappa}{2}\,\varepsilon^{\mu\tau\rho\sigma}\bar{\Psi}_\mu\gamma_\tau\,\partial_\rho\xi A\,\overleftrightarrow{\partial}_\sigma B$$
$$+\frac{\mathrm{i}\,\kappa}{2}\,\varepsilon^{\mu\tau\rho\sigma}\,\partial_\rho\bar{\Psi}_\mu\gamma_\tau\xi A\,\overleftrightarrow{\partial}_\sigma B \tag{5.15}$$

where, displaying only the A-dependent part,

$$T_{\mu\nu} \equiv \partial_\mu A\,\partial_\nu A - \tfrac{1}{2}\eta_{\mu\nu}\,\partial_\rho A\,\partial^\rho A + \cdots \tag{5.16}$$

which is just the energy–momentum tensor, and

$$A\,\overleftrightarrow{\partial}_\sigma B \equiv A\,\partial_\sigma B - B\,\partial_\sigma A. \tag{5.17}$$

To cancel the $T_{\mu\nu}$-dependent part of $\delta\mathscr{L}_1$ it is necessary to add to the Lagrangian the linearized gravity coupling of the graviton to the energy–momentum tensor

$$\mathscr{L}_2 = -\,\kappa h_{\mu\nu}T^{\mu\nu}. \tag{5.18}$$

With the aid of the transformation law for the graviton (4.53), it can be seen that $\delta\mathscr{L}_2$ cancels the $T_{\mu\nu}$ term in $\delta\mathscr{L}_1$ correct to order κ.

As a result of (5.12), the third term in (5.15) is cancelled by adding to the Lagrangian

$$\mathscr{L}_3 = -\frac{\mathrm{i}}{4}\kappa^2\varepsilon^{\mu\tau\rho\sigma}\bar{\Psi}_\mu\gamma_\tau\Psi_\rho A\overleftrightarrow{\partial}_\sigma B. \tag{5.19}$$

The last term in (5.15) is cancelled by the variation of the gravitino kinetic term of (4.21) if we modify the local supersymmetry transformation law of Ψ_μ by addition of a term to

$$\delta\Psi_\mu = 2\kappa^{-1}\,\partial_\mu\xi + \mathrm{i}\,\kappa\gamma_5\xi A\overleftrightarrow{\partial}_\mu B. \tag{5.20}$$

Of course, once the supersymmetry transformation law for the gravitino has been modified as in (5.20), the variation of the existing Lagrangian terms is correspondingly modified at order κ^2. For example, there is the extra term from $\mathscr{L}_1$,

$$\delta\mathscr{L}_1 = \frac{\mathrm{i}\,\kappa^2}{2}\bar{\xi}\gamma_5\not{\partial}(A - \mathrm{i}\,\gamma_5 B)\gamma^\mu\Psi A\overleftrightarrow{\partial}_\mu B + \cdots \tag{5.21}$$

which is cancelled by adding

$$\mathscr{L}_4 = -\frac{\kappa^2}{4}\bar{\Psi}\gamma_5\gamma^\tau\Psi A\overleftrightarrow{\partial}_\tau B. \tag{5.22}$$

There are also various four-fermion terms arising which we have not computed.

The process ends at order κ^2 apart from covariantization with respect to gravity, so we find that the coupling of a free massless chiral supermultiplet to supergravity is given by the Lagrangian

$$\begin{aligned}\mathscr{L} = &-\frac{1}{2\kappa^2}|\det e|R - \tfrac{1}{2}\varepsilon^{\mu\nu\rho\sigma}\bar{\Psi}_\mu\gamma_5\gamma_\nu\tilde{D}_\rho\Psi_\sigma + |\det e|\,\partial_\mu\varphi^*\,\partial^\mu\varphi \\ &+\frac{\mathrm{i}}{2}|\det e|\bar{\Psi}\gamma^\mu D_\mu\Psi - \frac{\kappa}{2}|\det e|\bar{\Psi}_\mu\not{\partial}\varphi(A - \mathrm{i}\,\gamma_5 B)\gamma^\mu\Psi \\ &-\frac{\mathrm{i}}{4}\kappa^2\varepsilon^{\mu\tau\rho\sigma}\bar{\Psi}_\mu\gamma_\tau\Psi_\rho A\overleftrightarrow{D}_\sigma B - \frac{\kappa^2}{4}|\det e|\bar{\Psi}\gamma_5\gamma^\mu\Psi A\overleftrightarrow{D}_\mu B \\ &+ \text{four-fermion terms}\end{aligned} \tag{5.23}$$

where all derivatives D_μ are covariantized with respect to gravity, the first two terms are as in (4.65), and $e_\mu{}^m$ is the vierbein. The local supersymmetry transformation laws are

$$\delta A = \bar{\xi}\Psi \tag{5.24}$$

$$\delta B = \mathrm{i}\,\bar{\xi}\gamma_5\Psi \tag{5.25}$$

$$\delta e_\mu{}^m = -\mathrm{i}\,\kappa\bar{\xi}\gamma^m\Psi_\mu \tag{5.26}$$

$$\delta\Psi_\mu = 2\kappa^{-1} D_\mu\xi + \mathrm{i}\,\kappa\xi A\,\overleftrightarrow{D}_\mu B + \text{two-fermion terms} \tag{5.27}$$

and

$$\delta\Psi = -\,\mathrm{i}\,\gamma^\mu D_\mu(A + \mathrm{i}\,\gamma_5 B)\xi + \text{two-fermion terms}. \tag{5.28}$$

A similar approach[1] may be used to derive the coupling of the vector supermultiplet to supergravity.

5.3 The general supergravity Lagrangian for chiral supermultiplets

In §5.2, the coupling to supergravity of free chiral supermultiplets was derived from the globally supersymmetric Lagrangians using the Noether procedure. For interacting (before coupling to supergravity) chiral supermultiplets, this same procedure has been used to determine the coupling to supergravity for some cases. However, the very laborious general result has in practice been derived using the less intuitively appealing but more efficient local tensor calculus technique[2],[3], and we now present the result. The most general globally supersymmetric Lagrangian for chiral superfields Φ_i describing complex scalar fields φ_i and Weyl spinor fields ψ_i, or equivalently Majorana spinor fields Ψ_i, is

$$\mathscr{L}_{\mathrm{GLOBAL}} = \int \mathrm{d}^4\theta\, K(\Phi^+, \Phi) + \int \mathrm{d}^2\theta\,(W(\Phi) + \mathrm{HC}) \tag{5.29}$$

where the first term in (2.82) has been generalized to allow a general function K of the superfields Φ_i^+ and Φ_i because non-renormalizable kinetic terms cannot be excluded in the presence of gravity. Correspondingly, the superpotential $W(\Phi)$ may contain arbitrary powers of the superfields Φ_i. The supergravity Lagrangian turns out to depend only on a single function of the scalar fields φ_i^* and φ_i, namely

$$G(\varphi^*, \varphi) = J(\varphi^*, \varphi) + \ln|W|^2 \tag{5.30}$$

where

$$J(\varphi^*, \varphi) = -\,3\ln(-K/3). \tag{5.31}$$

The function G is referred to as the Kähler potential. (Sometimes J is referred to in this way.) The same supergravity Lagrangians can be obtained for different choices of J and W, because G is invariant under the transformation

$$\begin{aligned} J &\rightarrow J + h(\varphi) + h^*(\varphi^*) \\ W &\rightarrow \mathrm{e}^{-h}W \end{aligned} \tag{5.32}$$

for an arbitrary function h.

It is convenient to write the supergravity Lagrangian in terms of the left and right chiral components of the Majorana spinors

$$\Psi_{iL} \equiv \tfrac{1}{2}(I - \gamma_5)\Psi_i \tag{5.33}$$

$$\Psi_{iR} = \tfrac{1}{2}(I + \gamma_5)\Psi_i \tag{5.34}$$

and their conjugates

$$\bar{\Psi}_{iL} = \bar{\Psi}_i \tfrac{1}{2}(I + \gamma_5) \tag{5.35}$$

$$\bar{\Psi}_{iR} = \bar{\Psi}_i \tfrac{1}{2}(I - \gamma_5). \tag{5.36}$$

This makes it easy to interpret the result in terms of Dirac spinors assembled from left and right chiral components of Majorana spinors.

The rather lengthy supergravity Lagrangian $\mathscr{L}$ may be split into terms as

$$\mathscr{L} = \mathscr{L}_B + \mathscr{L}_{FK} + \mathscr{L}_F \tag{5.37}$$

where $\mathscr{L}_B$ contains only bosonic fields, $\mathscr{L}_{FK}$ contains fermionic fields and covariant derivatives, and supplies the fermion kinetic energy terms, and $\mathscr{L}_F$ fermionic fields but *no* covariant derivatives. In detail[(3),(4)], with $|\det e|$ the determinant of the vierbein $e_\mu{}^m$, and R the curvature scalar, in units with κ^2 set equal to 1,

$$|\det e|^{-1}\mathscr{L}_B = -\tfrac{1}{2}R + G^i_j D_\mu\varphi_i D^\mu\varphi^{j*} + e^G\left(3 - G_i(G^{-1})^i_j G^j\right) \tag{5.38}$$

where the derivatives D_μ are covariantized with respect to gravity. To keep track of differentiation with respect to scalar fields and their adjoints, the scalar fields have been written as φ_i and their adjoints as φ^{i*}, and the derivatives of the Kähler potential as

$$G^i \equiv \frac{\partial G}{\partial \varphi_i} \qquad G_i \equiv \frac{\partial G}{\partial \varphi^{i*}} \tag{5.39}$$

and

$$G^i_j = \frac{\partial^2 G}{\partial \varphi_i\, \partial \varphi^{j*}}. \tag{5.40}$$

The inverse $(G^{-1})^i_j$ obeys

$$(G^{-1})^i_j G^j_k = \delta^i_k. \tag{5.41}$$

The kinetic terms and the tree level effective potential of the scalar fields are provided by $\mathscr{L}_B$:

$$|\det e|^{-1}\mathcal{L}_{\mathrm{FK}} = -\tfrac{1}{2}|\det e|^{-1}\varepsilon^{\mu\nu\rho\sigma}\bar{\Psi}_\mu\gamma_5\gamma_\nu D_\rho\Psi_\sigma$$
$$+\tfrac{1}{4}|\det e|^{-1}\varepsilon^{\mu\nu\rho\sigma}\bar{\Psi}_\mu\gamma_\nu\Psi_\rho(G^i D_\sigma\varphi_i - G_i D_\sigma\varphi^{i*})$$
$$+\left(\frac{\mathrm{i}}{2}G^i_j\bar{\Psi}_{i\mathrm{L}}\gamma^\mu D_\mu\Psi_{\mathrm{L}}{}^j + \frac{\mathrm{i}}{2}\bar{\Psi}_{i\mathrm{L}}\not{D}\varphi_j\Psi_{k\mathrm{L}}(-G^{ij}_k + \tfrac{1}{2}G^i_kG^j)\right.$$
$$\left.+\frac{1}{\sqrt{2}}G^i_j\bar{\Psi}_{\mu\mathrm{L}}\not{D}\varphi^{i*}\gamma^\mu\Psi_{j\mathrm{R}} + \mathrm{HC}\right). \tag{5.42}$$

This contains the kinetic terms for the fermions and some non-renormalizable interaction terms. Finally,

$$|\det e|^{-1}\mathcal{L}_{\mathrm{F}} = \frac{\mathrm{i}}{2}\mathrm{e}^{G/2}\bar{\Psi}_\mu\sigma^{\mu\nu}\Psi_\nu$$
$$+\left(\tfrac{1}{2}\mathrm{e}^{G/2}(-G^{ij} - G^iG^j + G^{ij}_k(G^{-1})^k_lG^l)\bar{\Psi}_{i\mathrm{L}}\Psi_{j\mathrm{R}}\right.$$
$$\left.+\frac{\mathrm{i}}{\sqrt{2}}\mathrm{e}^{G/2}G^i\bar{\Psi}_{\mu\mathrm{L}}\gamma^\mu\Psi_{i\mathrm{L}} + \mathrm{HC}\right) + \text{four-fermion terms} \tag{5.43}$$

with $\sigma^{\mu\nu}$ as in (4.35). This contribution to the Lagrangian contains the fermion Yukawa couplings, and numerous non-renormalizable terms which can be found in full in the original literature[3] or elsewhere[4]. (The zero-superpotential limit, corresponding to the supergravity-coupled Wess–Zumino model of §5.2, cannot be taken straightforwardly because the superpotential enters the Kähler potential logarithmically in (5.30).)

The local supersymmetry transformation laws (analogous to (5.24)–(5.28)) are

$$\delta\varphi_i = \sqrt{2}\bar{\xi}\Psi_i = \sqrt{2}\bar{\xi}_{\mathrm{R}}\Psi_{i\mathrm{L}} \tag{5.44}$$

$$\delta e_\mu{}^m = -\mathrm{i}\,\kappa\bar{\xi}\gamma^m\Psi_\mu \tag{5.45}$$

$$\delta\Psi_\mu = 2\kappa^{-1}D_\mu\xi + \kappa\xi(G^iD_\mu\varphi_i - G_iD_\mu\varphi^{i*})$$
$$+\mathrm{i}\,\mathrm{e}^{G/2}\gamma_\mu\xi + \text{two-fermion terms} \tag{5.46}$$

and

$$\delta\Psi_i = -\mathrm{i}\not{D}(A_i + \mathrm{i}\gamma_5B_i)\xi - \sqrt{2}e^{G/2}(G^{-1})^j_iG_j\xi + \text{two-fermion terms} \tag{5.47}$$

where φ_i is decomposed into real fields A_i and B_i as

$$\varphi_i = \frac{1}{\sqrt{2}}(A_i + \mathrm{i}B_i). \tag{5.48}$$

The renormalizable globally supersymmetric Lagrangian of (2.56) can be recovered for $m_{ij} = 0$ by taking the superpotential as in (2.45) (with $m_{ij} = 0$),

$$W(\Phi) = \tfrac{1}{3}\lambda_{ijk}\Phi_i\Phi_j\Phi_k \tag{5.49}$$

and the function K in (5.29) to be

$$K(\Phi^{\dagger}, \Phi) = \Phi_i^{\dagger}\Phi_i - 3 \tag{5.50}$$

which differs from the D-term in (2.44) by a constant, which has no significance in the globally supersymmetric theory but is of importance in the locally supersymmetric theory which is coupled to gravity. Neglecting non-renormalizable terms in (5.38), (5.42) and (5.43) then leads (Exercise 5.4) to (2.56), with $m_{ij} = 0$.

In general, in the presence of gravity, there is no requirement that the Lagrangian should be renormalizable, and, in particular, there is no reason for $K(\Phi^{\dagger}, \Phi)$ to correspond to only renormalizable kinetic terms in the globally supersymmetric Lagrangian. Accordingly, it is often convenient in model calculations to use the form of K that leads to minimal kinetic terms in the supergravity Lagrangian, namely

$$K = -3\exp(-\Phi_i\Phi^{i\dagger}/3). \tag{5.51}$$

Then,

$$G = \varphi_i\varphi^{i*} + \ln|W|^2 \tag{5.52}$$

and

$$G^i_j = \delta^i_j. \tag{5.53}$$

As a consequence of (5.53), the kinetic terms for the scalar fields (in (5.38)) are simply $\partial_\mu\varphi_i\,\partial^\mu\varphi_i^*$ and the kinetic terms for the fermion fields in (5.42) are

$$\frac{\mathrm{i}}{2}\bar{\Psi}_{i\mathrm{L}}\gamma^\mu\,\partial_\mu\Psi_{i\mathrm{L}} + \mathrm{HC}.$$

5.4 The general supergravity Lagrangian including vector supermultiplets

As observed in §5.2, the coupling of the vector supermultiplet to supergravity may be derived by the Noether procedure. However, in practice, the complete supergravity Lagrangian involving vector supermultiplets and chiral supermultiplets has been derived[5] by the local tensor calculus method[2]. To state the result we first need to write down the general (not necessarily renormalizable) globally supersymmetric Lagrangian for vector superfields V as defined in §3.6, and chiral superfields Φ_i:

$$\mathscr{L}_{\text{GLOBAL}} = \int \mathrm{d}^4\theta\, K(\Phi^\dagger\, \mathrm{e}^{2gV}, \Phi) + \int \mathrm{d}^2\theta\, (W(\Phi) + \text{HC})$$

$$+ \int \mathrm{d}^2\theta\, (f_{ab}(\Phi) W_a{}^\alpha W_{\alpha b} + \text{HC}) \tag{5.54}$$

where $W_a{}^\alpha$ is the gauge field strength superfield of (3.125) with spinor index α and gauge group index a, $f_{ab}(\Phi)$ is an arbitrary function of the chiral superfields which would be just δ_{ab} in the renormalizable case, $W(\Phi)$ is the superpotential, and, in the function K of (5.29), e^{2gV} has been introduced to couple the chiral supermultiplets to the gauge fields, as in §3.6.

The resulting supergravity Lagrangian $\mathscr{L}$ may be separated conveniently into several terms

$$\mathscr{L} = \tilde{\mathscr{L}}_{\mathrm{B}} + \tilde{\mathscr{L}}_{\mathrm{FK}} + \tilde{\mathscr{L}}_{\mathrm{F}} + \hat{\mathscr{L}}_{\mathrm{B}} + \hat{\mathscr{L}}_{\mathrm{FK}} + \hat{\mathscr{L}}_{\mathrm{F}} \tag{5.55}$$

where $\tilde{\mathscr{L}}_{\mathrm{B}}$, $\tilde{\mathscr{L}}_{\mathrm{FK}}$ and $\tilde{\mathscr{L}}_{\mathrm{F}}$ are identical to $\mathscr{L}_{\mathrm{B}}$, $\mathscr{L}_{\mathrm{FK}}$ and $\mathscr{L}_{\mathrm{F}}$ of (5.38), (5.42) and (5.43), except that the covariant derivatives are to be covariantized with respect to the gauge group in the usual way, as well as with respect to gravity. The terms $\hat{\mathscr{L}}_{\mathrm{B}}$, $\hat{\mathscr{L}}_{\mathrm{FK}}$ and $\hat{\mathscr{L}}_{\mathrm{F}}$ are as follows:

$$|\det e|^{-1}\hat{\mathscr{L}}_{\mathrm{B}} = -\tfrac{1}{4}(\operatorname{Re} f_{ab})(F_a)_{\mu\nu}F_b{}^{\mu\nu} + \frac{\mathrm{i}}{4}(\operatorname{Im} f_{ab})(F_a)_{\mu\nu}\tilde{F}_b{}^{\mu\nu}$$

$$-\frac{g^2}{2}(\operatorname{Re} f_{ab}^{-1})G^i(T_a)_{ij}\varphi_j G^k(T_b)_{kl}\varphi_l \tag{5.56}$$

where the gauge field strength $(F_a)_{\mu\nu}$ is

$$(F_a)_{\mu\nu} = \partial_\mu V_{\nu a} - \partial_\nu V_{\mu a} - g f_{abc} V_{\mu b} V_{\nu c} \tag{5.57}$$

and its dual $(\tilde{F}_a)_{\mu\nu}$ is

$$(\tilde{F}_a)_{\mu\nu} = \varepsilon_{\mu\nu\rho\sigma}(F_a)^{\rho\sigma}. \tag{5.58}$$

The last term in (5.56), which involves the generators $(T_a)_{ij}$ of the gauge group in the appropriate representation for the φ_i, is the D-term of (3.127) generalized to take account of $f_{ab}(\Phi)$ in (5.54),

$$|\det e|^{-1}\hat{\mathscr{L}}_{\mathrm{FK}} = \tfrac{1}{2}\operatorname{Re} f_{ab}(\varphi)\Big(\tfrac{1}{2}\bar{\lambda}_a \not{D}\lambda_b + \tfrac{1}{2}\bar{\lambda}_a\gamma^\mu\sigma^{\nu\rho}\Psi_\mu(F_b)_{\nu\rho}$$

$$+\tfrac{1}{2}G^i D^\mu\varphi_i\bar{\lambda}_{a\mathrm{L}}\gamma_\mu\lambda_{b\mathrm{L}}\Big) - \frac{\mathrm{i}}{8}\operatorname{Im} f_{ab}(\varphi)\, D_\mu(|\det e|\bar{\lambda}_a\gamma_5\gamma^\mu\lambda_b)$$

$$-\tfrac{1}{2}\frac{\partial f_{ab}(\varphi)}{\partial\varphi_i}\bar{\Psi}_{i\mathrm{R}}\sigma^{\mu\nu}(F_a)_{\mu\nu}\lambda_{b\mathrm{L}} + \text{HC} \tag{5.59}$$

where λ_a is the gaugino.

Finally,

$$|\det e|^{-1}\hat{\mathscr{L}}_F = \tfrac{1}{4}\mathrm{e}^{G/2}\frac{\partial f_{ab}{}^*}{\partial\varphi^{j*}}(G^{-1})^j_k G^k \lambda_a\lambda_b$$
$$-\frac{\mathrm{i}}{2}gG^i(T_a)_{ij}\varphi_j\bar{\Psi}_{\mu\mathrm{L}}\gamma^\mu\lambda_{a\mathrm{L}} + 2\mathrm{i}\,gG^i_j(T_a)_{ik}\varphi_k\bar{\lambda}_{a\mathrm{R}}\Psi_{i\mathrm{L}}$$
$$-\frac{\mathrm{i}}{2}g\,\mathrm{Re}\,f^{-1}_{ab}\frac{\partial f_{bc}}{\partial\varphi_k}G^i(T_a)_{ij}\varphi_j\bar{\Psi}_{k\mathrm{R}}\lambda_{c\mathrm{L}} + \mathrm{HC}$$
$$+ \text{four-fermion terms.} \qquad (5.60)$$

If some of the scalar fields φ_i develop expectation values, equation (5.60) can give rise to gaugino mass terms.

The corresponding local supersymmetry transformation laws are identical to (5.44)–(5.47) for $\delta\varphi_i$, $\delta e_\mu{}^m$, $\delta\Psi_\mu$ and $\delta\Psi_i$ except that the covariant derivative in (5.47) is also covariantized with respect to the gauge group and there are some extra two-fermion terms involving gauginos. In addition, there are the transformation laws for the gauge fields and gauginos,

$$\delta V^\mu_a = -\bar{\xi}_\mathrm{L}\gamma^\mu\lambda_{a\mathrm{L}} + \mathrm{HC} \qquad (5.61)$$

and

$$\delta\lambda_{a\mathrm{L}} = \sigma^{\mu\nu}(F_a)_{\mu\nu}\xi_\mathrm{L} + \frac{\mathrm{i}}{2}g\,\mathrm{Re}\,f^{-1}_{ab}G^i(T_b)_{ij}\varphi_j\xi_\mathrm{L} + \text{two-fermion terms} \qquad (5.62)$$

with $\sigma^{\mu\nu}$ as in (4.35). The complete expressions for the Lagrangian inclusive of four-fermion terms, and for the local supersymmetry transformation laws inclusive of two-fermion terms, can be found in the original literature[(5)] or elsewhere[(4)].

5.5 Spontaneous supersymmetry breaking in supergravity

In §2.9, it was found that for globally supersymmetric theories a supersymmetric vacuum state had zero energy, and if supersymmetry was spontaneously broken in the vacuum state it had positive energy. It was also found that spontaneous supersymmetry breaking occurs if one of the auxiliary fields F_i of a chiral supermultiplet Φ_i develops a non-zero VEV, or, as in §3.5, there is the alternative of the auxiliary field D_a of a vector supermultiplet V_a developing a VEV. A Goldstone fermion arises from the supermultiplet to which F_i or D_a belongs. For theories with local supersymmetry there are some differences and some similarities.

One difference is that the vacuum energy is no longer positive semi-definite. This can be seen by looking at the tree level effective potential V arising from equations (5.38) and (5.56). For simplicity, consider first the case of a single gauge-singlet chiral superfield Φ with minimal kinetic terms arising from

$$G = \varphi^*\varphi + \ln |W|^2 \tag{5.63}$$

as in (5.52). Then, after a little algebra, we find from (5.38) that

$$V = e^{\varphi^*\varphi}\left(\left|\frac{\partial W}{\partial \varphi} + \varphi^* W\right|^2 - 3|W|^2\right) \tag{5.64}$$

for supergravity, to be compared with

$$V = \left|\frac{\partial W}{\partial \varphi}\right|^2 \tag{5.65}$$

from (2.51) and (2.52) for the globally supersymmetric case. (We are using φ^* to denote the complex conjugate of the expectation value of φ.) It is clear from (5.64) that there is now the possibility of a ground state with negative energy.

For spontaneous supersymmetry breaking to occur, at least one of the fields in the theory must have a VEV that is *not* invariant under supersymmetry transformations. The only supersymmetry transformation laws amongst (5.44)–(5.48) and (5.61), (5.62) that can have a non-zero right-hand side without breaking Lorentz invariance are (5.47) and (5.62). If moreover we assume that there are *no* non-zero expectation values for the terms on the right-hand side involving fermionic fields, then (for non-spatially varying expectation values) we may simplify (5.47) and (5.62) to

$$\langle 0|\delta\Psi_i|0\rangle = -\, e^{G/2}(G^{-1})^j_i G_j \xi \tag{5.66}$$

and

$$\langle 0|\delta\lambda_a|0\rangle = \frac{i}{2} g \,\mathrm{Re}\, f^{-1}_{ab} G^i (T_b)_{ij} \varphi_j \xi \tag{5.67}$$

where the scalar fields on the right-hand sides of (5.66) and (5.67) are being used to denote their expectation values. (In §5.8, we shall discuss the case of supersymmetry breaking by a gaugino condensate, in which case we have to take account of a VEV for a product of two fermionic gaugino fields.)

For the case of a single gauge-singlet chiral superfield Φ with Kähler potential given by (5.63), equation (5.66) simplifies to

$$\langle 0|\delta\Psi|0\rangle = -\frac{\exp(\frac{1}{2}(\varphi^*\varphi + \ln |W|^2))}{W^*}\left(\frac{\partial W^*}{\partial \varphi^*} + \varphi W^*\right)\xi \tag{5.68}$$

and the right-hand side of (5.67) is zero. Then, the criterion for supersymmetry breaking is that the VEV of $\partial W/\partial\varphi + \varphi^* W$ should be non-zero. This is the generalization of F-term supersymmetry breaking to supergravity. Returning to (5.64), we see that

$$V = -3\,\mathrm{e}^{\varphi^*\varphi}|W|^2 \tag{5.69}$$

in a supersymmetric vacuum. Thus, the energy of a supersymmetric vacuum is negative. When supersymmetry is broken, the VEV of $\partial W/\partial\varphi + \varphi^* W$ is non-zero, and it is possible for a cancellation to occur in (5.64) to give a vacuum with zero energy. There is therefore the attractive possibility in supergravity theories of obtaining a vacuum state with zero cosmological constant when supersymmetry is broken.

Consider next the case of a gauge non-singlet chiral superfield Φ_i with minimal kinetic terms arising from

$$G = \varphi^{i^*}\varphi_i + \ln|W|^2 \tag{5.70}$$

and with the minimal choice of gauge field kinetic terms given by

$$f_{ab}(\varphi) = \delta_{ab}. \tag{5.71}$$

Then, from (5.38) and (5.56) we find the tree level effective potential

$$V = \mathrm{e}^{\varphi^{j^*}\varphi_j}\left(-3|W|^2 + \left|\frac{\partial W}{\partial\varphi_i} + \varphi^{i^*}W\right|^2\right) + \frac{g^2}{2}G^i(T_a)_{ij}\varphi_j G^k(T_a)_{kl}\varphi_l \tag{5.72}$$

where

$$G^i = \varphi^{i^*} + \frac{1}{W}\frac{\partial W}{\partial\varphi_i} \tag{5.73}$$

for supergravity, to be compared with the positive semi-definite

$$V = \left|\frac{\partial W}{\partial\varphi_i}\right|^2 + \frac{g^2}{2}\varphi^{i^*}(T_a)_{ij}\varphi_j\varphi^{k^*}(T_a)_{kl}\varphi_l \tag{5.74}$$

for the globally supersymmetric case. Assuming *no* expectation values on the right-hand sides of equations (5.47) and (5.62) involving fermionic fields, these equations may be simplified to

$$\langle 0|\delta\Psi_i|0\rangle = -\frac{\exp(\frac{1}{2}(\varphi^{j^*}\varphi_j + \ln|W|^2))}{W^*}\left(\frac{\partial W^*}{\partial\varphi_i^*} + \varphi^i W^*\right)\xi \tag{5.75}$$

and

$$\langle 0|\delta\lambda_a|0\rangle = \frac{\mathrm{i}}{2}gG^i(T_a)_{ij}\varphi_j\xi. \tag{5.76}$$

There are now two different mechanisms for supersymmetry breaking depending on whether one of the Ψ_i or one of the λ_a has a VEV that is not invariant under supersymmetry transformations. In the former case, the criterion for supersymmetry breaking is

$$\frac{\partial W}{\partial \varphi_i} + \varphi_i^* W \neq 0 \tag{5.77}$$

for some values of i, generalizing F-term supersymmetry breaking to supergravity. In the latter case, the criterion is

$$G^i (T_a)_{ij} \varphi_j \neq 0 \tag{5.78}$$

for some value of a, generalizing D-term supersymmetry breaking to supergravity. (In either case, a necessary condition for supersymmetry breaking is that G^i should be non-zero for at least one value of i.)

5.6 The super-Higgs mechanism and gravitino mass

In §2.9 and §2.10, we saw that in globally supersymmetric theories of chiral superfields when F-term supersymmetry breaking occurred with an auxiliary field developing an expectation value, the spinor in the supermultiplet of this auxiliary field was the Goldstone fermion. In locally supersymmetric theories, we might expect that the Goldstone fermion would be 'eaten' by the gauge field of local supersymmetry, namely the gravitino, and in this way the gravitino would acquire a mass. (The helicity $\pm\frac{1}{2}$ states of the Goldstone fermion are combined with the helicity $\pm\frac{3}{2}$ states of the massless gravitino to give the states of a massive spin-$\frac{3}{2}$ particle.)

That this happens[(3),(5)] can be seen by looking at the terms $\mathcal{L}_F$ of (5.43) quadratic in the spin-$\frac{1}{2}$ fields Ψ_i and the gravitino field Ψ_μ. In outline, when G^i is non-zero for some value of i (and we have seen in §5.5 that this is a necessary condition for supersymmetry breaking) there is a mixing of the would-be Goldstone fermion $G^i \Psi_i$ with the gravitino through the mass term $((\mathrm{i}/\sqrt{2})\, \mathrm{e}^{G/2} G^i \bar{\Psi}_{\mu \mathrm{L}} \gamma^\mu \Psi_{i\mathrm{L}} + \mathrm{HC})$. Once the Goldstone fermion has been 'eaten' by the gravitino (the super-Higgs mechanism) we are left with a massive gravitino Ψ'_μ with mass term $(\mathrm{i}/2)\, \mathrm{e}^{G_0/2} \bar{\Psi}'_\mu \sigma^{\mu\nu} \Psi'_\nu$, where G_0 is the expectation value of G in the physical vacuum. Recall that we have been working in units where κ^2 in (4.24) has been taken to be 1. Thus, all masses are in units of the Planck mass m_P defined by

$$8\pi G_\mathrm{N} m_\mathrm{P}^2 = \kappa^2 m_\mathrm{P}^2 = 1 \tag{5.79}$$

where G_N is the Newtonian gravitational constant. Explicitly restoring the mass unit, the gravitino mass $m_{3/2}$ is

$$m_{3/2} = \mathrm{e}^{G_0/2} m_\mathrm{P} \tag{5.80}$$

where

$$m_P \simeq 2.4 \times 10^{18}\,\text{GeV}. \tag{5.81}$$

It is instructive to work through the super-Higgs effect in a little more detail. For simplicity, we shall continue to use the minimal form of G as in (5.70). The appropriate choice of expectation values for G and its derivatives is that corresponding to the physical vacuum. First, the vacuum must be a minimum of the effective potential, V. Assuming, for simplicity, *no* D-terms, V is as in (5.38):

$$V = -\,e^G\left(3 - G_i(G^{-1})^i_j G^j\right). \tag{5.82}$$

To obtain

$$\frac{\partial V}{\partial \varphi_k} = 0 \tag{5.83}$$

when

$$G^i_j = \delta^i_j \tag{5.84}$$

we need

$$G^k + G_j G^{jk} = 0. \tag{5.85}$$

Second, we require the value of V in the physical vacuum to be zero to obtain a vanishing cosmological constant. Thus we also need

$$G_j G^j = 3. \tag{5.86}$$

For the minimal form (5.70) of G, and assuming that the expectation values of the scalar fields are real, the mass terms $\mathscr{L}^m_F$ derived from (5.43) simplify to

$$|\det e|^{-1}\mathscr{L}^m_F = \frac{i}{2}e^{G/2}\bar{\Psi}_\mu\sigma^{\mu\nu}\Psi_\nu + \frac{i}{\sqrt{2}}e^{G/2}G^i\bar{\Psi}_\mu\gamma^\mu\Psi_i - \tfrac{1}{2}e^{G/2}(G^{ij} + G^iG^j)\bar{\Psi}_i\Psi_j. \tag{5.87}$$

The contribution of the Goldstone fermion

$$\eta = G^i\Psi_i \tag{5.88}$$

may be separated from the contributions of the other chiral fermions by rewriting (5.87) in the form

$$|\det e|^{-1}\mathscr{L}^m_F = \frac{i}{2}e^{G/2}\bar{\Psi}_\mu\sigma^{\mu\nu}\Psi_\nu + \frac{i}{\sqrt{2}}e^{G/2}\bar{\Psi}_\mu\gamma^\mu\eta - \tfrac{1}{3}e^{G/2}\bar{\eta}\eta - \tfrac{1}{2}e^{G/2}(G^{ij} + \tfrac{1}{3}G^iG^j)\bar{\Psi}_i\Psi_j. \tag{5.89}$$

It is not difficult to verify using (5.85) and (5.86), as appropriate to the physical vacuum, that there is *no* contribution to the mass of η from the mass matrix

$$M^{ij} = G^{ij} + \tfrac{1}{3}G^iG^j. \tag{5.90}$$

It may be shown that all terms in the Lagrangian quadratic in the fermion fields may be written in terms of

$$\Psi'_\mu = \Psi_\mu - \frac{\mathrm{i}}{3\sqrt{2}}\gamma_\mu\eta - \frac{\sqrt{2}}{3}\mathrm{e}^{-G/2}\,\partial_\mu\eta \tag{5.91}$$

and the Ψ_i. In particular (Exercise 5.5) the mass terms of (5.89) may be cast in the form

$$|\det e|^{-1}\mathscr{L}_{\mathrm{F}}^{m} = \frac{\mathrm{i}}{2}\mathrm{e}^{G/2}\bar{\Psi}'_\mu\sigma^{\mu\nu}\Psi'_\nu - \tfrac{1}{2}\mathrm{e}^{G/2}(G^{ij} + \tfrac{1}{3}G^iG^j)\bar{\Psi}_i\Psi_j. \tag{5.92}$$

Whereas Ψ_μ was a massless spin-$\frac{3}{2}$ field, Ψ'_μ is a massive spin-$\frac{3}{2}$ field with the extra helicities deriving from the contribution of the Goldstone fermion η to (5.91). The mass of the gravitino is given by the value of $\mathrm{e}^{G/2}$ in the physical vacuum, as advertised in (5.80).

Further insight can be obtained by considering the local supersymmetry transformation laws (5.46) and (5.47). For the minimal form (5.70) of G, and in the physical vacuum, equation (5.47) yields

$$\delta\eta = -3\sqrt{2}\,\mathrm{e}^{G/2}\xi + \cdots. \tag{5.93}$$

To lowest order in κ, the Goldstone fermion can be gauged to zero by choosing

$$\xi = -\frac{1}{3\sqrt{2}}\mathrm{e}^{-G/2}\eta. \tag{5.94}$$

Then, the gauge-transformed gravitino field

$$\Psi'_\mu = \Psi_\mu + \delta\Psi_\mu \tag{5.95}$$

arising from (5.46) is given by (5.91). Thus, as expected in a Higgs phenomenon, there is a choice of (supersymmetric) gauge in which the Goldstone fermion disappears from the theory having been 'eaten' by the gauge-transformed gravitino Ψ'_μ.

For the case of D-term breaking of supersymmetry, the term $\mathscr{L}_{\mathrm{MIX}}$ in the Lagrangian mixing the gravitino with the would-be Goldstone fermion is given by

$$\begin{aligned}|\det e|^{-1}\mathscr{L}_{\mathrm{MIX}} = {} & \frac{\mathrm{i}}{\sqrt{2}}\mathrm{e}^{G/2}G^i\bar{\Psi}_{\mu\mathrm{L}}\gamma^\mu\Psi_{i\mathrm{L}} + \mathrm{HC} \\ & - \frac{\mathrm{i}}{2}gG^i(T_a)_{ij}\varphi_j\bar{\Psi}_{\mu\mathrm{L}}\gamma^\mu\lambda_{a\mathrm{L}} + \mathrm{HC}.\end{aligned} \tag{5.96}$$

(The latter term, arising from (5.60), is non-zero when (5.78) is satisfied, corresponding to D-term supersymmetry breaking.) For minimal G, as in (5.70), and real expectation values for the scalar fields, the Goldstone fermion is

$$\eta = G^i \Psi_i - \frac{g}{\sqrt{2}} \mathrm{e}^{-G/2} G^i (T_a)_{ij} \varphi_j \lambda_a \tag{5.97}$$

and similar arguments to those just given can be constructed (Exercise 5.6) to display the super-Higgs mechanism.

5.7 Hidden-sector supersymmetry breaking

The most successful applications of supergravity to the construction of supersymmetric grand unified models (see Chapter 6 and the review of Nilles[(4)]) have the supersymmetry breaking occurring in a 'hidden sector', by which is meant a sector of the theory that couples to the 'observable sector', of quark, leptons, gauge fields, Higgses and their supersymmetric partners, only through gravitational interactions. The simplest model of a supersymmetry-breaking hidden sector uses the Polonyi superpotential for a single gauge-singlet chiral superfield Φ,

$$W(\Phi) = m^2(\Phi + \beta) \tag{5.98}$$

where m and β are real parameters with dimensions of mass. Let us also for simplicity adopt minimal kinetic terms so that the Kähler potential is

$$G = \varphi^* \varphi + \ln |W|^2 \tag{5.99}$$

as in (5.63). Then, the tree level effective potential derived from (5.64) is

$$V = m^4 \mathrm{e}^{\varphi^* \varphi}(|1 + \varphi^*(\varphi + \beta)|^2 - 3|\varphi + \beta|^2). \tag{5.100}$$

For the special choice

$$\beta = 2 - \sqrt{3} \tag{5.101}$$

it may be shown (Exercise 5.7) that V has an absolute minimum at

$$\varphi = \varphi_0 = \sqrt{3} - 1 \tag{5.102}$$

with $V = 0$ (and so the desirable feature of a vanishing cosmological constant in the physical vacuum). At this minimum,

$$\frac{\partial W}{\partial \varphi} + \varphi^* W = \sqrt{3} m^2 \tag{5.103}$$

is non-zero and consequently supersymmetry is broken, as discussed in §5.6.

The Majorana fermion Ψ is the Goldstone fermion and is 'eaten' by the gravitino to give the gravitino a mass. Using (5.80), we then find

$$m_{3/2} = \frac{\exp(\sqrt{3}-1)^2}{2}\frac{m^2}{m_{\mathrm{P}}^2}m_{\mathrm{P}} \tag{5.104}$$

where the factors of m_{P} have been restored in $\mathrm{e}^{G_0/2}$. It is worth noticing that the gravitino mass can be much smaller than the Planck mass if m/m_{P} is small.

In (2.114), the supersymmetry-breaking scale M_S^2 for a globally supersymmetric theory was defined as the expectation value of the F-term responsible for supersymmetry breaking. For the case of supergravity, comparing (5.66) with (2.87), the supersymmetry-breaking scale is defined (almost) correspondingly by

$$M_S^2 = \mathrm{e}^{G/2}(G^{-1})_i^j G_j \tag{5.105}$$

where it is understood that the right-hand side denotes the expectation value taken at the absolute minimum of the effective potential. In the present case

$$M_S^2 = \mathrm{e}^{G/2}\left(\varphi^* + \frac{1}{W}\frac{\partial W}{\partial \varphi}\right) \tag{5.106}$$

and we find

$$M_S^2 = \sqrt{3}m_{3/2}m_{\mathrm{P}}. \tag{5.107}$$

(This is a general result for theories where the supersymmetry-breaking absolute minimum has $V = 0$.) Conversely,

$$m_{3/2} = M_S^2/\sqrt{3}m_{\mathrm{P}} \tag{5.108}$$

which means that the gravitino mass will be small compared with the supersymmetry-breaking scale whenever M_S is small on the Planck scale. For instance, a gravitino mass of the order of 100 GeV is obtained when M_S is of the order of 10^{10} GeV.

The masses of the fermions and scalars arising from the chiral supermultiplet Φ and the supergravity multiplet may be calculated in terms of $m_{3/2}$. For the fermions, equation (5.92) gives the mass matrix in the physical vacuum

$$|\det e|^{-1}\mathscr{L}_{\mathrm{F}}^m = \frac{\mathrm{i}}{2}m_{3/2}\bar{\Psi}'_\mu\sigma^{\mu\nu}\Psi'_\nu. \tag{5.109}$$

This is as expected because the fermion from the supermultiplet Φ is the Goldstone fermion that has been 'eaten' by the gravitino, and is in agreement with the remarks following (5.89). For the scalars, the mass terms $\mathscr{L}_{\mathrm{B}}^m$ arising from the (5.38) are given by (Exercise 5.8)

$$\mathscr{L}_{\mathrm{B}}^m = -2m_{3/2}^2\varphi'^*\varphi' - 2(\sqrt{3}-1)m_{3/2}^2(\varphi'\varphi' + \varphi'^*\varphi'^*) \tag{5.110}$$

where we have written

$$\varphi = \varphi_0 + \varphi' \tag{5.111}$$

with φ_0 the expectation value of φ in the physical vacuum as in (5.102). Defining real scalar fields A and B by

$$\varphi' = \frac{1}{\sqrt{2}}(A + \mathrm{i}\,B) \tag{5.112}$$

we then find

$$m_A^2 = 2\sqrt{3}m_{3/2}^2 \tag{5.113}$$

and

$$m_B^2 = 2(2 - \sqrt{3})m_{3/2}^2. \tag{5.114}$$

For globally supersymmetric theories with only chiral supermultiplets we found in §2.10 that the supertrace of the mass-squared matrix over real fields was zero. Here we find, with the inclusion of the supergravity multiplet, that

$$\mathrm{STr}\,M^2 = \sum_{J=0}^{3/2} (-1)^{2J}(2J+1)m_J^2 = -4m_{3/2}^2 + m_A^2 + m_B^2 = 0 \tag{5.115}$$

which is the same as for the globally supersymmetric case. However, for theories with the supergravity multiplet plus N chiral supermultiplets, this generalizes to

$$\mathrm{STr}\,M^2 = 2(N-1)m_{3/2}^2. \tag{5.116}$$

There is the attractive feature that the scalar particles are required by (5.116) to be more massive on the average than their fermionic superpartners. This accords with experience, in that, for instance, the squarks and sleptons must be more massive than the quarks and leptons. As we have remarked earlier, it is possible for $m_{3/2}$ to be much smaller than the supersymmetry-breaking scale and so, even when the supersymmetry-breaking scale is large, it is possible to have modest mass splittings within supermultiplets.

5.8 Supersymmetry breaking by gaugino condensates

As already mentioned following (5.67), another possible mechanism[6] for supersymmetry breaking in a supergravity theory would be for a product of two fermionic gaugino fields to develop a VEV. Including gaugino terms (5.47) is

$$\delta\Psi_i = -\mathrm{i}\,\not{D}(A_i + \mathrm{i}\gamma_5 B_i)\xi - \sqrt{2}\,\mathrm{e}^{G/2}(G^{-1})^i_j G_j \xi - \tfrac{1}{8} f_{abj}(G^{-1})^j_i \lambda_a \lambda_b$$
$$+ \text{ other two-fermion terms} \tag{5.117}$$

with

$$f_{abj} \equiv \frac{\partial f_{ab}}{\partial \varphi^{j*}} \tag{5.118}$$

where f_{ab} is the coefficient of the gauge field strength term as in (5.54). Thus, an expectation value for $\lambda^a\lambda^b$ can break supersymmetry by making the expectation value of Ψ_i non-invariant under a supersymmetry transformation. For this to occur it is necessary for some components of f_{abj} to be non-zero, which requires non-minimal gauge field kinetic terms.

It may be possible for gaugino condensation to occur in a hidden sector of the theory if the gauge group is a product of two factors one of which contains the gauge group of the standard model. For instance, in heterotic string theories, as we shall see in Chapter 9, it is possible for the gauge group to be $E_8 \times E_8$ with the first exceptional group factor containing the standard model gauge group and the second factor providing a hidden sector. It has been argued[6] that a gaugino condensate with

$$\langle \lambda^a \lambda^b \rangle \sim \mu^3 \tag{5.119}$$

should develop if the running gauge coupling constant for the hidden sector becomes strong at an energy scale μ.

In the presence of such a gaugino condensate, the four-fermion term from (5.60),

$$|\det e|^{-1}\mathscr{L}_{\mathrm{MIX}} = \tfrac{1}{2} f^i_{ab} \bar{\Psi}_{i\mathrm{L}} \sigma^{\mu\nu} \lambda_{a\mathrm{L}} \bar{\Psi}_{\nu\mathrm{L}} \gamma_\mu \lambda_{b\mathrm{R}} + \mathrm{HC} \tag{5.120}$$

will cause

$$\eta = f^i_{ab} \langle \lambda_a \lambda_b \rangle \Psi_i \tag{5.121}$$

to mix with the gravitino. Thus, η should be identified as the Goldstone fermion. The value of the gravitino mass $m_{3/2}$ arising from the supersymmetry breaking depends on the details of the Kähler potential because of (5.80). However, the order of magnitude $m_{3/2}$ may be determined by observing that supersymmetry breaking through gaugino condensates is known *not* to occur in the case of globally supersymmetric theories. It follows that the supersymmetry breaking tends to zero as $m_{\mathrm{P}} \to \infty$, and we expect the supersymmetry-breaking scale M_S^2 to be at most

$$M_S^2 \sim \mu^3/m_{\mathrm{P}} \tag{5.122}$$

with the possibility of it being suppressed by a further power of m_{P}. Correspondingly, the gravitino mass arising from (5.108) is at most

$$m_{3/2} \sim \mu^3/m_{\mathrm{P}}^2. \tag{5.123}$$

A small gravitino mass can result even for large values of μ provided that μ is small on the scale of the Planck mass.

5.9 Supersymmetry-breaking effects in the observable sector

In §5.7 and §5.8, mechanisms by which supersymmetry may be broken in the hidden sector have been discussed. The next question that must be addressed is of how the supersymmetry breaking that occurs in the hidden sector feeds through into the observable sector. In particular, we need an effective low-energy Lagrangian[(7),(8)] including supersymmetry-breaking effects that can be used at energies small compared with the Planck scale.

The chiral superfields may be divided into hidden-sector superfields Z_i and observable-sector superfields Y_r with corresponding scalars z_i and y_r, where we are using indices $i, j, k, \ldots$ for the hidden sector, and indices $r, s, t, \ldots$ for the observable sector. For simplicity, let us assume that the superpotential is additive in the hidden and observable sectors:

$$W(Z_i, Y_r) = \bar{W}(Z_i) + \tilde{W}(Y_r). \tag{5.124}$$

(This certainly avoids any couplings other than gravitational between the two sectors.) Let us also assume that the kinetic terms are minimal, corresponding to a Kähler potential

$$G = m_{\rm P}^{-2}(z^{i*}z_i + y^{r*}y_r) + \ln(|W|^2/m_{\rm P}^6) \tag{5.125}$$

where we have displayed the Planck mass, $m_{\rm P}$. Then, the tree level effective potential (excluding D-terms) deriving from (5.38) is

$$V = \exp((z^{i*}z_i + y^{r*}y_r)/m_{\rm P}^2)\left(\left|\frac{\partial\bar{W}}{\partial z_i} + \frac{z^{i*}}{m_{\rm P}^2}(\bar{W} + \tilde{W})\right|^2 + \left|\frac{\partial\tilde{W}}{\partial y_r} + \frac{y^{r*}}{m_{\rm P}^2}(\bar{W} + \tilde{W})\right|^2 - 3m_{\rm P}^{-2}|\bar{W} + \tilde{W}|^2\right). \tag{5.126}$$

A form for the effective potential appropriate to energies small compared with the Planck scale can be obtained by replacing z_i by the expectation value at the minimum of the effective potential and working to leading order in $m_{\rm P}^{-1}$. In general, the hidden-sector expectation values at the minimum of V may be written as

$$\langle z_i \rangle = a_i m_{\rm P} \tag{5.127}$$

$$\langle \bar{W} \rangle = \mu m_{\rm P}^2 \tag{5.128}$$

and

$$\left\langle \frac{\partial \tilde{W}}{\partial z_i} \right\rangle = c_i \mu m_{\rm P} \tag{5.129}$$

where a_i and c_i are dimensionless quantities, and μ is a mass scale characterizing the expectation value of the hidden-sector superpotential. (As we shall see in a moment it is convenient here to write this scale as $\mu m_{\rm P}^2$, rather than as $m^2 m_{\rm P}$, if we were to follow §5.7.) Using (5.80), the gravitino mass is then given by

$$m_{3/2} \approx e^{|a_i|^2/2} \mu . \tag{5.130}$$

In taking the low-energy limit, we need to hold $m_{3/2}$ fixed as $m_{\rm P} \to \infty$, and, from (5.130), this means working to leading order in $\mu/m_{\rm P}$. In the low-energy limit, equation (5.126) reduces (Exercise 5.9) to

$$V = e^{|a_i|^2} \left[\left| \frac{\partial \tilde{W}}{\partial y_r} \right|^2 + \mu^2 |y_r|^2 + \mu \left(y_r \frac{\partial \tilde{W}}{\partial y_r} + (A-3)\tilde{W} + \text{cc} \right) \right] \tag{5.131}$$

where

$$A = (c_i^* + a_i)a_i^* . \tag{5.132}$$

Using (5.130), and defining a modified superpotential

$$\hat{W} \equiv e^{|a_i|^2/2} \tilde{W} \tag{5.133}$$

then (5.131) may be rewritten as

$$V = \left| \frac{\partial \hat{W}}{\partial y_r} \right|^2 + m_{3/2}^2 |y_r|^2 + m_{3/2} \left(y_r \frac{\partial \hat{W}}{\partial y_r} + (A-3)\hat{W} + \text{cc} \right) \cdot \tag{5.134}$$

The first term is of the standard form (2.51) for unbroken global supersymmetry with superpotential $\hat{W}$, and the remaining terms are supersymmetry-breaking terms. The part of the low-energy Lagrangian involving chiral spin-$\frac{1}{2}$ fermions deriving from (5.43) is as for unbroken global supersymmetry with superpotential $\hat{W}$. Thus, $m_{3/2}$ is the supersymmetry-breaking mass splitting between bosons and fermions in the same chiral supermultiplet. In the special case where the hidden-sector superpotential is the Polonyi superpotential of §5.7, it is not difficult (Exercise 5.10) to check that

$$A = 3 - \sqrt{3} \qquad \text{Polonyi superpotential.} \tag{5.135}$$

The parameters A and $m_{3/2}$ in (5.134) should be understood as being defined at the Planck scale in the first instance. Thus, to employ the effective potential at the electroweak scale it will be necessary to run the parameters between the two energy scales by means of renormalization group equations, as will be discussed in Chapter 6.

An attractive possibility for the observable-sector superpotential $\hat{W}$ is that it should be trilinear in the chiral superfields Y_r, so as to avoid any small

(~100 GeV) adjustable mass parameters. Then, the effective potential simplifies to

$$V = \left|\frac{\partial \hat{W}}{\partial y_r}\right|^2 + m_{3/2}^2|y_r|^2 + Am_{3/2}(\hat{W} + \hat{W}^*). \tag{5.136}$$

A feature of (5.134) and (5.136) is that there is a universal supersymmetry-breaking mass $m_{3/2}$ for the scalars y_r from all chiral supermultiplets. It is possible to obtain effective potentials in which this universality is absent by taking the kinetic terms for the chiral superfields to be non-minimal. However, universal supersymmetry-breaking scalar masses may be desirable because flavour-changing neutral currents are then avoided in a natural way.

In general, the low-energy effective potential may contain D-terms. These are readily obtained from (5.56), and, for G of the form (5.125), the complete low-energy effective potential is

$$V = \left|\frac{\partial \hat{W}}{\partial y_r}\right|^2 + m_{3/2}^2|y_r|^2 + m_{3/2}\left(y_r\frac{\partial \hat{W}}{\partial y_r} + (A-3)\hat{W} + \mathrm{cc}\right)$$
$$+ \tfrac{1}{2}\operatorname{Re} f_{ab}^{-1} D_a D_b \tag{5.137}$$

with

$$D_a = g y^{r*}(T_a)_{rs} y_s. \tag{5.138}$$

Another possible source of supersymmetry breaking in the observable sector is the occurrence of gaugino masses while the corresponding gauge fields remain massless; so there are mass splittings within the vector supermultiplets. For this to be possible at tree level, it is necessary that non-minimal gauge kinetic terms should be present with f_{ab} in (5.54) a non-trivial function of the chiral superfields. Then, the first term in (5.60),

$$\mathcal{L}_{\mathrm{GM}} \equiv \tfrac{1}{4}\mathrm{e}^{G/2}\frac{\partial f_{ab}{}^*}{\partial \varphi^{j*}}(G^{-1})^j_k G^k \lambda_a \lambda_b \tag{5.139}$$

can induce gaugino masses provided that G^k is non-zero for some value of k, which is the case whenever supersymmetry is broken, as discussed in §5.5. Observing that $\partial f_{ab}{}^*/\partial \varphi^{j*}$ has dimensions of inverse mass, assuming a hidden-sector scalar field VEV of order m_{P} responsible for supersymmetry breaking, and defining a supersymmetry-breaking scale M_S as in (5.105), we see that the gaugino masses $m_{1/2}$ are of order of magnitude

$$m_{1/2} \sim M_S^2/m_{\mathrm{P}} \sim m_{3/2} \tag{5.140}$$

where (5.108) has been used. Thus, the gaugino masses are another possible supersymmetry-breaking effect on the scale of the gravitino mass.

5.10 No-scale supergravity

In order to solve the hierarchy problem of grand unified theories (see Chapter 6) it is necessary to ensure that the supersymmetry-breaking mass splittings in the observable sector are not greater than about 1 TeV. The discussion of §5.9 then suggests that we require a gravitino mass $m_{3/2}$ of not more than 1 TeV. In §5.7 such a gravitino mass was obtained by introducing by hand into the superpotential of the hidden sector a mass scale m that could be adjusted to obtain the required value of $m_{3/2}$ as in (5.104). This is a somewhat unnatural procedure and it might be more attractive to have a theory in which the only input mass scale is the Planck scale m_P. This can be achieved if (in the absence of observable-sector superfields) the effective potential for the hidden sector is flat (constant). Then, the expectation values of the hidden-sector scalars are undetermined at tree level and consequently the gravitino mass of (5.80) is undetermined at tree level. It may then be possible for non-gravitational radiative corrections to the effective potential to lift the degeneracy of the tree level effective potential, and so the gravitino mass $m_{3/2}$ may be hierarchically smaller than the Planck scale m_P. (When scalar vacuum expectation values arise from radiative corrections in this way they can be suppressed by an exponential factor relative to the natural mass scale because the radiative correction to the effective potential is logarithmic in the scalar VEV.) Such theories[9] are called no-scale supergravity theories.

Kähler potentials of the no-scale supergravity type occur naturally in the supergravity limit of string theories[10],[11], the simplest example being

$$G = -3\ln(z + z^*) \tag{5.141}$$

where z is a hidden-sector scalar. For this Kähler potential

$$\frac{\partial G}{\partial z} = \frac{\partial G}{\partial z^*} = -3(z + z^*)^{-1} \tag{5.142}$$

$$\frac{\partial^2 G}{\partial z\,\partial z^*} = 3(z + z^*)^{-2} \tag{5.143}$$

and consequently the effective potential

$$V = -e^G\left(3 - G_i(G^{-1})^i_j G^j\right) \tag{5.144}$$

of (5.82) gives

$$V = 0 \tag{5.145}$$

for *all* values of z.

The desired property of a gravitino mass undetermined at tree level may be retained when observable-sector superfields are incorporated provided that the tree level effective potential is positive semi-definite with a flat direction (a direction along which V is constant) along which the degeneracy in the value of z can be lifted by radiative corrections. A simple example is provided by the Kähler potential

$$G = -3\ln(z + z^* - ky^{r^*}y_r) + \ln|W|^2 \tag{5.146}$$

where k is a positive constant and the superpotential is

$$W = d_{pqr}y_p y_q y_r \tag{5.147}$$

with c a constant, and d_{pqr} numerical coefficients. For this Kähler potential

$$3(z + z^* - ky^{r^*}y_r)\mathbf{G}^{-1} = \begin{pmatrix} z + z^* & y_s \\ y^{r^*} & k^{-1}\delta^r_s \end{pmatrix} \tag{5.148}$$

where we have taken the upper index to be the row index as in (5.38). After a little algebra, equation (5.144) together with the D-terms of (5.56) gives the positive semi-definite effective potential (Exercise 5.11)

$$V = (3k)^{-1}(z + z^* - ky^{r^*}y_r)^{-2}\left|\frac{\partial W}{\partial y_r}\right|^2 + \frac{g^2}{2}\operatorname{Re} f^{-1}_{ab} D_a D_b \tag{5.149}$$

where

$$D_a = G^r(T_a)_{rs}y_s\,. \tag{5.150}$$

The minimum of the potential occurs for

$$\frac{\partial W}{\partial y_r} = 0 \qquad D_a = 0 \qquad \text{for all } r \text{ and } a \tag{5.151}$$

and the cosmological constant is then zero. The expectation value of z is undetermined at the minimum of the tree level effective potential, and so radiative corrections can again determine $m_{3/2}$.

Unlike (5.126), the form (5.149) of V yields *no* supersymmetry-breaking scalar masses or A-terms in the low-energy limit. Thus, at first sight there is no way that supersymmetry breaking can communicate itself to the observable sector. However, in no-scale supergravity theories deriving from string theory[10] there is often a second hidden-sector scalar field S which enters the gauge kinetic term f_{ab} in the form

$$f_{ab} = \delta_{ab}S\,. \tag{5.152}$$

Then, a supersymmetry-breaking gaugino mass can arise as in (5.139). Other supersymmetry-breaking effects (such as scalar masses) can then be induced by radiative corrections when the parameters of the theory are run to the electroweak scale using renormalization group equations.

Exercises

5.1 Rewrite the supersymmetry transformation laws (5.2) and (5.3) in terms of Majorana spinor and real scalar fields.

5.2 Derive (5.10) for the variation of the Lagrangian $\mathscr{L}_0$.

5.3 Derive the next-to-leading-order variation of the Lagrangian (5.15).

5.4 Show that the supergravity Lagrangian defined by (5.38), (5.42) and (5.43) reduces to the globally supersymmetric form (2.56) with $m_{ij} = 0$ when non-renormalizable terms are neglected.

5.5 Show that the mass term of (5.89) may be rewritten in the form (5.92).

5.6 Perform the calculations necessary to demonstrate explicitly the super-Higgs mechanism for the case of D-term supersymmetry breaking.

5.7 Show that the Polonyi model effective potential of (5.100) has an absolute minimum at $\varphi_0 = \sqrt{3} - 1$.

5.8 Derive the supersymmetry-breaking scalar mass terms of (5.110).

5.9 Derive the low-energy effective potential (5.131) for hidden-sector supersymmetry breaking of supergravity.

5.10 Show that the supersymmetry-breaking parameter A has the value $3 - \beta$ for the Polonyi superpotential.

5.11 Derive the effective potential (5.149) from the Kähler potential (5.146).

References

General references

Nath P, Arnowitt R and Chamseddine A H 1984 *Applied $N = 1$ Supergravity* (Singapore: World Scientific)

Nilles H P 1984 *Phys. Rep.* **110** 1

Srivastava P P 1986 *Supersymmetry, Superfields and Supergravity: An Introduction* (Bristol: Institute of Physics Publishing)

van Nieuwenhuizen P 1981 *Phys. Rep.* **68** 189

West P 1986 *Introduction to Supersymmetry and Supergravity* (Singapore: World Scientific)

References in the text

1 van Nieuwenhuizen P 1981 *Phys. Rep.* **68** 189 §1.11

2 West P 1986 *Introduction to Supersymmetry and Supergravity* (Singapore: World Scientific) Chapter 13 and references therein

3 Cremmer E, Julia B, Scherk J, Ferrara S, Girardelli L and van Nieuwenhuizen P 1979 *Nucl. Phys.* B **147** 105
4 Nilles H P 1984 *Phys. Rep.* **110** 1 §3
5 Cremmer E, Ferrara S, Girardello L and van Proeyen A 1983 *Nucl. Phys.* B **212** 413
6 Nilles H P 1982 *Phys. Lett.* **115B** 193
Ferrara S, Girardello L and Nilles H P 1983 *Phys. Lett.* **125B** 457
7 Nilles H P 1984 *Phys. Rep.* **110** 1 §6
8 Barbieri R, Ferrara S and Savoy C A 1982 *Phys. Lett.* **119B** 343
Chamseddine A H, Arnowitt R and Nath P 1982 *Phys. Rev. Lett.* **49** 970
9 Cremmer E, Ferrara S, Kounnas C and Nanopoulos D V 1983 *Phys. Lett.* **133B** 61
Ellis, J, Lahanas A B, Nanopoulos D V and Tamvakis K 1984 *Phys. Lett.* **134B** 429
Ellis J, Kounnas C and Nanopoulos D V 1984 *Nucl. Phys.* B **247** 373
10 Witten E 1985 *Phys. Lett.* **155B** 151
Bailin D, Love A and Thomas S 1986 *Nucl. Phys. B* **273** 537
Ferrara S, Kounnas C and Porrati M 1986 *Phys. Lett.* B **181** 263
11 Antoniadis I, Ellis J, Floratos E, Nanopoulos D V and Tomaras T 1987 *Phys. Lett.* **191B** 96
Ferrara S, Girardello L, Kounnas C and Porati M 1987 *Phys. Lett.* **192B** 368

6

SUPERGRAVITY GRAND UNIFIED THEORIES

6.1 The hierarchy problem

We noted in Chapter 3 that there is indirect evidence both for supersymmetry and for grand unification from the calculated evolution of the QCD and electroweak coupling constants (α_3, α_2, $\frac{5}{3}\alpha_1$) to a common value of $\alpha_G \simeq \frac{1}{26}$ at an energy scale m_X of order 10^{16} GeV. If these indications are confirmed experimentally, by the discovery of supersymmetric particles on LEP200 or LHC, for example, the most pressing questions will be to determine *what* supersymmetric grand unified theory (GUT) describes nature above this energy scale, and to understand how its gauge symmetry and supersymmetry are broken to the familiar gauge theories that we observe. We shall address these shortly.

First we discuss the generic 'hierarchy problem' that is inherent in all GUTS, namely the huge disparity between the (huge) energy scale at which the GUT symmetry is broken and the (TeV) energy scale characterizing the familiar breaking of the electroweak symmetry. The advantage of a supersymmetric theory is that these two scales may be built into the tree level effective potential, and the non-renormalization theorems discussed earlier then ensure that higher-order loop radiative corrections do not destroy the hierarchy. However, even supersymmetric theories do not explain the origin of the two scales in the parameters of the input potential. In a supergravity theory emerging from string theory the GUT scale may plausibly be near the string scale (related to the Planck scale of 10^{18} GeV), but the origin of the electroweak scale remains problematic. At the time of writing the favoured solution is that the electroweak symmetry breaking is driven by radiative corrections, in which the mass-squared term for the electroweak scalar 'runs' from a positive value at the GUT scale to a negative value at the TeV scale, thereby inducing the spontaneous breaking of the electroweak symmetry. To investigate such a scenario we first need to formulate the supergravity GUT. This is done in §6.2 for the minimal SU(5) theory. In §6.3 we take universal soft supersymmetry-breaking masses for the scalars (m_0) and gauginos ($m_{1/2}$), as well as trilinear scalar interactions with their associated A-parameters. We then formulate the (coupled) renormalization group equations for the Yukawa couplings, scalar masses and A-parameters and investigate the possibility of obtaining the required electroweak symmetry breakdown. The soft supersymmetry breaking also induces mix-

DOI: 10.1201/9780367805807-6

ing between the gauginos and Higgsinos. This is discussed in §6.4, and the current status of experimental searches for these and other sparticles is reviewed in §6.5. In §6.6 we consider proton decay in the context of supersymmetric GUTS. The principal generic difference compared with non-supersymmetric GUTS arises from the existence of additional mechanisms for inducing proton decay, some of them at an unacceptably high rate. We discuss the removal of the unwanted terms by the assumption of additional (discrete or continuous) symmetry.

The limited evidence for supersymmetry is most welcome for reasons that were briefly alluded to in Chapter 1. These relate to the 'hierarchy' problem that afflicts all non-supersymmetric GUTS. All GUTS require at least two levels of spontaneous symmetry breaking. The first is to break the GUT symmetry, and this is achieved by a scalar field Φ acquiring a vacuum expectation value (VEV)

$$\langle 0|\Phi|0\rangle = V = \mathrm{O}(10^{15}\,\mathrm{GeV}). \tag{6.1}$$

The second is the familiar electroweak breaking which is achieved by the neutral component φ of the Higgs doublet acquiring a VEV

$$\langle 0|\varphi|0\rangle = v \simeq 246\,\mathrm{GeV}. \tag{6.2}$$

The hierarchy problem derives from the vast difference between these two scales

$$V/v = \mathrm{O}(10^{13}) \tag{6.3}$$

and the difficulty of arranging this *in a natural way*. The tree level effective potential for these fields has the generic form

$$V_0(\Phi, \varphi) = -\tfrac{1}{2}A\Phi^2 + \tfrac{1}{4}B\Phi^4 - \tfrac{1}{2}a\varphi^2 + \tfrac{1}{4}b\varphi^4 + \tfrac{1}{2}\lambda\Phi^2\varphi^2. \tag{6.4}$$

The GUT symmetry breakdown (6.1) is achieved by choosing A, B such that

$$V^2 = A/B$$

and this sets the mass scale for the superheavy Higgs particle(s) associated with Φ. The problem arises at the electroweak symmetry breaking (6.2), since the superheavy scale V is communicated to the φ-sector by the term proportional to λ in (6.4). Such a term is always present, since a representation multiplied by its complex conjugate always includes the singlet representation. The superheavy Higgs fields decouple, and the required symmetry breaking is ensured by choosing

$$v^2 = (\mathrm{a} - \lambda V^2)/b. \tag{6.5}$$

Evidently to achieve the required value (6.2) it is necessary to fine-tune the value of a to one part in 10^{26}, and it is this fine-tuning that is thought to be so unnatural. It might not seem so distasteful if it only had to be done once. However, radiative corrections produce loop corrections to the effective potential and we have to retune at each order in perturbation theory to accommodate the induced corrections to a, λ, V, b. Even so it should be borne in mind that it *is* only a matter of taste; after all, we are by now quite accustomed to doing something that, in a way, is even worse. When we renormalize we have to choose counter-terms that cancel the *infinities* that arise from the divergent loop momentum integrations. It is (only) the further finite retunings that are now regarded as unaesthetic.

The advantage of supersymmetric theories is that it is only the tree level potential whose parameters have to be fine-tuned, as in (6.5). The non-renormalization theorems, discussed in §§2.6, 2.7, ensure that loop corrections do not destroy the hierarchy, so there is no necessity to retune. Of course, since we know that supersymmetry must be broken, the situation is not quite so clean. The non-renormalization theorems derive from a cancellation of radiative corrections between the contributions from supersymmetric partners. Thus if the average supersymmetry-breaking mass-squared splittings within supermultiplets are of order μ^2, we shall obtain finite radiative corrections from graphs like (2.71, 2, 3) with

$$\delta m_x^2 \sim \frac{\lambda^2}{8\pi^2}\mu^2 \tag{6.6}$$

with λ some Yukawa or gauge coupling constant. So the retuning problem will not arise provided the supersymmetry-breaking mass splitting μ is of order v:

$$\mu \sim v. \tag{6.7}$$

In other words the super-partners of the known quarks and leptons and gauge particles should all have masses less than or of order 1 TeV or so, if we insist that the hierarchy problem is to be solved by supersymmetry. This then suggests, but does not require, that we should attempt to relate supersymmetry breakdown to the electroweak symmetry-breaking mechanism.

We have seen in Chapter 5 that an attractive mechanism for achieving the required supersymmetry breakdown in supergravity theories is to assume the existence of a 'hidden' sector of the theory, involving particles that interact only gravitationally with the quarks, leptons, gauge particles of the 'observable' sector; such a situation is realized naturally in heterotic string theories. In the observable sector the supersymmetry breaking is manifested by the appearance of (soft) mass terms for the scalars and/or the gauginos. The masses of the scalars (m_0) and the masses of the gauginos ($m_{1/2}$) are of the order of the gravitino mass ($m_{3/2}$):

$$m_0 \sim m_{1/2} \sim m_{3/2} \sim M_S^2/m_P \tag{6.8}$$

where M_S measures the supersymmetry breaking scale. Then the required TeV scale for the particle–sparticle mass splitting is obtained when

$$M_S = \mathrm{O}(10^{11}\,\mathrm{GeV}). \tag{6.9}$$

In turn such a value for M_S might be understood as arising in a no-scale theory as a hierarchic suppression of the Planck scale m_{P}. However, this does not explain *why* the electroweak symmetry breaking scale (v) is of the same order of magnitude O(1 TeV) as the supersymmetry-breaking mass splittings (6.7).

Indeed, if we add to the supersymmetric standard model superpotential (3.139) the only renormalizable term involving the Higgs chiral superfields H_1, H_2 that is consistent with SU(2) × U(1) symmetry†

$$\tilde{W}_1 = mH_1^{\mathrm{T}}\,\mathrm{i}\,\tau^2 H_2 \tag{6.10}$$

then the F-terms generate mass terms

$$\sum_i \left|\frac{\partial \tilde{W}_1}{\partial H_i}\right|^2 = m^2(h_1^\dagger h_1 + h_2^\dagger h_2) \tag{6.11}$$

for the Higgs scalar components h_1, h_2, of H_1, H_2. We see that, since $m^2 > 0$, there is no possibility of breaking the SU(2) × U(1) symmetry spontaneously. Furthermore the additional terms in the low-energy effective potential (5.137), which arise in models possessing hidden-sector supergravity breaking, also do not generate the necessary negative mass-squared terms. Thus, far from relating the electroweak breaking to the supersymmetry breakdown, we cannot even achieve spontaneous symmetry breaking.

The simplest way out is to introduce a new SU(2) × U(1) singlet chiral superfield Y, and instead of $\tilde{W}_1$ add the term

$$\tilde{W}_2 = \lambda Y(H_1^{\mathrm{T}}\,\mathrm{i}\,\tau^2 H_2 - \mu^2) \tag{6.12}$$

to the standard model superpotential. Then the effective potential (5.137) for the Higgs sector is

$$\begin{aligned} V = {} & \lambda^2|h_1^{\mathrm{T}}\,\mathrm{i}\,\tau^2 h_2 - \mu^2|^2 + \lambda^2 y^* y(h_1^\dagger h_1 + h_2^\dagger h_2) \\ & + m_{3/2}^2(y^* y + h_1^\dagger h_1 + h_2^\dagger h_2) + m_{3/2}\lambda A(y h_1^{\mathrm{T}}\,\mathrm{i}\,\tau^2 h_2 + \mathrm{HC}) \\ & + (2 - A)\lambda\mu^2(y + y^*) + \tfrac{1}{8}g'^2(h_1^\dagger h_1 - h_2^\dagger h_2)^2 \\ & + \tfrac{1}{8}y^2(h_1^\dagger \tau^i h_1 - h_2^\dagger \tau^i h_2)^2 \end{aligned} \tag{6.13}$$

where the last two terms are the D-terms. In writing (6.13) we have suppressed the factors $\exp(\frac{1}{2}|a_i|^2)$ in the definitions (5.130) and (5.133) of $m_{3/2}$ and W. They may be restored at the end by rescaling all masses with this

†We ignore until §6.6 the possibility of adding terms that do not conserve baryon number and/or lepton number.

factor. It is easy to verify (Exercise 6.1) that (6.13) does lead to the spontaneous breaking of SU(2) × U(1) to $U(1)_{em}$ provided

$$(\lambda\mu)^2 > m_{3/2}^2. \qquad (6.14)$$

We shall not proceed further with the minimization of V. Suffice it to say that the magnitude of (the common value of) the VEVS of h_1 and h_2 is determined by the parameter μ that we had to introduce in order to achieve the SU(2) × U(1) breaking. Thus the electroweak breaking scale is effectively input via μ and is not related to the supersymmetry-breaking scale. The solution to this problem that is favoured at present is that the electroweak symmetry breaking is induced by radiative corrections. In this scenario the mass-squared term for the electroweak Higgs fields is positive (or zero) at the GUT scale (6.1), but as the couplings evolve, because of radiative corrections, the renormalization group equations drive the mass-squared negative at the TeV scale (6.2), thereby causing the electroweak transition. We shall see later how this can happen. First we formulate the minimal SU(5) supergravity GUT.

6.2 The minimal SU(5) supergravity GUT

We have described in the previous chapter (§5.4) how to write down the supergravity Lagrangian for general (gauge) vector superfields coupled to general chiral supermultiplets. The procedure starts with a version of the global supersymmetric theory, generalized, by the introduction of the Kähler potential (G or K) and the function f_{ab}, to include non-renormalizable terms. The original (renormalizable) theory is fixed once the superpotential $W(\Phi)$ is known. In the case of the SU(5) GUT, the chiral superfields must include those associated with the three generations of chiral fermions. These are denoted by $\psi^{(f)i}$ ($i = 1, 2, \ldots, 5$) transforming as the $\bar{\mathbf{5}}$ representation of SU(5), and $\chi^{(f)}_{[ij]}$, which transforms as the **10** representation; $f = 1, 2, 3$ labels the generations. In addition there is the adjoint of scalars Σ^a ($a = 1, \ldots, 24$) necessary to break the SU(5) symmetry to SU(3) × SU(2) × U(1), and *two* further Higgs scalar multiplets, denoted $H_i, \bar{H}^i$ ($i = 1, \ldots, 5$) associated with the electroweak symmetry breakdown; H transforms as a **5**, and $\bar{H}$ as a $\bar{\mathbf{5}}$. The couplings required to generate fermion masses derive from the superpotential

$$\tilde{W}_m = \sum_{f,g} [m_{fg}^{(d)} \psi^{(f)i} \chi^{(g)}_{[ij]} \bar{H}^j + m_{fg}^{(u)} \epsilon^{ijklm} \chi^{(f)}_{[ij]} \chi^{(g)}_{[kl]} H_m]. \qquad (6.15)$$

In addition we need the terms that lead to SU(5) breaking. It is convenient to represent the adjoint scalars by a (traceless Hermitian 5 × 5) matrix

$$\Sigma^i_j = \Sigma^a (t^a)^i_j \qquad (6.16)$$

where t^a are the 5×5 matrices representing SU(5).

Then the required breaking can arise from

$$\widetilde{W}_\Sigma = \lambda_1(\tfrac{1}{3}\,\mathrm{tr}\,\Sigma^3 + \tfrac{1}{2}M_1\,\mathrm{tr}\,\Sigma^2). \tag{6.17}$$

Ignoring the supergravity contributions for the moment, the effective potential is given by

$$\begin{aligned} V = \sum_a \left|\frac{\partial \widetilde{W}_\Sigma}{\partial \Sigma^a}\right|^2 &= \mathrm{tr}\left|\frac{\partial \widetilde{W}_\Sigma}{\partial \Sigma^i_j} - \tfrac{1}{5}\delta^i_j \frac{\partial \widetilde{W}_\Sigma}{\partial \Sigma^k_k}\right|^2 \\ &= \lambda_1^2\,\mathrm{tr}\,|\Sigma^2 + M_1\Sigma - \tfrac{1}{5}\,\mathrm{tr}\,\Sigma^2|^2 \\ &= \lambda_1^2[\mathrm{tr}\,\Sigma^4 - \tfrac{1}{5}(\mathrm{tr}\,\Sigma^2)^2 + M_1^2\,\mathrm{tr}\,\Sigma^2 + 2M_1\,\mathrm{tr}\,\Sigma^3]. \end{aligned} \tag{6.18}$$

It is easy to verify that the F-term

$$F_\Sigma \equiv \Sigma^2 + M_1\Sigma - \tfrac{1}{5}\,\mathrm{tr}\,\Sigma^2 \tag{6.19}$$

is zero for three independence choices of Σ. These are

$$\Sigma = 0 \tag{6.20a}$$

$$\Sigma = \tfrac{1}{3}M_1\,\mathrm{diag}(1, 1, 1, 1, -4) \tag{6.20b}$$

$$\Sigma = M_1\,\mathrm{diag}(2, 2, 2, -3, -3). \tag{6.20c}$$

Thus V has three degenerate minima at which $V = 0$, so supersymmetry is unbroken. Clearly, at the first SU(5) is unbroken, at the second it is broken to SU(4) × U(1), and at the third to SU(3) × SU(2) × U(1). This is a generic feature of globally supersymmetric theories. However, in the case of supergravity GUTs we have seen how the supersymmetry may be broken via gravitational interactions with a hidden sector and so we might anticipate that the supergravity contributions in (5.137) will lift the degeneracy and, we hope, select the SU(3) × SU(2) × U(1) minimum as the preferred phase. In the presence of the supergravity corrections, the stationary points of the effective potential occur when

$$F_\Sigma + m_P^{-2}\Sigma(\widetilde{W} + \langle\widetilde{W}\rangle) = 0 \tag{6.21}$$

where $\langle\widetilde{W}\rangle$ is given by (5.128). In the limit $m_P \to \infty$ this gives

$$\Sigma^2 + (M_1 + \mu/\lambda_1)\Sigma - \tfrac{1}{5}\,\mathrm{tr}\,\Sigma^2 = 0. \tag{6.22}$$

Thus, as in (6.20), there are three solutions, but the positions are shifted by the replacement

$$M_1 \to M_1 + \mu/\lambda \tag{6.23}$$

which is small if $m_{3/2}$ is of TeV order, while M_1 is on the GUT scale. These (supersymmetric) vacua are no longer degenerate, since V has the value

$$V = e^{|a_i|^2} 2\mu(A-3)\tilde{W} \tag{6.24}$$

at such stationary points, and $\tilde{W}$ has different values for each of the three minima. Allowing for the shift (6.23), we find, corresponding to the three cases in (6.20),

$$e^{|a_i|^2}\tilde{W} = 0 \tag{6.25a}$$

$$e^{|a_i|^2}\tilde{W} = \tfrac{10}{27}\lambda_1(M_1 + \mu/\lambda_1)^2(M_1 - 2\mu/\lambda_1) \tag{6.25b}$$

$$e^{|a_i|^2}\tilde{W} = 5\lambda_1(M_1 + \mu/\lambda_1)^2(M_1 - 2\mu/\lambda_1). \tag{6.25c}$$

Thus if $A < 3$, as it is for the Polonyi potential, the global minimum is the SU(3) × SU(2) × U(1) phase, as required.

Besides the terms (6.15) and (6.17) of the superpotential, we must have in addition the terms

$$\tilde{W}_H = \lambda_2 \bar{H}(\Sigma + 3M_2)H. \tag{6.26}$$

At the SU(3) × SU(2) × U(1) minimum we may replace Σ by its VEV to obtain an effective superpotential that generates mass terms for the colour triplet and doublet scalars in H and $\bar{H}$. Explicitly we see

$$m_H(\mathbf{3}) = m_{\bar{H}}(\bar{\mathbf{3}}) = \lambda_2[3M_2 + 2(M_1 + \mu/\lambda_1)] \tag{6.27a}$$

$$m_H(\mathbf{2}) = m_{\bar{H}}(\mathbf{2}) = 3\lambda_2[M_2 - (M_1 + \mu/\lambda_1)]. \tag{6.27b}$$

We have seen that M_1 must be of order 10^{16} GeV in order to break the SU(5) symmetry at the 'observed' scale. This is welcome as a mass scale for the Higgs colour triplet scalars, since they can mediate proton decay, and would do so at an unacceptably high rate if their masses were less than O(10^{15} GeV). However, we also require light electroweak Higgs doublet scalar fields h_1, h_2, and this requires

$$M_1 = M_2 \tag{6.28}$$

to TeV accuracy. Dropping the massive colour triplet states then gives an effective SU(2) × U(1) superpotential of the form (6.10) with

$$m = -3\lambda_2 m_{3/2}/\lambda_1 \tag{6.29}$$

and we noted before that this cannot generate spontaneous breaking of the electroweak gauge group. However, we must also include the supergravity-induced contributions to the effective potential, besides the global supersymmetric contribution (6.11). Including such terms, we find, using (5.137),

$$V_{\text{eff}} = (m^2 + m_{3/2}^2)(h_1^\dagger h_1 + h_2^\dagger h_2) + m m_{3/2}[(A-1)h_1^{\text{T}}\,\mathrm{i}\,\tau^2 h_2 + \text{HC}] \tag{6.30}$$

which yields the required symmetry breaking, provided that

$$m^2 - m m_{3/2}|A-1| + m_{3/2}^2 < 0. \tag{6.31}$$

This is satisfied for certain values of $m/m_{3/2}$ provided that

$$|A-1|>2. \tag{6.32}$$

Of course the required TeV scale for m is a result of the fine-tuning (6.28), which we have input; nothing in the model requires it to be so, and in this sense we are no better off than in the original model (6.12).

What we really need is a model that has no mass scale input, and in which all low-energy mass scales derive from the supersymmetry breakdown which is transmitted to the observable sector by the parameter $m_{3/2}$. Since there are to be no explicit mass parameters in the superpotential, it must be trilinear in the fields.

We discussed just this possibility in §5.9. It leads to the effective potential given in (5.136) since

$$y_r \frac{\partial \hat{W}}{\partial y_r} = 3\hat{W}. \tag{6.33}$$

Then we can also write

$$V_{\text{eff}} = \left|\frac{\partial \hat{W}}{\partial y_r}\right|^2 + m_{3/2}^2 |y_r|^2 + \tfrac{1}{3} m_{3/2}\left[A y_r \frac{\partial \hat{W}}{\partial y_r} + \text{cc}\right]. \tag{6.34}$$

Since $\hat{W}$ is homogeneous, the effective potential is zero when all fields have zero VEVS. Thus the only way to obtain symmetry breakdown, which requires non-zero VEVS, is when V_{eff} is negative. However, it is easy to see that this can happen only if

$$|A|>3. \tag{6.35}$$

To illustrate this consider a simplified model with three singlet (chiral superfields) and

$$\hat{W} = \lambda XYZ. \tag{6.36}$$

Then

$$V_{\text{eff}} = \lambda^2[|XY|^2 + |YZ|^2 + |ZX|^2] + m_{3/2}^2[|X|^2 + |Y|^2 + |Z|^2] + m_{3/2}\lambda(AXYZ + \text{HC}) \tag{6.37}$$

where we are now using the same symbol to denote the chiral superfield and its scalar component. The symmetry requires that V_{eff} attains its minimum when

$$|\langle X\rangle| = |\langle Y\rangle| = |\langle Z\rangle| \equiv v \tag{6.38}$$

and

$$\arg A + \arg\langle X\rangle + \arg\langle Y\rangle + \arg\langle Z\rangle = \pi. \tag{6.39}$$

Then the stationary points are given by (Exercise 6.2)

$$v = 0,\ \tfrac{1}{4}\frac{m_{3/2}}{\lambda}[|A| \pm \sqrt{|A|^2 - 8}] \tag{6.40}$$

at which the potential has the value

$$V_{\text{eff}} = v^2(m_{3/2}^2 - \lambda^2 v^2). \tag{6.41}$$

When $|A| > 3$,

$$\lambda_v = \tfrac{1}{4}m_{3/2}[|A| + \sqrt{|A|^2 - 8}] > m_{3/2} \tag{6.42}$$

so the non-trivial minimum is the global minimum, as claimed in (6.35).

To realize this mechanism in electroweak theory it is essential to enlarge the Higgs sector, since we cannot make an SU(2) singlet using three doublets. The 'cheapest' enlargement is to introduce a *singlet* superfield Y and take[1]

$$\tilde{W}_H = \lambda H_1^{\mathrm{T}}\,\mathrm{i}\,\tau^2 H_2 Y + \tfrac{1}{3}\sigma Y^3. \tag{6.43}$$

Then, as above, when $|A| > 3$ there is an absolute minimum of the effective potential at which the SU(2) × U(1) symmetry is broken. Unfortunately, when we include the further terms in the superpotential that are required to generate the fermion masses, the absolute minimum typically occurs at a place where the slepton and squark fields have a non-zero VEV: in other words the $SU(3) \times U(1)_{\text{em}}$ gauge symmetry is preferentially broken. This is already pretty clear from (6.38) since the global minimum occurs when *all* fields have a non-zero VEV, including in our case coloured and charged scalars.

The best bet at this juncture is to arrange that

$$A = 3 \tag{6.44}$$

since then all of the local minima occur at V_{eff}, and so are degenerate. Radiative corrections might perhaps spoil this degeneracy, and, we hope, select the charge-conserving SU(2) × U(1)-breaking solution as the absolute minimum.

Actually the introduction of singlet fields, such as Y, introduces further unsatisfactory features for supergravity GUTs. If Y in (6.43) is an SU(5) singlet, then it can couple to the heavy colour triplet Higgs, and this threatens the stability of the TeV mass scale $m_{3/2}$ which transmits the supersymmetry breaking to the observable sector[2]. We shall not pursue this point further, but instead consider the possibility that we can sustain just the minimal Higgs sector (with *no* singlets Y). The breaking of the electroweak symmetry at tree level is then impossible, as we have seen, but it might be driven by radiative corrections.

6.3 Renormalization group equations[3]

At the unification scale (the Planck scale?) we shall assume that the scalar masses have a universal value m_0. To see whether electroweak breaking can arise from radiative corrections we need to write down the renormalization group equation for the squared mass of the electroweak Higgs scalars (h_1, h_2). Of course this equation involves other masses and coupling constants so it is necessary to consider the evolution of all of the parameters of the theory as we run the energy scale from 10^{16} GeV down to electroweak scale.

We have already written down in (3.139) the part of the observable-sector superpotential that is necessary to induce mass terms for the quarks and leptons at the electroweak symmetry breakdown. The only other term consistent with the gauge symmetry (and renormalizability) is that given in (6.10). It is, as we have said, unaesthetic to input any small mass scale (such as m) other than the gravitino mass $m_{3/2}$ deriving from supersymmetry breaking in the hidden sector. However, without such a term the theory possesses axions, as we shall see later, so we shall include it without further ado. Assuming, as in §5.9, that the full superpotential is the sum of the observable-sector piece plus a hidden sector piece $\overline{W}$, the effective potential for the scalar particles is given by

$$V_{\text{eff}} = \left|\frac{\partial \hat{W}}{\partial y_r}\right|^2 + m_r^2|y_r|^2 + m_{3/2}\left\{\sum_l A_l G^l L^{l\mathrm{T}}\,\mathrm{i}\,\tau^2 H_2 l^c \right.$$
$$+ \sum_{f,g}\left[A_d^{fg} G_{fg}^d Q^{f\mathrm{T}}\,\mathrm{i}\,\tau^2 H_2 D^{cg} + A_u^{fg} G_{fg}^u Q^{f\mathrm{T}}\,\mathrm{i}\,\tau^2 H_1 U^{cg}\right]$$
$$\left. + B\mu H_1^{\mathrm{T}}\,\mathrm{i}\,\tau^2 H_2 + \mathrm{HC}\right\} + \tfrac{1}{2}D^2 \tag{6.45}$$

where we are now using the same symbol to denote the chiral superfield and its scalar component. The scalar masses m_r^2 have the common values

$$m_r^2 = m_0^2 \tag{6.46}$$

(perhaps equal to $m_{3/2}^2$) at the unification scale. Similarly at this scale the various A-parameters have a common value:

$$A_d^{fg} = A_u^{fg} = A_l = A \tag{6.47}$$

and, as in (6.30),

$$B = A - 1. \tag{6.48}$$

Allowing for possibly non-minimal gauge kinetic functions f_{ab}^i, where $i = 1, 2, 3$ refers to the three gauge groups U(1), $\mathrm{SU_L}(2)$, $\mathrm{SU}_c(3)$, the D-terms in (6.45) are given by

$$\tfrac{1}{2}D^2 = \tfrac{1}{2}\sum_i \operatorname{Re} f^{i\,-1}_{ab} D^i_a D^i_b \tag{6.49a}$$

where

$$D^i_a = \tfrac{1}{2} g_i y^\dagger_r (T_a)_{rs} y_s. \tag{6.49b}$$

The non-minimal gauge kinetic functions can also give rise to non-zero gaugino masses, as discussed in §5.9. Thus besides the soft supersymmetry-breaking masses in (6.45) we also include gaugino (Majorana) mass terms

$$\tfrac{1}{2}\sum_i \tilde{m}_i \bar{\Lambda}^a_i \Lambda^a_i \tag{6.50}$$

where as before $i = 1, 2, 3$ labels the three gauge groups.

At the Planck scale these too have a common value

$$\tilde{m}_i = m_{1/2} \tag{6.51}$$

of order $m_{3/2}$.

Before formulating the renormalization group equations that determine whether the electroweak symmetry is broken by radiative corrections, let us first try to get some sort of feel for whether it is even feasible. The terms in (6.45) involving just the electroweak Higgs scalars lead to an effective potential having the general form

$$V_H = m_1^2 H_1^\dagger H_1 + m_2^2 H_2^\dagger H_2 - m_3^2 (H_1^{\mathrm{T}}\, \mathrm{i}\tau^2 H_2 + \mathrm{HC}) + \tfrac{1}{8} g_1^2 (H_1^\dagger H_1 - H_2^\dagger H_2)^2 + \tfrac{1}{8} g_2^2 (H_1^\dagger \tau^i H_1 + H_2^\dagger \tau^i H_2)^2 \tag{6.52}$$

and electroweak symmetry breaking occurs when this potential has a non-trivial minimum. This will happen in particular if $m_1^2 < 0$, as we shall see. The generic structure of the radiative contributions to a scalar squared mass is

$$\delta m_\varphi^2 \sim g^2 (m_F^2 - m_B^2) \tag{6.53}$$

where g measures the coupling strength, m_F^2 is the contribution from a virtual-fermion loop, and m_B is from a virtual-boson loop; the opposite sign derives from the usual -1 factor from closed fermion loops. It follows that the contributions from the gauge interactions is typically positive, since the gauge vector particles are massless, while the gauginos have a non-zero mass from the supersymmetry breaking. However, besides their electroweak gauge interactions the Higgs scalars have Yukawa couplings to the quarks and leptons, as displayed in (6.45). The largest of these is the coupling of H_1 to the top quark, since it is the most massive. Its supersymmetric partner, the top squark (or stop), is even more massive, so the top/stop contribution is negative and can in principle drive m_1^2 negative. This gives a non-zero VEV to H_1, and in fact H_2. Furthermore the non-zero VEV for H_2 leaves $\mathrm{U}(1)_{\mathrm{em}}$ unbroken, as required (Exercise 6.3).

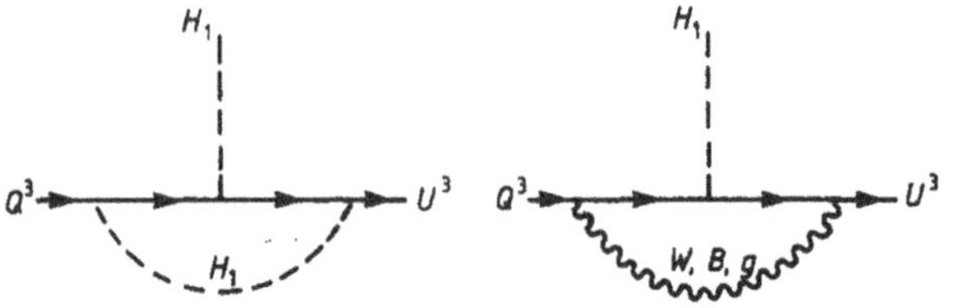

Figure 6.1 Diagrams contributing to the G^u_{33} renormalization group equation.

We denote the VEVs of the neutral components of $H_{1,2}$ by $v_{1,2}$, and these are non-zero when the minimum of

$$V_H(v_1, v_2) = m_1^2 v_1^2 + m_2^2 v_2^2 + 2m_3^2 v_1 v_2 + \tfrac{1}{8}(g_1^2 + g_2^2)(v_1^2 - v_2^2)^2 \qquad (6.54)$$

is away from the origin. Note that the quartic terms derive entirely from the D-terms, and that they vanish if $|v_1| = |v_2|$. Stability against $v_{1,2}$ running to infinity in this case therefore requires

$$m_1^2 + m_2^2 \geqslant 2|m_3^2|. \qquad (6.55)$$

Also, the breakdown of SU(2) × U(1) requires that the quadratic part is negative in some direction of the (v_1, v_2) plane, and for this it is necessary that

$$m_1^2 m_2^2 < m_3^4 \qquad (6.56)$$

which is satisfied, in particular, in the case where m_1^2 is negative, as we envisage.

We have already noted the renormalization group equations for the gauge coupling constants in §3.8. For three generations of chiral fermions ($n_{\rm G} = 3$) and the minimal Higgs content in a supersymmetric theory ($n_{\rm H} = 2$), these become

$$M\frac{d\alpha_1}{dM} = \frac{11}{2\pi}\alpha_1^2 \qquad (6.57a)$$

$$M\frac{d\alpha_2}{dM} = \frac{1}{2\pi}\alpha_2^2 \qquad (6.57b)$$

$$M\frac{d\alpha_3}{dM} = -\frac{3}{2\pi}\alpha_3^2 \qquad (6.57c)$$

at single-loop order. The corresponding equations for the Yukawa coupling constants may be computed in a similar manner. We are especially concerned with the largest of these, namely G^u_{33}, which eventually generates the top quark mass. It is customary to neglect the contributions to all diagrams from all except this Yukawa coupling. Its renormalization group equation can be computed from the diagrams shown in figure 6.1.

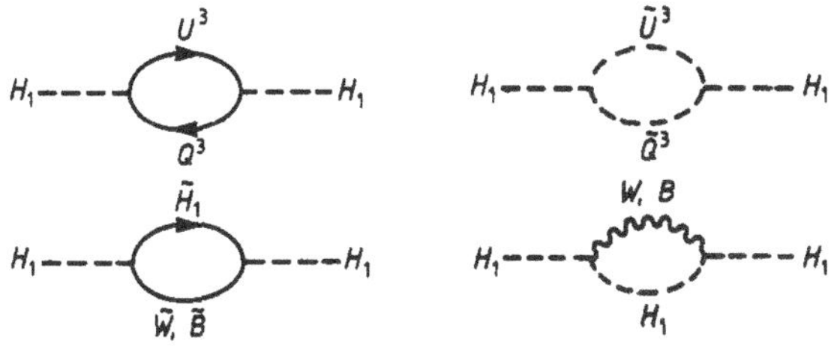

Figure 6.2 Diagrams contributing to the m_1^2 renormalization group equation.

They give

$$M\frac{dG_{33}^u}{dM} = \frac{3}{8\pi^2}(G_{33}^u)^3 - \frac{G_{33}^u}{2\pi}\left(\tfrac{8}{3}\alpha_3 + \tfrac{3}{2}\alpha_2 + \tfrac{13}{18}\alpha_1\right). \tag{6.58}$$

Similarly the corresponding three-scalar vertex given in (6.45) generates an equation for the A-parameter:

$$M\frac{d}{dM}(A_{33}^u m_{3/2}) = \frac{3}{4\pi^2}(A_{33}^u m_{3/2})(G_{33}^u)^2 - \frac{1}{\pi}\left(\tfrac{8}{3}\alpha_3\tilde{m}_3 + \tfrac{3}{2}\alpha_2\tilde{m}_2 + \tfrac{13}{18}\alpha_1\tilde{m}_1\right) \tag{6.59}$$

where $\tilde{m}_i$ are the gaugino masses, whose evolution is given by

$$\alpha_i/\tilde{m}_i = \text{constant}. \tag{6.60}$$

The spontaneous symmetry breakdown is driven by the running of m_1^2, as we have already seen. Its evolution is calculated from the diagrams in figure 6.2:

$$M\frac{dm_1^2}{dM} = \frac{3(G_{33}^u)^2}{8\pi^2}\left[m_1^2 + m_{U^3}^2 + m_{Q^3}^2 + (m_{3/2}A_{33}^u)^2\right] - \frac{2}{\pi}\left[\tfrac{3}{4}\alpha_2\tilde{m}_2^2 + \tfrac{1}{4}\alpha_1\tilde{m}_1^2\right]. \tag{6.61}$$

Evidently we also need the equations for $m_{U^3}^2$ and $m_{Q^3}^2$ which may be calculated in a similar manner:

$$M\frac{dm_{U^3}^2}{dM} = \frac{(G_{33}^u)^2}{4\pi^2}\left[m_1^2 + m_{U^3}^2 + m_{Q^3}^2 + (m_{3/2}A_{33}^u)^2\right] - \frac{2}{\pi}\left(\tfrac{4}{3}\alpha_3\tilde{m}_3^2 + \tfrac{4}{9}\alpha_1\tilde{m}_1^2\right) \tag{6.62}$$

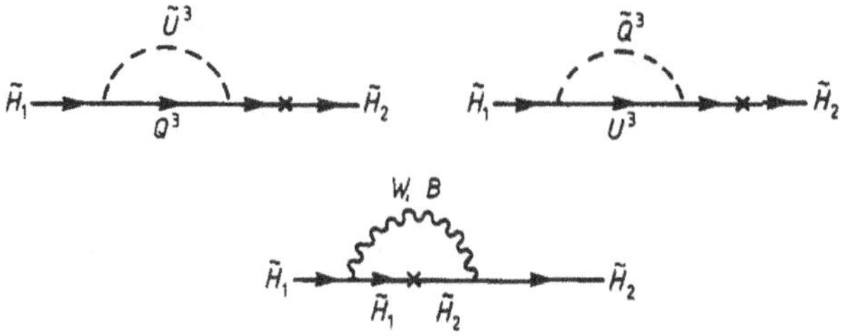

Figure 6.3 Diagrams contributing to the μ renormalization group equation.

$$M\frac{\mathrm{d}m_{Q^3}^2}{\mathrm{d}M} = \frac{(G_{33}^u)^2}{8\pi^2}[m_1^2 + m_{U^3}^2 + m_{Q^3}^2 + (m_{3/2}A_{33}^u)^2]$$
$$-\frac{2}{\pi}\left(\tfrac{4}{3}\alpha_3\tilde{m}_3^2 + \tfrac{3}{4}\alpha_2\tilde{m}_2^2 + \tfrac{1}{36}\alpha_1\tilde{m}_1^2\right). \tag{6.63}$$

The parameter m_3^2 in (6.52) is given by

$$m_3^2 = B\mu m_{3/2} \tag{6.64}$$

where the evolution of μ is determined from the diagrams in figure 6.3.
They give

$$\frac{M}{\mu}\frac{\mathrm{d}\mu}{\mathrm{d}M} = \frac{3}{16\pi^2}(G_{33}^u)^2 - \frac{1}{4\pi}(\alpha_1 + 3\alpha_2). \tag{6.65}$$

The corresponding diagrams with external scalars determine the evolution of the B-parameter

$$M\frac{\mathrm{d}}{\mathrm{d}M}(Bm_{3/2}) = -\frac{3}{8\pi^2}(A_{33}^u m_{3/2})(G_{33}^u)^2 - \frac{2}{\pi}\left(\tfrac{3}{4}\alpha_2\tilde{m}_2 + \tfrac{1}{4}\alpha_1\tilde{m}_1\right). \tag{6.66}$$

If we neglect all Yukawa couplings except for G_{33}^u then the running of all other scalar masses is determined just by the gauge couplings. In particular the mass m_2^2 in (6.52) satisfies

$$M\frac{\mathrm{d}m_2^2}{\mathrm{d}M} = -\frac{2}{\pi}\left(\tfrac{3}{4}\alpha_2\tilde{m}_2^2 + \tfrac{1}{4}\alpha_1\tilde{m}_1^2\right). \tag{6.67}$$

It is now straightforward, in principle, to perform the numerical integration of these equations for given input values of the parameters A, $m_{1/2}$, μ, $m_{3/2}$, m_0, G_{33}^u at the unification scale m_X at which the gauge coupling constants have the common value

$$\alpha_3(m_X) = \alpha_2(m_X) = \tfrac{5}{3}\alpha_1(m_X) \equiv \alpha_{\mathrm{GUT}}(m_X). \tag{3.153}$$

The values of these last two parameters are those required by the data and given in (3.164) and (3.167).

To get a feel for what actually happens we note first that the Yukawa coupling contributions to the running of the masses m_1^2, $m_{U^3}^2$, $m_{Q^3}^2$ in (6.61),

(6.62) and (6.63) have the effect of reducing the squared mass as the scale M is reduced. Further, since the coefficients of $(G^u_{33})^2$ are in the ratio 3:2:1, $m_{\bar{1}}^2$ decreases at a faster rate than the other masses squared; thus for suitable values of the parameters $m_{\bar{1}}^2$ becomes negative before the squark mass-squared terms do. By subtracting the equations it is clear that this remains true even allowing for the contributions of the gauge couplings, which oppose the decrease.

For this to work the coupling $(G^u_{33})^2$ must be large enough to drive $m_{\bar{1}}^2$ negative, and this leads to a lower bound on the mass of the top quark, since it is fixed when G^u_{33} is known:

$$m_t = G^u_{33} v_1 . \tag{6.68}$$

The actual value of the lower bound depends on the other parameters, notably A. However, we can also see from (6.58) that for large G^u_{33} the first term dominates and this *reduces* G^u_{33} as M is reduced; if G^u_{33} is too large at the unification scale, it becomes too small at lower energies to drive $m_{\bar{1}}^2$ negative. In this way an upper bound on G^u_{33}, and therefore m_t, is found. Altogether the allowed range is[4]

$$100\ \mathrm{GeV} \lesssim m_t \lesssim 200\ \mathrm{GeV} \tag{6.69}$$

with the renormalization group fixed point of (6.58) tending to attract G^u_{33} to the value

$$(G^u_{33})^2 = \frac{4\pi}{3}\left[\tfrac{8}{3}\alpha_3(m_z) + \tfrac{3}{2}\alpha_2(m_z) + \tfrac{13}{18}\alpha_1(m_z)\right] \tag{6.70}$$

corresponding to $m_t \simeq 210$ GeV. The values of $m_{3/2}$ that allow this are typically in the range

$$80\ \mathrm{GeV} \lesssim m_{3/2} \lesssim 300\ \mathrm{GeV}. \tag{6.71}$$

The renormalization group equation (6.61) shows that the running of $m_{\bar{1}}^2$ to the required negative value is assisted by the $(m_{3/2}A^u_{33})^2$ term. A larger A-value permits the symmetry breaking with a smaller value of G^u_{33}, and therefore of m_t. However, we have already noted that A cannot be too large, or else the global minimum of the scalar potential will not be that which preserves $SU(3)_c \times U(1)_{em}$. Allowing for different masses, as induced by radiative corrections, the A-parameter A_λ associated with the term λXYZ in the superpotential is constrained by[5]

$$|A_\lambda|^2 < 3(m_X^2 + m_Y^2 + m_Z^2)/m_{3/2}^2 \tag{6.72}$$

in order to avoid the unwanted minimum. (Note that, as required, this reduces to $|A| < 3$ when all masses are $m_{3/2}$.)

Incidentally this model also illustrates the source of the axion problem to which we have alluded. Evidently such a superpotential possesses a *global* U(1) symmetry in which

$$X \to e^{i\alpha_X} X \qquad Y \to e^{i\alpha_Y} Y \qquad Z \to e^{-i(\alpha_X + \alpha_Y)} Z \tag{6.73}$$

with α_X, α_Y arbitrary. One choice of α_X, α_Y is associated with the (gauged) $U(1)_{em}$ symmetry, but there is clearly an independent U(1) that is *not* gauged. The occurrence of the spontaneous symmetry breaking will violate the global symmetry and generate an unwanted Peccei–Quinn axion[(6)]. It is for this reason that the μ-term (6.10) was introduced into the potential (6.45)—it excludes any such global symmetry.

The upshot of all of this is that for a range of unification scale parameters the renormalization group equations *do* generate the required symmetry breaking, and all low-energy parameters are determined by a few basic input parameters.

In order to test the consistency of the theory with the data, we need to establish the relationship of the low-energy parameters determined by the renormalization group to the experimentally measured quantities. The low-energy effective potential is determined by the parameters m_i^2 ($i = 1, 2, 3$), α_j ($j = 1, 2$), and using (6.64) these are all known at low energies, for given input values of the parameters. Using these we may determine the VEVS $v_{1,2}$. It is convenient to use instead the related quantities v, θ defined by

$$v^2 \equiv v_1^2 + v_2^2 \tag{6.74a}$$

$$\tan\theta = v_1/v_2. \tag{6.74b}$$

Then θ is given by

$$\sin 2\theta = -\frac{2m_3^2}{m_1^2 + m_2^2} \tag{6.75}$$

(provided that the constraint (6.55) is satisfied) and v^2 by

$$2\pi(\alpha_1 + \alpha_2)v^2 = \frac{m_1^2 - m_2^2}{\cos 2\theta} + \frac{2m_3^2}{\sin 2\theta}. \tag{6.76}$$

The gauge boson masses are

$$m_W^2 = 2\pi\alpha_2 v^2 \tag{6.77a}$$

$$m_Z^2 = 2\pi(\alpha_1 + \alpha_2)v^2. \tag{6.77b}$$

The physical Higgs scalar particles are found by combining the fields in the two scalar doublets

$$H_1 = \begin{pmatrix} H_1^+ \\ H_1^0 \end{pmatrix} \qquad H_2 = \begin{pmatrix} H_2^0 \\ H_2^- \end{pmatrix}. \tag{6.78}$$

It is convenient to introduce

$$\psi_2 = -\mathrm{i}\,\tau^2 H_2^* = \begin{pmatrix} -H_2^+ \\ \bar{H}_2^0 \end{pmatrix} \tag{6.79}$$

and then define

$$\chi = (\cos\theta)H_1 - (\sin\theta)\psi_2 \tag{6.80a}$$

$$\varphi = (\sin\theta)H_1 + (\cos\theta)\psi_2 \tag{6.80b}$$

so that

$$\langle 0|\chi|0\rangle = 0 \qquad \langle 0|\varphi|0\rangle = \begin{pmatrix} 0 \\ v \end{pmatrix}. \tag{6.81}$$

Evidently χ is an 'ordinary' doublet while φ supplies the three Goldstone boson fields. This is to say, the fields χ^+, χ^0 and φ_1 in

$$\chi = \begin{pmatrix} \chi^+ \\ \chi^0 \end{pmatrix} \qquad \varphi = \begin{pmatrix} \varphi^+ \\ v + (1/\sqrt{2})(\varphi_1 + \mathrm{i}\,\varphi_2) \end{pmatrix} \tag{6.82}$$

are physical fields, and the Goldstone bosons φ^+, φ_2 are eaten in the Higgs mechanism. Expressing the low-energy potential in terms of these fields then enables us to determine the masses of the physical states. The charged scalars $\chi^\pm$ have

$$m_{\chi^\pm}^2 = m_1^2 + m_2^2 + m_\mathrm{W}^2 \tag{6.83}$$

and, writing

$$\chi^0 = \frac{1}{\sqrt{2}}(\chi_1 + \mathrm{i}\,\chi_2) \tag{6.84}$$

the field χ_2 is a pseudoscalar, and has a mass given by

$$m_{\chi_2}^2 = m_1^2 + m_2^2. \tag{6.85}$$

The two fields φ_1 and χ_1 are scalars and mix and produce eigenstates with mass eigenvalues given by

$$m_a^2, m_b^2 = \tfrac{1}{2}(m_{\chi_2}^2 + m_\mathrm{Z}^2) \pm \tfrac{1}{2}[(m_{\chi_2}^2 + m_\mathrm{Z}^2)^2 - 4m_{\chi_2}^2 m_\mathrm{Z}^2 \cos^2 2\theta]^{1/2} \tag{6.86}$$

which requires that there is one scalar *lighter* than the Z boson.

The mass parameters for the first- and second-generation squarks and sleptons satisfy renormalization group equations of a form similar to (6.67), and these can be integrated analytically; the generic structure is

$$M\frac{\mathrm{d}m^2}{\mathrm{d}M} = \frac{-1}{2\pi}\sum_i c_i \alpha_i \tilde{m}_i^2 \tag{6.87}$$

which is solved by

$$m^2(m_X) - m^2(m_Z) = \sum_i \frac{c_i}{2b_i}[\tilde{m}_i^2(m_X) - \tilde{m}_i^2(m_Z)] \tag{6.88}$$

where the b_i are given in (6.57) as

$$M\frac{\mathrm{d}\alpha_i}{\mathrm{d}M} = -b_i\alpha_i^2. \tag{6.89}$$

We shall not give details of the parameter c_i, but refer the interested reader to reference (3). However, it is important to recognize that these mass parameters are not the physical masses of the squarks and sleptons under discussion. This is because the D-terms and the trilinear scalar terms in (6.45) generate additional contributions to the squared masses when the scalars acquire non-zero VEVS, besides the explicit mass terms $m_{\tilde{f}}^2$. The determination of these additional contributions is controlled by the evolution of the various A-parameters, as well as that of the Yukawa couplings. We omit details of these too.

Because of the non-negligible Yukawa couplings the situation is even more complicated for the top squarks (the 'stops'). The two scalars $(\tilde{t}_\mathrm{L}, \tilde{t}_\mathrm{R})$ have a non-diagonal mass matrix[7] which has two eigenvalues

$$m_{\tilde{t}_1,\tilde{t}_2}^2 = \tfrac{1}{2}(m_\mathrm{LL}^2 + m_\mathrm{RR}^2) \mp [(m_\mathrm{LL}^2 - m_\mathrm{RR}^2)^2 + 4m_\mathrm{LR}^2]^{1/2} \tag{6.90a}$$

where

$$m_\mathrm{LL}^2 \equiv m_{Q^3}^2 + m_t^2 + \pi(\alpha_2 - \tfrac{1}{3}\alpha_1)v^2\cos 2\theta \tag{6.90b}$$

$$m_\mathrm{RR}^2 \equiv m_{U^3}^2 + m_t^2 + \frac{4\pi}{3}\alpha_1 v^2 \cos 2\theta \tag{6.90c}$$

$$m_\mathrm{LR}^2 \equiv vG_{33}^u[A_{33}^u m_{3/2}\sin\theta + \mu\cos\theta] \tag{6.90d}$$

are determined by integrating the renormalization equations given earlier. The important point here is that the lighter state $(\tilde{t}_1)$ can in principle be lighter than the top quark itself.

6.4 Charginos and neutralinos

We have just seen that because of the existence of the soft supersymmetry-breaking trilinear scalar interactions, as well as the inclusion of the μ-term in the superpotential, there is a non-zero coupling of the $\tilde{t}_\mathrm{L}$ squark to the $\tilde{t}_\mathrm{R}$ squark, and consequently a non-diagonal stop mass matrix, when the electroweak symmetry is spontaneously broken. As a result the mass eigenstates are superpositions of $\tilde{t}_\mathrm{L}$ and $\tilde{t}_\mathrm{R}$. A similar mixing effect occurs in the fermionic sector between the electroweak gauginos and the Higgsinos when the electroweak symmetry is broken. Of course, since charge is conserved (because $U(1)_{em}$ is unbroken) there is only mixing between states of the same charge.

We start with the mixing between the charged gauginos, the Winos $\widetilde{W}^{\pm}$, and the Higgsinos $\widetilde{H}_i^{\pm}$ $(i = 1, 2)$. We have input non-zero gaugino masses in (6.50), deriving from possibly non-minimal gauge kinetic functions in the hidden sector. The Wino mass terms come from the $a = 1, 2$ generators of the $SU_L(2)$ gauge group and can be written as

$$\tfrac{1}{2}\tilde{m}_2(\bar{\Lambda}_2^1\Lambda_2^1 + \bar{\Lambda}_2^2\Lambda_2^2) = \tilde{m}_2(\widetilde{W}^+\widetilde{W}^- + \bar{\widetilde{W}}{}^+\bar{\widetilde{W}}{}^-) \tag{6.91a}$$

where

$$\widetilde{W}^{\pm} = \frac{1}{\sqrt{2}}(\lambda_2^1 \mp \mathrm{i}\,\lambda_2^2) \tag{6.91b}$$

are Weyl spinors constructed from the Weyl spinors λ_2^a of the SU(2) gaugino Majorana spinor fields Λ_2^a. Similarly the Higgsino mass terms derive from the $\mu H_1\,\mathrm{i}\,\tau_2 H_2$ terms introduced into the superpotential for the reasons discussed earlier. The charged terms come from the $\mu H_1^+ H_2^-$ piece, which, as in (3.56), generates the mass term

$$\mu(\widetilde{H}_1^+\widetilde{H}_2^- + \bar{\widetilde{H}}{}_1^+\bar{\widetilde{H}}{}_2^-) \tag{6.92}$$

where $\widetilde{H}_1^+$, $\widetilde{H}_2^-$ are the spinor components of the superfields H_1^+, H_2^-. The bilinear coupling of these fields arises from the gauge terms displayed in (3.128*a*) when the neutral components h_1^0, h_2^0 of the scalar doublets H_1, H_2 develop VEVs. Then we find

$$g_2\sqrt{2}\sum_{\substack{a=1,2\\ i=1,2}} H_i^\dagger t^a \lambda^a \widetilde{H}_i = g_2(v_1\widetilde{W}^-\widetilde{H}_1^+ + v_2\widetilde{W}^+\widetilde{H}_2^-) \tag{6.93}$$

after spontaneous symmetry breaking. All of these mass terms may be combined using the 'chargino' mass matrix (C) as

$$(\widetilde{W}^+ \quad \mathrm{i}\,\widetilde{H}_1^+)\begin{pmatrix} \tilde{m}_2 & g_2v_2 \\ g_2v_1 & -\mu \end{pmatrix}\begin{pmatrix} \widetilde{W}^- \\ \mathrm{i}\,\widetilde{H}_2^- \end{pmatrix}. \tag{6.94}$$

We diagonalize C by forming the linear combinations

$$\begin{pmatrix} \cos\theta_\pm & -\sin\theta_\pm \\ \sin\theta_\pm & \cos\theta_\pm \end{pmatrix}\begin{pmatrix} \widetilde{W}^{\pm} \\ \mathrm{i}\,\widetilde{H}^{\pm} \end{pmatrix} \tag{6.95}$$

and choosing the angles $\theta_\pm$ appropriately. Then the mass eigenvalues are

$$\tilde{m}_2\cos\theta_+\cos\theta_- + g_2v_1\cos\theta_-\sin\theta_+ + g_2v_2\sin\theta_-\cos\theta_+ - \mu\sin\theta_+\sin\theta_- \tag{6.96a}$$

$$-\tilde{m}_2\sin\theta_+\sin\theta_- + g_2v_1\sin\theta_-\cos\theta_+ + g_2v_2\cos\theta_-\sin\theta_+ + \mu\cos\theta_+\cos\theta_-. \tag{6.96b}$$

A similar, but more complicated, treatment applies to the mixing of the neutral gauginos with the neutral Higgsinos. The two neutral electroweak gauginos are the SU(2) Wino $\tilde{W}^3$ and the U(1) Bino $\tilde{B}$. These mix with the two neutral Higgsinos after spontaneous symmetry breaking and yield the 'neutralino' mass matrix (N)

$$\tfrac{1}{2}(\tilde{W}^3 \quad \tilde{B} \quad \tilde{H}_1^0 \quad \tilde{H}_2^0)$$

$$\times \begin{bmatrix} \tilde{m}_2 & 0 & -\mathrm{i}\,g_2 v_1/\sqrt{2} & \mathrm{i}\,g_2 v_2/\sqrt{2} \\ 0 & \tilde{m}_1 & \mathrm{i}\,g_1 v_1/\sqrt{2} & -\mathrm{i}\,g_1 v_2/\sqrt{2} \\ -\mathrm{i}\,g_2 v_1/\sqrt{2} & \mathrm{i}\,g_1 v_1/\sqrt{2} & 0 & -\mu \\ \mathrm{i}\,g_2 v_2/\sqrt{2} & -\mathrm{i}\,g_1 v_2/\sqrt{2} & -\mu & 0 \end{bmatrix} \begin{bmatrix} \tilde{W}^3 \\ \tilde{B} \\ \tilde{H}_1^0 \\ \tilde{H}_2^0 \end{bmatrix}. \tag{6.97}$$

This too can be diagonalized, but we shall give no further details here. We merely note that the masses $\tilde{m}_1$ and $\tilde{m}_2$ are related by virtue of the renormalization group equation (6.60), the unification condition (3.153) and the common gaugino mass ($m_{1/2}$) at the unification scale:

$$\frac{\alpha_{\mathrm{GUT}}(m_X)}{m_{1/2}} = \frac{\alpha_2}{\tilde{m}_2} = \frac{5\alpha_1}{3\tilde{m}_1} \left(= \frac{\alpha_3}{\tilde{m}_3} \right). \tag{6.98}$$

Note also that when μ and $\tilde{m}_2$ (and hence $\tilde{m}_1$) are zero, the combinations photino

$$\tilde{\gamma} \equiv (g_1 \tilde{W}^3 + g_2 \tilde{B})/(g_1^2 + g_2^2)^{1/2} \tag{6.99}$$

and

$$\tilde{S} \equiv (v_1 \tilde{H}_2^0 + v_2 \tilde{H}_1^0)/v \tag{6.100}$$

are massless, while the two orthogonal combinations Zino

$$\tilde{Z} \equiv (g_1 \tilde{B} - g_2 \tilde{W}^3)/(g_1^2 + g_2^2)^{1/2} \tag{6.101}$$

and

$$\tilde{A} \equiv (v_1 \tilde{H}_1^0 - v_2 \tilde{H}_2^0)/v \tag{6.102}$$

give degenerate states $(1/\sqrt{2})(\tilde{Z} \pm \tilde{A})$ with mass

$$m_{\tilde{Z}} = \frac{1}{\sqrt{2}} (g_1^2 + g_2^2)^{1/2} v = m_Z. \tag{6.103}$$

In the same limit the chargino mass eigenstates are the Dirac spinors $(\tilde{W}^+, \bar{\tilde{H}}_2^-)$ and $(\tilde{H}_1^+, \bar{\tilde{W}}^-)$ with masses $g_2 v_2$, $g_2 v_1$ respectively, both of which are less than $2m_W$.

6.5 Experimental signatures

The experimental searches for supersymmetric particles are all constrained by the assumption that 'R-parity' is conserved[8]. This imposes a discrete symmetry on the allowed interactions, which requires, in particular that the (so far unobserved) super-partners of the known particles can only be produced in pairs. Then the lightest supersymmetric particle (LSP) is stable. This assumption of R-parity conservation is introduced primarily to simplify the otherwise prodigiously complicated production and decay mechanisms. However, we shall see in §6.6 that the known absence of fast proton decay requires some such symmetry in the supersymmetric theories that we are considering.

LEP experiments give lower bounds in the masses of sparticles[9]: the absence of Z decays that are not in accord with the predictions of the standard model bound the charged slepton, squark and chargino:

$$m_{\tilde{q}}, m_{\tilde{l}}, m_{\tilde{W}} \gtrsim \tfrac{1}{2} m_Z \tag{6.104}$$

while the 'invisible' width of the Z gives a sneutrino bound

$$m_{\tilde{\nu}} > 42 \text{ GeV}. \tag{6.105}$$

In principle, better limits on the squark and gluino masses can be obtained at $\bar{\text{p}}\text{p}$ colliders, if they are light enough to be pair produced. Ultimately the sparticles decay to the LSP which escapes detection, thereby giving an imbalance of energy transverse to the beam. However, to derive mass bounds further assumptions must be made. All analyses assume that the lightest neutralino ($\tilde{\chi}_1^0$) is the LSP. If the next lightest sparticle is a squark, it is assumed to decay *only* via

$$\tilde{q} \to \text{q}\tilde{\chi}_1^0 \tag{6.106}$$

so the experimental signal from squark pair production will be *two* jets plus missing transverse energy. On the other hand, if the gluinos are the next lightest sparticles, it is assumed that the only decay is

$$\tilde{g} \to \text{q}\bar{\text{q}}\tilde{\chi}_1^0 \tag{6.107}$$

(via a virtual squark), so the signal from gluino pair production is *four* jets plus missing transverse energy. Then with the further assumptions that the LSP is massless and that all squarks have the same mass, the 1991 CDF data give

$$\tilde{m}_3 \equiv m_{\tilde{g}} > 150 \text{ GeV} \qquad m_{\tilde{q}} > 170 \text{ GeV}. \tag{6.108}$$

Given the numerous assumptions needed it is difficult to assess how reliable these bounds are. If the squarks and gluinos are sufficiently massive that chargino and other neutralino decays are kinematically allowed, then these dominate the direct decays of $\tilde{g}$, $\tilde{q}$, and the LSP is produced at the end of a

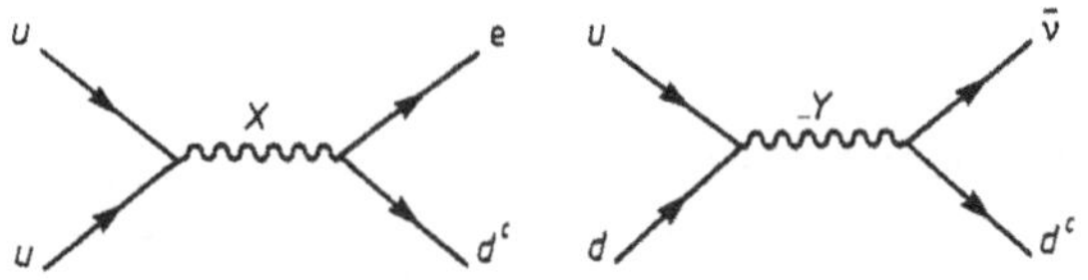

Figure 6.4 B- and L-non-conserving diagrams in the SU(5) GUT.

cascade of decays. Allowing for the cascade decays weakens the lower bounds[10] to

$$m_{\tilde{g}} \gtrsim 135\ \text{GeV} \qquad m_{\tilde{q}} \gtrsim 130\ \text{GeV}. \tag{6.109}$$

Then using (6.60) we can bound the SU(2) gaugino mass parameter:

$$\tilde{m}_2 = \frac{\alpha_2}{\alpha_3}\tilde{m}_3 \gtrsim 40\ \text{GeV} \tag{6.110}$$

using (3.158), (3.159), (3.161). Combining this with the absence of observed neutralinos in Z-decays gives a lower bound of

$$m_{\tilde{\chi}_1^0} \gtrsim 20\ \text{GeV} \tag{6.111}$$

for the lightest neutralino[11].

6.6 Proton decay

The minimal (non-supersymmetric) SU(5) GUT has 12 diquark and lepto-quark gauge bosons $X_i^{\mp 4/3}$, $Y_i^{\pm 1/3}$ which couple to baryon-number- (B-) and lepton-number- (L-) non-conserving currents. These gauge bosons acquire masses $m_X = m_Y = \mathrm{O}(10^{15}\ \text{GeV})$ when the SU(5) gauge symmetry is broken, and the massive bosons can mediate low-energy B- and L-non-conserving processes. In particular the Feynman diagrams shown in figure 6.4 generate a (colour singlet) effective Lagrangian for nucleon decay ($\mathrm{p} \to \mathrm{e}^+\pi^0, \bar{\nu}_\mathrm{e}\pi^+$).

They give

$$\begin{aligned}\mathcal{L}_{\text{eff}} = \frac{g_5^2}{8m_X^2}\epsilon_{ijk}\{&[\bar{u}_i^c\gamma^\mu(1-\gamma_5)u_j][\bar{\mathrm{e}}^c\gamma_\mu(1-\gamma_5)d_k + \bar{d}_k^c\gamma_\mu(1-\gamma_5)\mathrm{e}] \\ &+ [\bar{u}_i^c\gamma^\mu(1-\gamma_5)d_j][\bar{d}_k^c\gamma_\mu(1-\gamma_5)\nu_\mathrm{e}]\}\end{aligned} \tag{6.112}$$

using just the quark and lepton fields of the first generation; g_5 is the gauge coupling constant of the SU(5) GUT, evaluated at the scale m_X. This leads to a decay width

$$\Gamma(\mathrm{p} \to \mathrm{e}^+\pi^0) = \mathrm{O}\left(\frac{\alpha_5^2}{m_X^4}m_\mathrm{P}^5\right). \tag{6.113}$$

With the renormalization group equations predicting the GUT 'fine-structure constant' $\alpha_5 = g_5^2/4\pi \simeq 1/42$ and the unification scale $m_X \sim 10^{15}$ GeV, this gives a lifetime

$$\Gamma^{-1}(p \to e^+\pi^0) = O(10^{33}\,\text{yr}). \tag{6.114}$$

More sophisticated treatments[12], including an enhancement factor due to gluon radiative corrections, reduce this to

$$\Gamma^{-1}(p \to e^+\pi^0) = 4.5 \times 10^{29 \pm 1.7} \tag{6.115}$$

well below the measured lower bound[13] of 6×10^{32} yr, which is why the non-supersymmetric theory is dead. The supersymmetric SU(5) GUT of course allows proton decay via the same diagrams but the larger values of (α_G and) m_X, given in (3.164) and (3.167), increase the predicted value of τ_p way above the measured lower bound, principally because of the proportionality of the width to m_X^{-4}.

Another way to see that the matrix element is proportional to m_X^{-2}, and hence that $\Gamma \propto m_X^{-4}$, is to note that the mass dimension of the four-quark field operator in $\mathcal{L}_{\text{eff}}$ is $[M^6]$. Then to ensure that $\mathcal{L}_{\text{eff}}$ has the required dimension $[M^4]$, it is necessary to supply $[M^{-2}]$, and the unification scale m_X provides this in the gauge-boson-mediated processes shown in figure 6.4. In fact it is clear that something like this must be true of *any* (non-supersymmetric) GUT: to construct a colour singlet baryon-number-non-conserving operator the only possibility is to use *three* quark fields; then to make a Lorentz invariant effective Lagrangian a further fermion (lepton) field must also occur. The missing $[M^{-2}]$ is supplied by whatever mass can mediate the baryon number non-conservation. In the minimal SU(5) GUT $p \to e^+\pi^0$ can also be mediated by the colour triplet Higgs in the **5** representation, which naturally has a mass of order m_X (but is not compelled to do so).

In a supersymmetric GUT, however, there are additional operators that might arise as an effective Lagrangian. In the first place, there are additional SU(3) × SU(2) × U(1)-invariant terms that can be added to the superpotential (3.139):

$$\delta W = \sum_{l,m,n} \lambda^{(1)}_{lmn} L^{(n)\mathrm{T}} \mathrm{i}\tau^2 L^{(m)} l^c + \sum_{l,f,g} \lambda^{(2)}_{lfg} L^{(l)\mathrm{T}}\,\mathrm{i}\tau^2 Q^{(f)} D^{c(g)}$$
$$+ \sum_{f,g,h} \lambda^{(3)}_{fgh} U^{c(f)} D^{c(g)} D^{c(h)} \tag{6.116}$$

where $l, m, n = e, \mu, \tau$ label the leptons and $f, g, h = 1, 2, 3$ label the three quark generations. Each of these violates L- and/or B-conservation, and taking the F-part gives a dimension 4-operator, just as it does for the terms in (3.139). Thus these operators are unsuppressed by any power of the superheavy mass scale, and, if they are all present, they generate an amplitude for

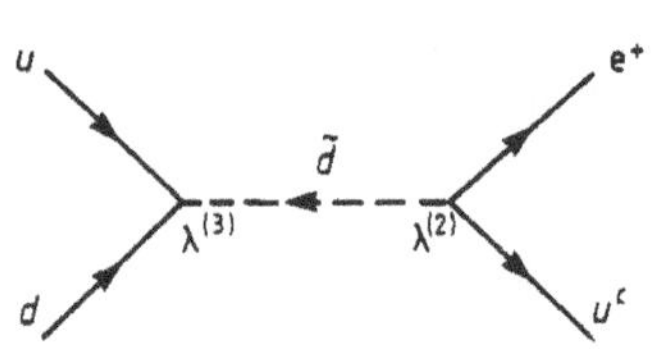

Figure 6.5 *d*-squark-mediated *B*- and *L*-non-conserving amplitude.

proton decay suppressed only by the supersymmetry mass breaking scale[14], as shown in figure 6.5.

It is for this reason that an additional (discrete) symmetry must be invoked to forbid them: estimating the proton decay rate using the diagram of figure 6.5, and comparing with the measured bound leads to the constraint[15]

$$\lambda^{(2)}\lambda^{(3)} \lesssim m_{\mathrm{SUSY}}^2/m_X^2 \sim 10^{-26}. \tag{6.117}$$

The simplest way to expunge the offending terms is to require a 'family reflection symmetry'[15] under which

$$L^{(l)} \to -L^{(l)} \qquad l^c \to -l^c \qquad Q^{(f)} \to -Q^{(f)}$$

$$U^{c(f)} \to -U^{c(f)} \qquad D^{c(f)} \to -D^{c(f)}$$

$$H_1 \to H_1 \qquad H_2 \to H_2. \tag{6.118}$$

Then, all of the terms (3.139) needed to generate masses are allowed, as is the term (6.10) needed to ensure $v_1 \neq v_2$, but all of the *B*- and *L*-non-conserving terms (6.116) are forbidden.

We shall discuss later alternative ways of removing these dimension-4 operators. Assuming that they are absent for some reason, there remains the possibility of dimension-5 *B*- and *L*-non-conserving operators[15], which will only be suppressed by a single power of the superheavy scale, and which may well be the dominant contributors to proton decay in supersymmetric theories.

The dimension-5 *B*- and *L*-non-conserving operators allowed by the SU(3) × SU(2) × U(1) symmetry have the structure

$$O_1 = [QQQL]_F \tag{6.119a}$$

$$O_2 = [U^cU^cD^cl^c]_F \tag{6.119b}$$

$$O_3 = [QQQH_2]_F \tag{6.119c}$$

$$O_4 = [QU^cl^cL]_F \tag{6.119d}$$

$$O_5 = [LLH_1H_1]_F \tag{6.119e}$$

$$O_6 = [LH_2H_1H_1]_F \tag{6.119f}$$

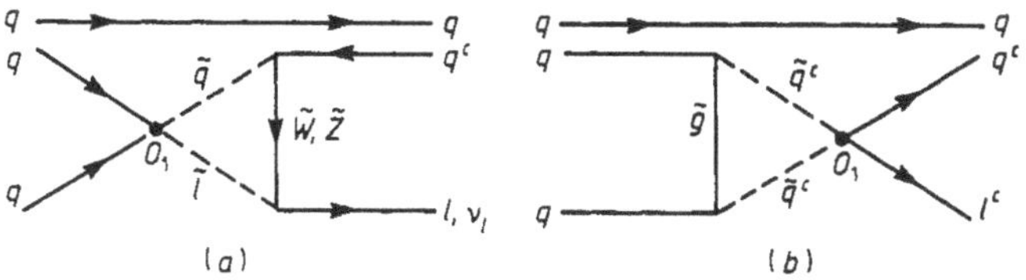

Figure 6.6 Proton decay induced by O_1.

$$O_7 = [H_1 H_1 l^{c\dagger}]_D \tag{6.119g}$$

$$O_8 = [H_1^\dagger H_2 l^c]_D \tag{6.119h}$$

$$O_9 = [Q U^c L^\dagger]_D \tag{6.119i}$$

$$O_{10} = [U^c D^{c\dagger} l^c]_D \tag{6.119j}$$

where we have omitted the generation labels, as well as all of the group labels. Each chiral superfield has dimension 1. Since the superspace coordinate θ has dimension $-\frac{1}{2}$, as in (3.21), the dimension of the F-part of a product of n superfields is

$$[\Phi_1 \cdots \Phi_n]_F\colon \quad [M^{n+1}] \tag{6.120}$$

while the dimension of the D-part of a product is

$$[\Phi_1 \cdots \Phi_n^\dagger]_D\colon \quad [M^{n+2}]. \tag{6.121}$$

To generate proton decay, which does not involve any sparticles, the above operators must be dressed by radiative corrections involving only (light) particles and sparticles. Thus O_1 generates the diagrams shown in figure 6.6. In figure 6.6(a) two incoming quarks annihilate to produce a squark and a slepton—this is the dimension-5 B- and L-violation—and these convert to a quark and lepton via electroweak gaugino exchange. As in the previous examples, proton decay is achieved by adding a spectator quark.

Evidently not all of the above operators can by themselves generate proton decay, since some of them ($O_{4,5,6,7,8,9,10}$) do not violate conservation of B, and O_3 does not have L-non-conservation. Of course, if *some* of the dimension-4 operators (6.116) survived, the missing ingredient might be supplied this way. Thus O_3 together with the term proportional to $\lambda^{(2)}$ will generate proton decay, as in figure 6.5, when the electroweak symmetry is broken, so H_2 is replaced by $\langle O|H_2|O\rangle \sim v_2$. At any rate, if the family reflection symmetry (6.118) is invoked, only $O_{1,2,4}$ survive, and O_4 cannot generate proton decay in the absence of the dimension-4 operators which are also deleted by the symmetry.

We therefore consider just $O_{1,2}$, remembering that the chiral superfields obey Bose statistics. Consider first O_2:

$$O_2 = \epsilon_{ijk} U_i^{c(f)} U_j^{c(g)} D_k^{c(h)} l^c \tag{6.122}$$

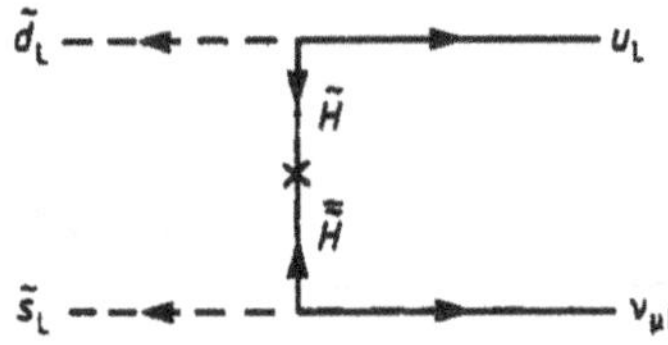

Figure 6.7 Higgsino contribution to O_1.

where $i, j, k = 1, 2, 3$ are the SU(3) labels. Bose statistics therefore require that the generation labels f, g are different, $f \neq g$, so one of the U^c fields must be the c-quark (c^c) superfield (or else t^c). Since this is an SU(2) singlet, neither the electroweak nor the strong dressings, shown in figure 6.6, can change the flavour. It follows that O_2 does not contribute to nucleon decay, since t, c are too massive. Finally, then, there is O_1:

$$O_1 = \epsilon_{ijk}(Q_i^{(f)\mathrm{T}}\, \mathrm{i}\,\tau^2 Q_j^{(g)})(Q_k^{(h)\mathrm{T}}\, \mathrm{i}\,\tau^2 L^{(l)})$$

$$= \epsilon_{ijk}(u_i^{(f)} d_j^{(g)} - d_i^{(f)} u_j^{(g)})(u_k^{(h)} l - d_k^{(h)} \nu_l) \tag{6.123}$$

and the Bose statistics again requires the use of at least *two* quark generations. On this occasion, however, we can use Wino exchange in figure 6.6 to convert the second- or third-generation up-like flavour to a down-like flavour (d, s); in the case of a neutrino decay, however, we can use strong dressing, via gluino exchange, which leaves the (down-like) flavour unchanged. It follows that the dominant decay mode will be

$$\mathrm{p} \rightarrow \mathrm{K}^+ \bar{\nu} \tag{6.124}$$

in a supersymmetric theory, provided, of course that in the supersymmetric GUT the operator O_1 actually arises. The expression (6.123) for O_1 gives a two-fermion–two-sfermion vertex in which both incoming fermions are left chiral, and both incoming sfermions are the partners of left chiral fermions. Such a vertex therefore cannot arise via gaugino exchange, since gauge particles conserve helicity. It can, however, arise in the supersymmetric SU(5) theory via colour triplet Higgsino exchange, as shown in figure 6.7, in which the cross denotes the Dirac mass of the Higgsino.

Using the vertices in (6.15), we can estimate the effective strength of this vertex

$$G_1 = G_{11}^{(u)} G_{22}^{(d)} / m_{\tilde{H}}(\mathbf{3}) \tag{6.125}$$

where $m_{\tilde{H}}(\mathbf{3})$ is the colour triplet Higgsino mass. The dressing via strong interactions, as shown in figure 6.6, then gives an effective Lagrangian of the form

$$\mathscr{L}_{\text{eff}} \simeq G_{\text{eff}} \epsilon_{ijk} (u_i d_j)(s_k \nu_\mu) \tag{6.126}$$

(omitting the γ-matrices) with

$$G_{\text{eff}} = G_1 \frac{\alpha_3}{2\pi} \frac{m_g}{m_s^2} \tag{6.127}$$

where m_g is the gluino mass and m_s the squark mass. Reasonable estimates of the parameters then yield

$$G_{\text{eff}} \sim g_5^2/8m_X^2. \tag{6.128}$$

Thus the dominant decay mode (into kaons) has a lifetime of order 10^{30} yr, as before.

All of this is dependent upon the assumed family reflection symmetry. There are other symmetries that forbid the dimension-4 operators. In particular, if the offending terms are dropped, the remaining theory has a larger (global) symmetry called R-symmetry. This is a U(1) continuous symmetry, parametrized by α, under which the superspace coordinate θ transforms as

$$\theta \rightarrow e^{i\alpha}\theta \tag{6.129}$$

and a general chiral superfield Φ as

$$\Phi \rightarrow e^{iR\alpha}\Phi. \tag{6.130}$$

So we can say that θ and Φ have R-charges:

$$R(\theta) = 1 \qquad R(\Phi) = R. \tag{6.131}$$

Evidently, since the superpotential W contributes $\int d^2\theta\, W$ to the Lagrangian, the theory is R-invariant if

$$R(W) = 2. \tag{6.132}$$

Referring now to the terms that interest us, namely (3.139), (6.10) and (6.116), we see that there is an R-invariance in which the superfields have charges

$$R(Q^{(f)}, L^{(l)}, U^{(f)c}, D^{(f)c}, l^c) = \tfrac{1}{2} \tag{6.133a}$$

$$R(H_1, H_2) = 1 \tag{6.133b}$$

and that *excludes* the (offending) terms (6.116).

The transformation properties (6.129), (6.130) show that the scalar and spinor components of Φ have charges

$$R(\varphi) = R \qquad R(\psi) = R - 1 \tag{6.134}$$

and according to (6.133)

$$R(q, l) = -\tfrac{1}{2} \qquad R(\tilde{q}, \tilde{l}) = \tfrac{1}{2} \tag{6.135a}$$

$$R(h_1, h_2) = 1 \qquad R(\tilde{h}_1, \tilde{h}_2) = 0. \tag{6.135b}$$

Since it is real, the vector superfield V has zero charge:

$$R(V) = 0. \tag{6.136}$$

Referring to (3.4), we deduce that the gauge field component V_μ has

$$R(V_\mu) = 0 \tag{6.137}$$

since θ, $\bar{\theta}$ have opposite charges. It follows too from (3.4) that the gaugino fields transform non-trivially:

$$R(\lambda) = 1 \tag{6.138}$$

and this means that gaugino Majorana mass terms such as (6.50) are forbidden by R-symmetry. Since there is experimental evidence that the gluinos have non-zero masses, as we have seen in (6.108), it seems that the R-invariance must be (spontaneously) broken. This in turn generates an unwanted Goldstone boson (actually an axion, since the U(1) is anomalous). Thus it appears that the continuous symmetry cannot be used to exclude the unwanted terms in the superpotential. This has led to a fuller investigation of the discrete R-parities that may be used to exclude some, but not all, of the B- and L-non-conserving operators, but in a way that is consistent with present data on proton decay[(16)].

Exercises

6.1 Show that the potential (6.13) does generate the spontaneous breaking of SU(2) × U(1) provided that $|\lambda\mu| > m_{3/2}$.

6.2 Verify (6.40) and (6.41).

6.3 Show that the minimum of the Higgs potential (6.52) leaves $U(1)_{em}$ unbroken.

6.4 Calculate the R-charges of the chiral superfields in the supersymmetric standard model that are needed to give all 'particles' zero R-charge, and all 'sparticles' non-zero R-charge.

References

General references

The following articles have been most useful in preparing this chapter.

Bartl A, Majoretto W and Mösslacher B 1991 *Proc. Joint Int. Lepton–Photon Symp. and Europhysics Conf. on High Energy Physics (Geneva, 1991)* (Singapore: World Scientific)
Nilles H P 1984 *Phys. Rep.* C **110** 1
—— 1991 *Lectures at the Phenomenological Aspects of Supersymmetry Workshop (Munich, 1991)*; *Preprint* MPI-Ph/91-106 (MPI, Munich)

References in the text

1 Frere J-M, Jones D R T and Raby S 1983 *Nucl. Phys.* B **222** 11
2 Polchinski J and Susskind L 1982 *Phys. Rev.* D **26** 3661
Nilles H P, Srednicki M and Wyler D 1983 *Phys. Lett.* **124B** 337
Lahunnas A B 1983 *Phys. Lett.* **124B** 341
3 Inoue K, Kakuto A, Komatsu H and Takeshita S 1982 *Prog. Theor. Phys.* **67** 1889; 1982 *Prog. Theor. Phys.* **68** 927
4 Alvarez-Gaume L, Polchinski J and Wise M B 1983 *Nucl. Phys.* B **221** 495
Nilles H P 1984 *Phys. Rep.* C **110** 1
Drees M and Nojiri M M 1992 *Nucl. Phys.* B **369** 54
Inoue I, Kawasaki M, Yamaguchi M, Yanagida T, Kelley S *et al* 1991 *Phys. Lett.* **273B** 423
5 Derendinger J P and Savoy C 1984 *Nucl. Phys.* B **237** 307
6 Peccei R D and Quinn H R 1977 *Phys. Rev. Lett.* **38** 1440; 1977 *Phys. Rev.* D **16** 1791
7 Ellis J and Zwirner F 1990 *Nucl. Phys.* B **338** 317
8 Fayet P 1975 *Nucl. Phys.* B **90** 104
Salam A and Strathdee J 1975 *Nucl. Phys.* B **87** 85
9 Davier M 1991 *Proc. Joint Int. Lepton–Photon Symp. and Europhysics Conf. on High Energy Physics (Geneva, 1991)* (Singapore: World Scientific)
Decamp D *et al* 1990 *Phys. Lett.* **244B** 541
Akrawy M *et al* 1990 *Phys. Lett.* **252B** 290
Abren D A *et al* 1990 *Phys. Lett.* **247B** 157
10 Baer H, Tata X and Woodside J 1990 *Phys. Rev.* D **42** 1568
11 Roszkowski L 1990 *Phys. Lett.* **252B** 471
12 Tomozawa Y 1981 *Phys. Rev. Lett.* **46** 463
Berezenskii V, Joffe B and Kogan Ya 1981 *Phys. Lett.* **105B** 33
Donoghue J and Golowich E 1982 *Phys. Rev.* D **26** 2888
Lucha W 1983 *Phys. Lett.* **122B** 381
Isgur N and Wise M 1982 *Phys. Lett.* **117** 179
13 Becker-Stendy *et al* 1990 *Phys. Rev.* D **42** 2974
14 Dimpoulos S and Georgi H 1981 *Nucl. Phys.* B **193** 150
15 Weinberg S 1982 *Phys. Rev.* D **26** 287
Sakai N and Yanagida T 1982 *Nucl. Phys.* B **197** 533
16 Ibanez L and Ross G G 1992 *Nucl. Phys.* B **368** 3

7

THE BOSONIC STRING

7.1 Introduction

Historically, relativistic string theory was developed as a possible theory of strong interactions[1] before the role was filled by quantum chromodynamics. Subsequently, the occurrence of a massless spin-2 particle in the spectrum of states of the string suggested an alternative use for string theory as a possible framework for gravitation[2] that might succeed in overcoming the problems that had been encountered in trying to obtain a renormalizable theory of quantum gravity. In particular, there is an intuitive argument, which we shall explain in the next section, as to why string theory might not suffer from the ultraviolet divergences of quantum field theory of point particles. This argument is borne out by detailed calculations of string loop diagrams. In the case of the heterotic string, to be discussed in Chapter 9, not only does the theory provide a satisfactory framework for quantum gravity but it also appears able to unify gravity with the strong, weak and electromagnetic interactions. These remarkable properties make string theory a candidate theory of all physics (a 'theory of everything'). At the present time, the theory is usually formulated in the form of relativistic quantum mechanics, a relativistic quantum field theory of strings not yet having been fully developed (if indeed this is the appropriate framework for a complete string theory).

7.2 The bosonic string action

Whereas the position of a point particle in D dimensions may be described by degrees of freedom $X^\mu(\tau)$, $\mu = 0, 1, \ldots, D-1$, depending only on a time-like coordinate τ, to describe a string we need in addition a space-like coordinate σ. Then, the string degrees of freedom $X^\mu(\tau, \sigma)$, $\mu = 0, 1, \ldots, D-1$, can trace out a curve as σ varies at fixed τ. The curve may be open or closed (open or closed strings as in figure 7.1) and it is convenient to take

$$0 \leqslant \sigma \leqslant \pi \tag{7.1}$$

to be the range over which σ varies as the string is traced out from one end to the other for an open string, or once round the string for a closed string. As τ varies the string sweeps out a world sheet (figure 7.2) just as a point particle sweeps out a world line, and it is appropriate to refer to the coordinates τ and

DOI: 10.1201/9780367805807-7

Figure 7.1 Open and closed strings.

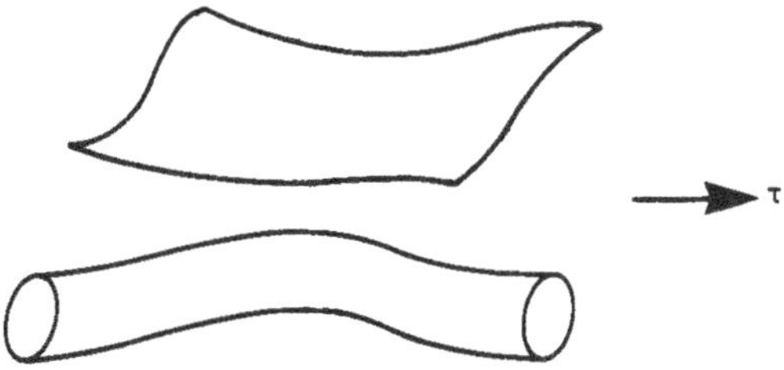

Figure 7.2 Open- and closed-string world sheets.

σ as world sheet coordinates. We shall see in §7.8 that consistency of the theory requires D to be 26.

An action is required to embody the dynamics of the bosonic string. For a massless, non-relativistic string of length L the action of classical mechanics is

$$S = -\frac{T}{2}\int_{t_i}^{t_f} \mathrm{d}t \int_0^L \mathrm{d}x \left(\frac{\mathrm{d}y}{\mathrm{d}x}\right)^2 \tag{7.2}$$

where y is the displacement of the string at position x, T is the tension in the string, and t_i and t_f are some initial and final times. When writing down an analogous action for the relativistic bosonic string propagating in D-dimensional flat space we wish to ensure that it is not only Lorentz covariant in form but also that the physics does not depend on the particular choice of world sheet coordinates τ and σ. To this end we introduce a space-time metric

$$\eta_{\mu\nu} = \mathrm{diag}(1, -1, \cdots, -1) \qquad \mu, \nu = 0, 1, \cdots, D-1 \tag{7.3}$$

and a world sheet metric $h_{\alpha\beta}(\tau, \sigma)$ of signature $(+, -)$ where $\alpha = 0$ and 1 refers to τ and σ, respectively, and adopt the action for the relativistic bosonic string

$$S = -\frac{T}{2}\int_{\tau_i}^{\tau_f} \mathrm{d}\tau \int_0^{\pi} \mathrm{d}\sigma\, (-\det h)^{1/2} h^{\alpha\beta}\eta^{\mu\nu}\, \partial_\alpha X_\mu\, \partial_\beta X_\nu. \tag{7.4}$$

Figure 7.3 Point particle and string interactions.

The required world sheet coordinate covariance then follows in the same way as the space-time coordinate covariance of general relativity. Thus, the action (7.4) has by construction local world sheet reparametrization invariance, and global D-dimensional space-time Poincaré invariance. In detail, the infinitesimal transformation for world sheet reparametrization invariance is

$$\delta X^\mu = \xi^\alpha \partial_\alpha X^\mu \qquad \delta h^{\alpha\beta} = \xi^\gamma \partial_\gamma h^{\alpha\beta} + \partial_\gamma \xi^\alpha h^{\gamma\beta} + \partial_\gamma \xi^\beta h^{\alpha\gamma} \tag{7.5}$$

where ξ^α is the infinitesimal shift in the coordinates (τ, σ). The transformation for global D-dimensional space-time Poincaré invariance is

$$\delta X^\mu = l^\mu{}_\nu X^\nu + a^\mu \qquad \delta h_{\alpha\beta} = 0 \tag{7.6}$$

where $l^\mu{}_\nu$ and a^μ are constants with

$$l_{\mu\nu} = \eta_{\mu\rho} l^\rho{}_\nu \tag{7.7}$$

antisymmetric. Additionally, at least at the classical level, the action has a local Weyl scaling or conformal invariance under the coordinate-dependent rescaling of the world sheet metric

$$\delta h_{\alpha\beta} = \Lambda(\tau, \sigma) h_{\alpha\beta} \qquad \delta X^\mu = 0. \tag{7.8}$$

It should be noted that this is *not* included in the world sheet reparametrization invariance, because those transformations have $\delta X^\mu \neq 0$ as a result of the τ- and σ-dependence of X^μ. The existence of this accidental conformal invariance is peculiar to strings and is not shared by reparametrization-invariant objects with more dimensions such as membranes.

As mentioned in the introduction, string theory is expected to be free from the ultraviolet divergences that occur in the quantum field theory of point particles, for the following reason. In the quantum field theory of point particles, interactions are associated with vertices where world lines meet, and, correspondingly, string interactions should be associated with world sheets joining, as in figure 7.3 for closed strings. An important difference is that whereas for a point particle theory there is a well defined point where the interaction occurs, for a string theory there is no well defined point at which the two strings merge into one. If we consider two observers corresponding to different choices of world sheet coordinates their lines of

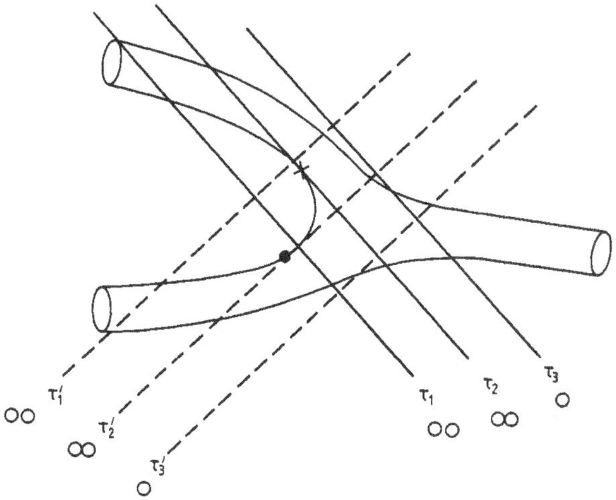

Figure 7.4 Strings merging as seen by different observers. The solid lines are lines of constant τ and the dashed lines are lines of constant τ'. What the observers see at various values of τ and τ' is depicted alongside.

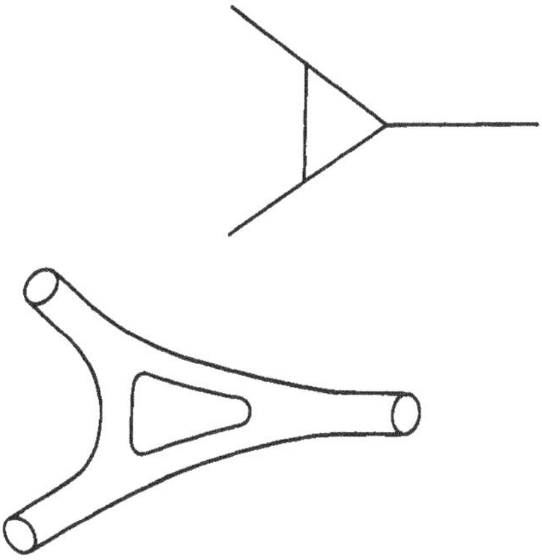

Figure 7.5 Loop corrections to a vertex in point particle field theory and in string theory.

constant τ differ† and, as in figure 7.4, one observer will see the two strings merging at the point marked by the solid dot, and the other at the point marked by the cross. If we now consider loop corrections to the vertex as in figure 7.5, in the point particle field theory diagram ultraviolet divergences

†As will be seen in §7.7, it is possible in the light cone gauge to choose the world sheet coordinates such that τ is identified with any specified combination of X^+ and X^i, and it is this freedom which is exploited here.

arise from propagators meeting at a well defined interaction point. On the other hand, we do not expect such problems in the string case because there is *no* well defined point at which the interaction occurs. This expectation is confirmed by detailed calculations in string loop perturbation theory[(3)].

7.3 Equations of motion and covariant gauges

String equations of motion may be obtained as Euler–Lagrange equations by varying the action with respect to X^μ and with respect to $h_{\alpha\beta}$. The former equations of motion are

$$\partial_\alpha\left((-\det h)^{1/2}h^{\alpha\beta}\,\partial_\beta X^\mu\right) = 0 \tag{7.9}$$

where it has been assumed that the surface terms vanish, a point to which we return shortly. The latter equations of motion are

$$T_{\alpha\beta} \equiv -\,\partial_\alpha X^\mu\,\partial_\beta X_\mu + \tfrac{1}{2}h_{\alpha\beta}h^{\gamma\delta}\,\partial_\gamma X^\mu\,\partial_\delta X_\mu = 0 \tag{7.10}$$

where $T_{\alpha\beta}$ may be interpreted as the energy–momentum tensor obtained by regarding the string action as the action for a two-dimensional field theory of D free scalar fields X^μ. In deriving (7.10), the identities

$$\delta(\det h) = \det h\; h^{\gamma\delta}\,\delta h_{\gamma\delta} \tag{7.11}$$

and

$$\delta h^{\alpha\beta} = -\,h^{\alpha\gamma}h^{\beta\delta}\,\delta h_{\gamma\delta} \tag{7.12}$$

are useful. For any solution of the equations of motion (7.10), the integrand of the string action (7.4) may be cast in the form $|\det\partial_\alpha X^\mu\,\partial_\beta X_\mu|^{1/2}$, which shows that the action has a geometrical interpretation in terms of the area of the string world sheet.

Just as in gauge field theory or general relativity, there are fewer independent dynamical degrees of freedom than appear explicitly in the action, and the number of degrees of freedom may be reduced by a suitable choice of gauge. In the present case, the gauge symmetries are the two-dimensional world sheet reparametrization invariances (7.5), and the conformal invariance (7.8). The reparametrization invariance may be used to reduce the metric $h_{\alpha\beta}$ of signature $(+,-)$, which in the first instance has three independent components, to the form

$$h_{\alpha\beta} = \mathrm{e}^{\varphi(\tau,\,\sigma)}\eta_{\alpha\beta} \tag{7.13}$$

where

$$\eta_{\alpha\beta} \equiv \mathrm{diag}(1, -1). \tag{7.14}$$

This choice of gauge is usually referred to as conformal gauge. If we also exploit the conformal invariance (7.8), the metric may be further simplified to

$$h_{\alpha\beta} = \eta_{\alpha\beta}. \tag{7.15}$$

When we refer to covariant gauges in what follows we shall have (7.15) in mind. There is some further gauge freedom, which we do not employ for the time being, that can be used to reduce the number of components of X^μ. This will be used in §7.7 to obtain the non-covariant light cone gauge which does not put all the components of X^μ on the same footing.

In the covariant gauge (7.15), the equation of motion (7.9) simplifies to the one-dimensional wave equation

$$\partial_\alpha \partial^\alpha X^\mu = \left(\frac{\partial^2}{\partial\tau^2} - \frac{\partial^2}{\partial\sigma^2}\right) X^\mu = 0. \tag{7.16}$$

The string degrees of freedom X^μ are further constrained by (7.10) which may now be written as

$$T_{00} = T_{11} = -\frac{1}{2}\left(\frac{\partial X^\mu}{\partial\tau}\frac{\partial X_\mu}{\partial\tau} + \frac{\partial X^\mu}{\partial\sigma}\frac{\partial X_\mu}{\partial\sigma}\right) = 0 \tag{7.17}$$

and

$$T_{01} = T_{10} = -\frac{\partial X^\mu}{\partial\tau}\frac{\partial X_\mu}{\partial\sigma} = 0 \tag{7.18}$$

where we are using indices 0 and 1 to refer to τ and σ, respectively. In particular,

$$h^{\alpha\beta}T_{\alpha\beta} = T_{00} - T_{11} = 0 \tag{7.19}$$

i.e. the two-dimensional energy–momentum tensor is traceless.

As remarked earlier, the validity of (7.16) depends on the vanishing of the surface terms when the variation of the action is made. In the case of a closed string, the boundary conditions

$$X^\mu(\tau, \sigma + \pi) = X^\mu(\tau, \sigma) \qquad \text{closed string} \tag{7.20}$$

ensure that the surface terms are zero. For an open string, and in covariant gauge, the surface terms again vanish provided that we impose the boundary conditions

$$\frac{\partial X^\mu}{\partial\sigma} = 0 \qquad \text{when } \sigma = 0 \text{ and } \sigma = \pi \qquad \text{open string.} \tag{7.21}$$

7.4 Mode expansion and quantization

For a closed string, the general solution of the wave equation (7.16) consistent with the boundary conditions (7.20) is (Exercise 7.1)

$$X^\mu = x^\mu + l^2 p^\mu \tau$$
$$+ \frac{i}{2} l \sum_{n \neq 0} \left(\frac{1}{n} \alpha_n^\mu e^{-2in(\tau - \sigma)} + \frac{1}{n} \tilde{\alpha}_n^\mu e^{-2in(\tau + \sigma)} \right)$$

closed string (7.22)

where the zero-frequency part (zero mode) has been treated separately, and a fundamental length

$$l = (\pi T)^{-1/2} \tag{7.23}$$

has been introduced for later convenience. We shall often work in units where l is 1. Hermicity of the operator X^μ requires the centre-of-mass coordinates and momenta x^μ and p^μ to be Hermitian, and imposes the conditions on the Hermitian adjoints of the oscillator coefficients

$$(\alpha_n^\mu)^\dagger = \alpha_{-n}^\mu \qquad (\tilde{\alpha}_n^\mu)^\dagger = \tilde{\alpha}_{-n}^\mu . \tag{7.24}$$

As always for solutions of the one-dimensional wave equation, the solution may be separated into the sum of a term depending on $\tau - \sigma$ (the right-mover part) and a term depending on $\tau + \sigma$ (the left-mover part). Thus, for the closed string, there is the separation into right mover X_R^μ and left mover X_L^μ,

$$X^\mu(\tau, \sigma) = X_R^\mu(\tau - \sigma) + X_L^\mu(\tau + \sigma) \tag{7.25}$$

where

$$X_R^\mu(\tau - \sigma) = \tfrac{1}{2} x^\mu + \tfrac{1}{2} l^2 p^\mu (\tau - \sigma) + \frac{i}{2} l \sum_{n \neq 0} \frac{1}{n} \alpha_n^\mu e^{-2in(\tau - \sigma)} \tag{7.26}$$

and

$$X_L^\mu(\tau + \sigma) = \tfrac{1}{2} x^\mu + \tfrac{1}{2} l^2 p^\mu (\tau + \sigma) + \frac{i}{2} l \sum_{n \neq 0} \frac{1}{n} \tilde{\alpha}_n^\mu e^{-2in(\tau + \sigma)} \tag{7.27}$$

Quantization of the closed string requires us to identify the canonical momentum conjugate to X^μ,

$$P_\mu(\tau, \sigma) = -\frac{\partial \mathcal{L}}{\partial \dot{X}^\mu} \tag{7.28}$$

where

$$\dot{X}^\mu \equiv \frac{\partial X^\mu}{\partial \tau} \tag{7.29}$$

and $\mathcal{L}$ is the Lagrangian density defined by writing the action as

$$S = \int_{\tau_i}^{\tau_f} d\tau \int_0^{\pi} d\sigma \, \mathscr{L}. \tag{7.30}$$

From the action (7.4) in covariant gauge,

$$P^{\mu}(\tau, \sigma) = T \frac{\partial X^{\mu}}{\partial \tau} = T\dot{X}^{\mu}. \tag{7.31}$$

Equal-τ canonical commutation relations may now be imposed, namely

$$[X^{\mu}(\tau, \sigma), X^{\nu}(\tau, \sigma')] = [\dot{X}^{\mu}(\tau, \sigma), \dot{X}^{\nu}(\tau, \sigma')] = 0 \tag{7.32}$$

and

$$[P^{\mu}(\tau, \sigma), X^{\nu}(\tau, \sigma')] = T[\dot{X}^{\mu}(\tau, \sigma), X^{\nu}(\tau, \sigma')] = \mathrm{i}\, \delta(\sigma - \sigma')\eta^{\mu\nu}. \tag{7.33}$$

Substituting (7.26) and (7.27) into (7.32) and (7.33) it can be seen that (Exercise 7.2) the corresponding commutator for the centre-of-mass coordinate x^{μ} and the centre-of-mass momentum p^{μ} of the string is

$$[x^{\mu}, p^{\nu}] = -\mathrm{i}\, \eta^{\mu\nu} \tag{7.34}$$

and the corresponding commutators for the oscillators α_n^{μ} and $\tilde{\alpha}_n^{\mu}$ are

$$[\alpha_m^{\mu}, \alpha_n^{\nu}] = -m\delta_{m+n,0}\eta^{\mu\nu} \tag{7.35}$$

$$[\tilde{\alpha}_m^{\mu}, \tilde{\alpha}_n^{\nu}] = -m\delta_{m+n,0}\eta^{\mu\nu} \tag{7.36}$$

and

$$[\alpha_m^{\mu}, \tilde{\alpha}_n^{\nu}] = 0. \tag{7.37}$$

The operators

$$a_n^{\mu\dagger} = \frac{1}{\sqrt{n}} \alpha_{-n}^{\mu} \qquad a_n^{\mu} = \frac{1}{\sqrt{n}} \alpha_n^{\mu} \qquad n > 0 \tag{7.38}$$

and the operators

$$\tilde{a}_n^{\mu\dagger} = \frac{1}{\sqrt{n}} \tilde{\alpha}_{-n}^{\mu} \qquad \tilde{a}_n^{\mu} = \frac{1}{\sqrt{n}} \tilde{\alpha}_n^{\mu} \qquad n > 0 \tag{7.39}$$

have standard commutation relations for harmonic oscillator creation and annihilation operators, with a factor of $\eta^{\mu\nu}$,

$$[a_m^{\mu}, a_n^{\nu\dagger}] = -\delta_{mn}\eta^{\mu\nu}. \tag{7.40}$$

It remains to implement the constraints (7.17) and (7.18) on X^{μ}, and we return to this in §7.5.

The Hamiltonian in covariant gauge

$$H = \int_0^{\pi} d\sigma \left(-P_{\mu} \frac{\partial X^{\mu}}{\partial \tau} - \mathscr{L} \right) = -\frac{T}{2} \int_0^{\pi} d\sigma \left(\frac{\partial X^{\mu}}{\partial \tau} \frac{\partial X_{\mu}}{\partial \tau} + \frac{\partial X^{\mu}}{\partial \sigma} \frac{\partial X_{\mu}}{\partial \sigma} \right) \tag{7.41}$$

may be written in terms of the oscillators α_n^μ and $\tilde{\alpha}_n^\mu$ in the mode expansion (7.22) as (Exercise 7.3)

$$H = -\sum_{n \neq 0} (\alpha^\mu_{-n}\alpha_{\mu n} + \tilde{\alpha}^\mu_{-n}\tilde{\alpha}_{\mu n}) - \frac{l^2}{2}p^\mu p_\mu . \tag{7.42}$$

For an open string, the general solution of the wave equation (7.16) consistent with the boundary conditions (7.21) is (Exercise 7.4)

$$X^\mu = x^\mu + l^2 p^\mu \tau + \mathrm{i}\, l \sum_{n \neq 0} \frac{1}{n} \alpha_n^\mu \mathrm{e}^{-\mathrm{i}\, n\tau} \cos(n\sigma) \qquad \text{open string.} \tag{7.43}$$

In the open-string case, the left- and right-mover oscillator terms are *not* independent, having been linked by the boundary conditions (7.21), and a separation into left and right movers is not particularly useful. Quantizing in the same way as for the closed string using (7.32) and (7.33) leads to the non-zero commutators for the centre-of-mass coordinates

$$[x^\mu, p^\nu] = -\mathrm{i}\,\eta^{\mu\nu} \tag{7.45}$$

and for the oscillator coefficients

$$[\alpha_n^\mu, \alpha_n^\nu] = -m\delta_{m+n,0}\eta^{\mu\nu}. \tag{7.46}$$

The constraints (7.17) and (7.18) on X^μ will be discussed in §7.7. For the open string, the Hamiltonian takes the form (Exercise 7.5)

$$H = -\tfrac{1}{2}\sum_{n \neq 0} \alpha^\mu_{-n}\alpha_{\mu n} - \frac{l^2}{2}p^\mu p_\mu . \tag{7.47}$$

7.5 Virasoro algebra and masses of states for the closed string

The string degrees of freedom in covariant gauge must not only satisfy the wave equation (7.16), leading to the mode expansion (7.22), but must also satisfy the constraint equations (7.17) and (7.18) which originate from the $h_{\alpha\beta}$ equation of motion and are a consequence of the reparametrization invariance of the string action. To analyse these constraints it is convenient to define combinations T_{++} and T_{--} of the components of the two-dimensional energy–momentum tensor $T_{\alpha\beta}$ by

$$T_{++} \equiv \tfrac{1}{2}(T_{00} + T_{01}) = -\tfrac{1}{4}\left(\frac{\partial X^\mu}{\partial \tau} + \frac{\partial X^\mu}{\partial \sigma}\right)\left(\frac{\partial X_\mu}{\partial \tau} + \frac{\partial X_\mu}{\partial \sigma}\right) \tag{7.48}$$

and

$$T_{--} \equiv \tfrac{1}{2}(T_{00} - T_{01}) = -\tfrac{1}{4}\left(\frac{\partial X^\mu}{\partial \tau} - \frac{\partial X^\mu}{\partial \sigma}\right)\left(\frac{\partial X_\mu}{\partial \tau} - \frac{\partial X_\mu}{\partial \sigma}\right). \tag{7.49}$$

Noticing that T_{++} can only depend on $X^\mu_{\mathrm{L}}(\tau + \sigma)$, that T_{--} can only depend on $X^\mu_{\mathrm{R}}(\tau - \sigma)$ and that

$$\frac{\partial X^\mu_{\mathrm{L}}}{\partial \tau} = \frac{\partial X^\mu_{\mathrm{L}}}{\partial \sigma} \qquad \frac{\partial X^\mu_{\mathrm{R}}}{\partial \tau} = -\frac{\partial X^\mu_{\mathrm{R}}}{\partial \sigma} \tag{7.50}$$

we can write (7.48) and (7.49) more succinctly as

$$T_{++} = -\frac{\partial X^\mu_{\mathrm{L}}}{\partial \tau}\frac{\partial X_{\mu\mathrm{L}}}{\partial \tau} \tag{7.51}$$

and

$$T_{--} = -\frac{\partial X^\mu_{\mathrm{R}}}{\partial \tau}\frac{\partial X_{\mu\mathrm{R}}}{\partial \tau}. \tag{7.52}$$

The constraint equations (7.17) and (7.18) may be written as

$$T_{++} = T_{--} = 0. \tag{7.53}$$

Classically, we are free to implement these constraints straightforwardly. However, quantum mechanically we shall see that it is not possible to impose the full content of (7.53) without inconsistency, and in that case it is useful to work with the Fourier components

$$\begin{aligned} L_m &\equiv \frac{T}{2}\int_0^\pi \mathrm{d}\sigma\, \mathrm{e}^{2\mathrm{i}m(\tau-\sigma)} T_{--} \\ &= -\frac{T}{2}\int_0^\pi \mathrm{d}\sigma\, \mathrm{e}^{2\mathrm{i}m(\tau-\sigma)} \left(\frac{\partial X^\mu_{\mathrm{R}}}{\partial \tau}\right)^2 \qquad m \neq 0 \end{aligned} \tag{7.54}$$

and

$$\begin{aligned} \tilde{L}_m &\equiv \frac{T}{2}\int_0^\pi \mathrm{d}\sigma\, \mathrm{e}^{2\mathrm{i}m(\tau+\sigma)} T_{++} \\ &= -\frac{T}{2}\int_0^\pi \mathrm{d}\sigma\, \mathrm{e}^{2\mathrm{i}m(\tau+\sigma)} \left(\frac{\partial X^\mu_{\mathrm{L}}}{\partial \tau}\right)^2 \qquad m \neq 0 \end{aligned} \tag{7.55}$$

which are referred to as the Virasoro operators. Substituting the expansions (7.26) and (7.27) into (7.54) and (7.55) gives the expressions for the Virasoro operators:

$$L_m = -\tfrac{1}{2}\sum_{n=-\infty}^{\infty} \alpha^\mu_{m-n}\alpha_{\mu n} \qquad m \neq 0 \tag{7.56}$$

and

$$\tilde{L}_m = -\tfrac{1}{2} \sum_{n=-\infty}^{\infty} \tilde{\alpha}^{\mu}_{m-n}\tilde{\alpha}_{\mu n} \qquad m \neq 0 \tag{7.57}$$

where we have defined

$$\alpha^{\mu}_0 = \tilde{\alpha}^{\mu}_0 = \frac{l}{2} p^{\mu} \tag{7.58}$$

with l as in (7.23), so as to be able to cast the result in this neat form. Strictly, if we were to use (7.54) and (7.55) to define L_0 and $\tilde{L}_0$ then (7.56) and (7.57) would also apply for $m = 0$. However, for reasons that we discuss in a moment we shall denote the expressions obtained in this way by L_0' and $\tilde{L}_0'$. Thus,

$$L_0' \equiv \frac{T}{2}\int_0^{\pi} \mathrm{d}\sigma\, T_{--} = -\tfrac{1}{2} \sum_{n=-\infty}^{\infty} \alpha^{\mu}_{-n}\alpha_{\mu n} \tag{7.59}$$

and

$$\tilde{L}_0' \equiv \frac{T}{2}\int_0^{\pi} \mathrm{d}\sigma\, T_{++} = -\tfrac{1}{2} \sum_{n=-\infty}^{\infty} \tilde{\alpha}^{\mu}_{-n}\tilde{\alpha}_{\mu n}. \tag{7.60}$$

If we arrange the oscillators in L_0' and $\tilde{L}_0'$ in normal-ordered form with all annihilation operators α^{μ}_n, $n > 0$, to the right of all creation operators α^{μ}_{-n}, $n > 0$, then we have

$$L_0' = -\tfrac{1}{2}\alpha^{\mu}_0\alpha_{\mu 0} - \sum_{n=1}^{\infty} \alpha^{\mu}_{-n}\alpha_{\mu n} - a \tag{7.61}$$

and

$$\tilde{L}_0' = -\tfrac{1}{2}\tilde{\alpha}^{\mu}_0\tilde{\alpha}_{\mu 0} - \sum_{n=1}^{\infty} \tilde{\alpha}^{\mu}_{-n}\tilde{\alpha}_{\mu n} - a \tag{7.62}$$

where a is a formally infinite constant arising from the commutators. This constant needs to be regularized in some way, and we shall show in §7.8 that we should take a to be 1. (No such problem arises for L_m and $\tilde{L}_m$ when $m \neq 0$.) It is conventional to define L_0 and $\tilde{L}_0$ to be just the normal-ordered products

$$L_0 \equiv -\tfrac{1}{2} : \sum_{n=-\infty}^{\infty} \alpha^{\mu}_{-n}\alpha_{\mu n} : \tag{7.63}$$

and

$$\tilde{L}_0 \equiv -\tfrac{1}{2}\!: \sum_{n=-\infty}^{\infty} \tilde{\alpha}^{\mu}_{-n}\tilde{\alpha}_{\mu n}\!: . \tag{7.64}$$

With this definition,

$$L_0 = -\tfrac{1}{2}\alpha^{\mu}_0\alpha_{\mu 0} - \sum_{n=1}^{\infty} \alpha^{\mu}_{-n}\alpha_{\mu n} = L_0' + a \tag{7.65}$$

and

$$\tilde{L}_0 = -\tfrac{1}{2}\tilde{\alpha}^{\mu}_0\tilde{\alpha}_{\mu 0} - \sum_{n=1}^{\infty} \tilde{\alpha}^{\mu}_{-n}\tilde{\alpha}_{\mu n} = \tilde{L}_0' + a. \tag{7.66}$$

The Hamiltonian (7.41) is related to the Virasoro operators by

$$H = 2(L_0' + \tilde{L}_0') = 2(L_0 + \tilde{L}_0 - 2a). \tag{7.67}$$

The constraint equations (7.53) may be formulated in terms of the Virasoro operators, which are the Fourier components of T_{++} and T_{--}, as the conditions for physical states $|\varphi\rangle$,

$$L_m|\varphi\rangle = \tilde{L}_m|\varphi\rangle = 0 \qquad m > 0 \tag{7.68}$$

and

$$L_0'|\varphi\rangle = \tilde{L}_0'|\varphi\rangle = 0 \tag{7.69}$$

or equivalently,

$$L_0|\varphi\rangle = \tilde{L}_0|\varphi\rangle = a|\varphi\rangle. \tag{7.70}$$

(We shall see in §7.8 that a should be taken to be 1.) The conditions (7.68) are applied only for $m > 0$ and not for $m < 0$ for reasons to do with the algebra of the Virasoro operators.

The algebra may be derived from the explicit expressions (7.56), (7.57), (7.65) and (7.66) for the Virasoro operators using the commutation relations (7.35)–(7.37) which are also valid for m or n equal to zero with the definition (7.58) of α^{μ}_0 and $\tilde{\alpha}^{\mu}_0$. Let us concentrate for the moment on the algebra of the L_m. When $m + n \neq 0$, $[L_m, L_n]$ can be calculated straightforwardly (Exercise 7.6). In particular, when m and n are both non-zero we obtain

$$\begin{aligned}[L_m, L_n] &= -\tfrac{1}{2}\sum_p p\alpha^{\mu}_{m-p}\alpha_{\mu,\,p+n} - \tfrac{1}{2}\sum_p (m-p)\alpha^{\mu}_{m+n-p}\alpha_{\mu p} \\ &= -\sum_p (m-p)\alpha^{\mu}_p\alpha_{\mu,\,m+n-p}.\end{aligned} \tag{7.71}$$

Using the antisymmetry of the commutator in m and n, and comparing with (7.56), this may be cast in the form

$$[L_m, L_n] = (m-n)L_{m+n} \qquad m, n \neq 0. \tag{7.72}$$

A similar calculation shows that the same result applies when one of m or n is zero, but not both.

It remains to consider the case m and n both non-zero but $m + n = 0$. (The case m and n both zero is trivial.) Then, both the infinite sums in the first line of (7.71) have normal-ordering problems and adding the two terms is very delicate. We therefore write the general formula

$$[L_m, L_n] = (m-n)L_{m+n} + b(m)\delta_{m+n,0} \tag{7.73}$$

where $b(m)$ is a constant deriving from the proper treatment of normal ordering when $m + n$ is zero. This constant, referred to as the conformal anomaly, may be evaluated by considering $[L_m, L_{-m}]$. As a preliminary to this evaluation recall that the $a_n^{\mu\dagger}$ and a_n^μ of (7.38) have harmonic oscillator commutation relations and a Fock space may therefore be defined by applying creation operators $a_n^{\mu\dagger}$ (or α_{-n}^μ) to a ground state that is annihilated by all the a_n^μ (or α_n^μ) where n is positive. A ground state and a corresponding Fock space of the oscillator algebra may be constructed for any value of the centre-of-mass momentum p of the string and all possible ground states are conveniently labelled as $|0; p\rangle$. In the case of $|0; 0\rangle$, the Fock space ground state is also annihilated by α_0^μ of (7.58). (Notice that this is *not* a physical state obeying (7.70).) The constant $b(m)$ may now be evaluated (Exercise 7.7) as the expectation value of $[L_m, L_{-m}]$ with $m > 0$, in the Fock space ground state $|0; 0\rangle$ with zero centre-of-mass momentum.

$$\langle 0; 0|[L_m, L_{-m}]|0; 0\rangle = \frac{D}{2}\sum_{p=1}^{m-1} p(m-p) = \frac{D}{12}m(m^2-1). \tag{7.74}$$

In deriving (7.74) oscillators have been moved past other oscillators so as to act on this ground state with the aid of the commutators (7.35). Thus, the right-mover part of the Virasoro algebra for general m and n is

$$[L_m, L_n] = (m-n)L_{m+n} + \frac{D}{12}m(m^2-1)\delta_{m+n,0}. \tag{7.75}$$

The left-mover part of the Virasoro algebra

$$[\tilde{L}_m, \tilde{L}_n] = (m-n)\tilde{L}_{m+n} + \frac{D}{12}m(m^2-1)\delta_{m+n,0} \tag{7.76}$$

is derived in exactly the same fashion, and because the oscillators α_n^μ and $\tilde{\alpha}_n^\mu$ always commute, the algebra is completed by

$$[L_m, \tilde{L}_n] = 0. \tag{7.77}$$

The presence of the constant term (the conformal anomaly) in (7.75) and (7.76) can now be seen to prevent us from extending the conditions (7.68) and (7.70) for physical states to negative values of m. If we consider the expectation value of $[L_m, L_{-m}]$, with $m > 0$, in a normalized physical state $|\varphi\rangle$, we find with the aid of (7.68), (7.75) and (7.70) that

$$\langle\varphi|L_m L_{-m}|\varphi\rangle = 2ma + \frac{D}{12}m(m^2 - 1). \tag{7.78}$$

If we were to require $L_{-m}|\varphi\rangle$ to be zero for two or more values of m with $m \geqslant 1$ we would have an inconsistency. We shall see later that in physically acceptable theories the appropriate choices of D and a are $D = 26$ and $a = 1$. In that case, we reach the stronger conclusion that we cannot consistently demand $L_{-m}|\varphi\rangle$ to be zero for any positive m. However, condition (7.68) is sufficient to ensure that $\langle\varphi|L_n|\varphi\rangle$ is zero for all non-zero n, because when n is negative we can use the fact that

$$L_n = L_{-n}^{\dagger}. \tag{7.79}$$

The physical state conditions (7.68)–(7.70) have an important role in the theory in preventing the occurrence of ghosts, by which we mean, in the present context, states with negative norm. That such states could arise can be seen by considering the commutation relations (7.40) of the creation and annihilation operators. For the time components, we have

$$[a_m^0, a_m^{0\,\dagger}] = -1. \tag{7.80}$$

Thus, the Fock space state $a_m^{0\,\dagger}|0; 0\rangle$ has negative norm because

$$\langle 0; 0|a_m^0 a_m^{0\,\dagger}|0; 0\rangle = \langle 0; 0|[a_m^0, a_m^{0\,\dagger}]|0; 0\rangle = -1. \tag{7.81}$$

We should hope that the physical state conditions might forbid the occurrence of such states. There are certainly enough conditions because (7.68) applies for every positive value of n, and indeed it can be shown that for $D = 26$ and $a = 1$ all ghosts are removed from the theory by these conditions[4]. Zero-norm states occur in the theory for $D = 26$ but these decouple from scattering amplitudes of positive-norm physical states.

The masses M of physical states in the closed-string theory may be obtained from (7.70). In units where $l = 1$, equation (7.70) yields

$$\tfrac{1}{8}M^2 = \tfrac{1}{8}p^\mu p_\mu = -\sum_{n=1}^{\infty} \alpha_{-n}^{\mu}\alpha_{\mu n} - a = -\sum_{n=1}^{\infty} \tilde{\alpha}_{-n}^{\mu}\tilde{\alpha}_{\mu n} - a. \tag{7.82}$$

This may be written as

$$M^2 = M_{\mathrm{R}}^2 + M_{\mathrm{L}}^2 \tag{7.83}$$

where

$$\tfrac{1}{4}M_{\mathrm{R}}^2 = -\sum_{n=1}^{\infty} \alpha_{-n}^{\mu}\alpha_{\mu n} - a \tag{7.84}$$

$$\tfrac{1}{4}M_{\mathrm{L}}^2 = -\sum_{n=1}^{\infty} \tilde{\alpha}_{-n}^{\mu}\tilde{\alpha}_{\mu n} - a \tag{7.85}$$

and

$$M_{\mathrm{R}}^2 = M_{\mathrm{L}}^2. \tag{7.86}$$

Thus we may think of the squared mass as having equal contributions from the left and right movers. This last equation is a consequence of

$$L_0|\varphi\rangle = \tilde{L}_0|\varphi\rangle \tag{7.87}$$

for all physical states $|\varphi\rangle$. Because of the commutators (7.35) and (7.36), acting with α_{-m}^{μ} $(m > 0)$ on a state increases the value of $\tfrac{1}{4}M_{\mathrm{R}}^2$ by m, and acting with $\tilde{\alpha}_{-m}^{\mu}$ increases the value of $\tfrac{1}{4}M_{\mathrm{L}}^2$ by m.

7.6 Virasoro algebra and masses of states for the open string

As discussed in §7.4, the boundary conditions for the open string mean that a separation into left and right movers is not useful because the left- and right-mover oscillator terms are not independent. Consequently, the open-string expansion (7.43) only involves one set of oscillators α_n^{μ}, and, correspondingly, we would only expect to define a single Virasoro algebra, rather than independent algebras for the left and right movers. The constraint equations (7.17) and (7.18) may as before be written as the pair of equations

$$T_{++} = T_{--} = 0 \tag{7.88}$$

with T_{++} and T_{--} as in (7.48) and (7.49). Generators L_m of a Virasoro algebra may then be defined by taking a combination of Fourier components of T_{++} and T_{--} in the following way

$$L_m = T\int_0^{\pi} \mathrm{d}\sigma\, \mathrm{e}^{\mathrm{i}m(\tau+\sigma)}T_{++} + T\int_0^{\pi} \mathrm{d}\sigma\, \mathrm{e}^{\mathrm{i}m(\tau-\sigma)}T_{--}. \tag{7.89}$$

It is not too difficult to check that these generators may be expressed in terms of the oscillators α_n^{μ} of (7.43) as

$$L_m = -\tfrac{1}{2}\sum_{n=-\infty}^{\infty} \alpha_{m-n}^{\mu}\alpha_{\mu n} \qquad m \neq 0 \tag{7.90}$$

where we have defined

$$\alpha_0^\mu = lp^\mu \tag{7.91}$$

with l as in (7.23). Notice that this differs by a factor of two from the definition (7.58) for the closed string. Just as for the right movers of the closed string we define

$$L_0 \equiv -\tfrac{1}{2}\!:\sum_{n=-\infty}^{\infty} \alpha_{-n}^\mu \alpha_{\mu n}\!: \tag{7.92}$$

and impose the physical state conditions

$$L_m|\varphi\rangle = 0 \qquad m > 0 \tag{7.93}$$

and

$$L_0|\varphi\rangle = a|\varphi\rangle \tag{7.94}$$

where

$$L_0 = -\tfrac{1}{2}\sum_{n=-\infty}^{\infty} \alpha_{-n}^\mu \alpha_{\mu n} + a. \tag{7.95}$$

The same Virasoro algebra as before is obtained because we are dealing with oscillators with the same commutation relations.

$$[L_m, L_n] = (m-n)L_{m+n} + \frac{D}{12} m(m^2-1)\delta_{m+n,0}. \tag{7.96}$$

The Hamiltonian of (7.47) is related to L_0 by

$$H = L_0 - a. \tag{7.97}$$

The masses of physical states are obtained from (7.94), and (7.91). In units where $l = 1$,

$$\tfrac{1}{2}M^2 = -\sum_{n=1}^{\infty} \alpha_{-n}^\mu \alpha_{\mu n} - a. \tag{7.98}$$

This is of almost the same form as the corresponding expression (7.82) for the closed string but differs by a factor of four on the left-hand side.

7.7 The light cone gauge

In §7.3, we employed the reparametrization and local Weyl scaling invariance of the action of the string to choose the covariant gauge defined by

(7.15). In choosing this gauge not all the gauge freedom has been removed and it is possible to impose further gauge conditions that reduce the number of non-trivial components of X^μ and leave only physical dynamical degrees of freedom[(5)]. Let us first define light cone coordinates for a string in D dimensions

$$X^{\pm} \equiv \frac{1}{\sqrt{2}}(X^0 \pm X^{D-1}). \tag{7.99}$$

A reparametrization (7.5) followed by a local Weyl scaling (7.8) is consistent with the gauge conditions (7.15), provided that we choose

$$\Lambda \eta^{\alpha\beta} = -(\partial^\alpha \xi^\beta + \partial^\beta \xi^\alpha). \tag{7.100}$$

This residual gauge invariance allows us to make the essentially trivial choice for X^+

$$X^+(\tau, \sigma) = x^+ + l^2 p^+ \tau \tag{7.101}$$

where x^+ and p^+ are constants, and l is as in (7.23). To show this we first notice that (7.100) implies that $\xi^0 \pm \xi^1$ may be taken to be arbitrary functions of $\tau \pm \sigma$, respectively. In particular, ξ^0 is the sum of an arbitrary function of $\tau + \sigma$ and an arbitrary function of $\tau - \sigma$. Recalling that ξ^α is an infinitesimal shift on (τ, σ) we see that we may combine a reparametrization and a local Weyl scaling to obtain a new τ, say τ', which is the sum of an arbitrary function of $\tau + \sigma$ and an arbitrary function of $\tau - \sigma$. This means that τ' may be identified with any chosen solution u of the wave equation

$$\left(\frac{\partial^2}{\partial \tau^2} - \frac{\partial^2}{\partial \sigma^2}\right) u = 0. \tag{7.102}$$

Since X^+ obeys this equation, and consequently so does $aX^+ + b$ for any constants a and b, we may make the identification (7.101), where the prime on the variable τ' for the chosen gauge has now been dropped.

The coordinate X^- may now be calculated in terms of the transverse degrees of freedom X^i, $i = 1, \ldots, D-2$, in the following way. The constraint equations (7.17) and (7.18) written in terms of light cone variables become

$$\frac{\partial X^+}{\partial \tau}\frac{\partial X^-}{\partial \tau} + \frac{\partial X^+}{\partial \sigma}\frac{\partial X^-}{\partial \sigma} = \frac{1}{2}\left(\left(\frac{\partial X^i}{\partial \tau}\right)^2 + \left(\frac{\partial X^i}{\partial \sigma}\right)^2\right) \tag{7.103}$$

and

$$\frac{\partial X^+}{\partial \tau}\frac{\partial X^-}{\partial \sigma} + \frac{\partial X^-}{\partial \tau}\frac{\partial X^+}{\partial \sigma} = \frac{\partial X^i}{\partial \tau}\frac{\partial X^i}{\partial \sigma}. \tag{7.104}$$

Using (7.101) for X^+ gives

$$l^2 p^+ \frac{\partial X^-}{\partial \tau} = \frac{1}{2}\left(\left(\frac{\partial X^i}{\partial \tau}\right)^2 + \left(\frac{\partial X^i}{\partial \sigma}\right)^2\right) \tag{7.105}$$

and

$$l^2 p^+ \frac{\partial X^-}{\partial \sigma} = \frac{\partial X^i}{\partial \tau}\frac{\partial X^i}{\partial \sigma} \tag{7.106}$$

which may be solved to obtain X^- in terms of the transverse degrees of freedom X apart from an integration constant x^-. Thus, in the light cone gauge the only dynamical degrees of freedom are the transverse degrees of freedom X^i. We need only solve the string equations of motion (7.16) for the transverse degrees of freedom, using closed- or open-string boundary combinations as appropriate. There will be the usual mode expansions (7.22) or (7.26) and (7.27) for the closed string, and (7.43) for the open string, for the transverse degrees of freedom.

In the light cone gauge, the derivation of the masses of physical states proceeds somewhat differently to the derivation of §7.5 for the covariant gauge. In the case of the closed string, equation (7.105) may be written with the aid of (7.31) and (7.101) as

$$l^2 p^+ T^{-1} p^- = \tfrac{1}{2}\left(\left(\frac{\partial X^i}{\partial \tau}\right)^2 + \left(\frac{\partial X^i}{\partial \sigma}\right)^2\right). \tag{7.107}$$

Using the expansion (7.22), and integrating from 0 to π with respect to σ, gives (Exercise 7.8)

$$\tfrac{1}{8}M^2 = \tfrac{1}{8}p^2 = \tfrac{1}{8}(2p^+p^- - (p^i)^2) = \tfrac{1}{2}\sum_{n=1}^{\infty}(\alpha^i_{-n}\alpha^i_n + \tilde{\alpha}^i_{-n}\tilde{\alpha}^i_n) - a \tag{7.108}$$

where we have chosen $l = 1$, or equivalently, from (7.23), have taken $T = \pi^{-1}$. The normal-ordering constant a may be determined because it arises from reordering some of the terms in the original sum over n from $-\infty$ to ∞ using the commutators (7.35) and (7.36). The result is

$$a = -\frac{(D-2)}{2}\sum_{n=1}^{\infty} n. \tag{7.109}$$

If we regularize this divergent expression using ζ-function regularization with $\zeta(s) = \Sigma_{n=1}^{\infty} n^{-s}$ analytically continued to $s = -1$, then

$$a = -\frac{(D-2)}{2}\zeta(-1) = \frac{(D-2)}{24}. \tag{7.110}$$

(A discussion of the use of ζ-function regularization in the field theory context can be found in Ramond[6].)

For the closed-string case, equation (7.106) may be integrated from 0 to π with respect to σ and the periodicity X^- in the interval 0 to π may be used to deduce that

$$\int_0^\pi \frac{\partial X^i}{\partial \tau}\frac{\partial X^i}{\partial \sigma}\,\mathrm{d}\sigma = 0. \tag{7.111}$$

Using the expansion (7.22) then yields the further condition

$$\sum_{n=1}^{\infty} \tilde{\alpha}^i_{-n}\tilde{\alpha}^i_n = \sum_{n=1}^{\infty} \alpha^i_{-n}\alpha^i_n \tag{7.112}$$

where the normal-ordering constants, which are the same on the left- and right-hand sides, have cancelled. Combining (7.108) and (7.112) we may write

$$M^2 = M^2_{\mathrm{R}} + M^2_{\mathrm{L}} \tag{7.113}$$

where

$$\tfrac{1}{4}M^2_{\mathrm{R}} = \sum_{n=1}^{\infty} \alpha^i_{-n}\alpha^i_n - a \tag{7.114}$$

$$\tfrac{1}{4}M^2_{\mathrm{L}} = \sum_{n=1}^{\infty} \tilde{\alpha}^i_{-n}\tilde{\alpha}^i_n - a \tag{7.115}$$

$$M^2_{\mathrm{R}} = M^2_{\mathrm{L}} \tag{7.116}$$

and a is given by (7.110).

A similar treatment may be used for the open-string case except that there is no condition like (7.112) because the string degrees of freedom are not in this case periodic in σ in the interval 0 to π. For the open string we obtain (Exercise 7.9)

$$\tfrac{1}{2}M^2 = \sum_{n=1}^{\infty} \alpha^i_{-n}\alpha^i_n - a \tag{7.117}$$

with a again given by (7.110).

7.8 Low-lying string states

Bosonic string states may be constructed in the light cone gauge by acting with products of oscillators α^i_{-n} and $\tilde{\alpha}^i_{-n}$, $i = 1, \ldots, D-2$, on the ground states $|0\rangle_{\mathrm{R}}$ and $|0\rangle_{\mathrm{L}}$ for the right and left movers, respectively, and forming the product of the right- and left-mover states obtained. To calculate the masses of these states we need the value of the normal-ordering constant a in (7.114) and (7.115). This constant may be determined by constructing the Lorentz generators for the D-dimensional string theory, using the Noether currents for the Lorentz group, and requiring that the commutators of the Lorentz generators do not have any quantum anomalies spoiling the Lorentz invariance[(5)]. The vanishing of these anomalies requires

$$D = 26 \tag{7.118}$$

and (without using (7.110), which is a check on the result)

$$a = 1. \tag{7.119}$$

We shall see shortly that there is a simple argument from the masses of the string states that leads to the same conclusion. In later chapters it will be seen how theories in 4 dimensions rather than 26 dimensions may be constructed.

For the closed bosonic string, the ground state $|0\rangle_R$ for the right movers is obtained when the oscillator contribution to (7.114) is zero, and so $|0\rangle_R$ has $M_R^2 = -4$, and similarly $|0\rangle_L$ has $M_L^2 = -4$. Thus, the closed-string ground state has $M^2 = -8$, in units where l of (7.23) is 1. This tachyonic ground state constitutes a problem for the bosonic string which can be avoided for the superstring of Chapter 8, and for the heterotic string of Chapter 9 which employs superstring right movers and closed-bosonic-string left movers.

To obtain massless states consistently with (7.116) it is necessary for M_R^2 and M_L^2 to be separately zero. Thus, the massless closed string states are

$$\begin{aligned}\alpha^i_{-1}|0\rangle_R\tilde{\alpha}^j_{-1}|0\rangle_L = {} & \tfrac{1}{2}(\alpha^i_{-1}|0\rangle_R\tilde{\alpha}^j_{-1}|0\rangle_L + \alpha^j_{-1}|0\rangle_R\tilde{\alpha}^i_{-1}|0\rangle_L \\ & - 2\delta^{ij}(D-2)^{-1}\alpha^k_{-1}|0\rangle_R\tilde{\alpha}^k_{-1}|0\rangle_L) \\ & + \delta^{ij}(D-2)^{-1}\alpha^k_{-1}|0\rangle_R\tilde{\alpha}^k_{-1}|0\rangle_L \\ & + \tfrac{1}{2}(\alpha^i_{-1}|0\rangle_R\tilde{\alpha}^j_{-1}|0\rangle_L - \alpha^j_{-1}|0\rangle_R\tilde{\alpha}^i_{-1}|0\rangle_L).\end{aligned} \tag{7.120}$$

The three terms in (7.120) correspond to a traceless symmetric tensor, which may be identified with the graviton for $D = 26$ dimensions, a scalar referred to as the dilaton, and an antisymmetric tensor. These last two types of massless particle accompanying the graviton are characteristic of string theory.

Since we are working in the light cone gauge where only the transverse degrees of freedom are present, we should expect massive states to fall into irreducible representations of SO(25) and massless states into irreducible representations of SO(24). The states of (7.120) do fall neatly into SO(24) representations but there are no other states of the same mass, no matter what the value of a, that can complete SO(25) multiplets. This is a further argument that the states of (7.120) should be massless, and that the choice $a = 1$ is the correct one. Consistency with (7.110) then requires $D = 26$.

7.9 Path integral quantization

In quantum field theory, there are two basic approaches to quantization. The first is canonical quantization, in which the classical field φ is replaced by an operator field $\hat{\varphi}$ upon which canonical commutation relations are imposed. This is analogous to the procedures that we have been using up to this point in the quantization of string theory. The second is the path integral approach, in which the generating functional $W[J]$ is constructed as a path integral over the exponential of the action in the presence of source terms, and the Green functions of the theory are obtained as functional derivatives of the generating functional. We shall now develop this approach to the bosonic string[7].

Because the bosonic string possesses gauge symmetries, namely world sheet reparametrization invariance (7.5), the procedure required for path integral quantization of the string is very similar to the Faddeev–Popov method required for the quantization of gauge field theories[8],[6],[9]. It is convenient to introduce world sheet 'light cone' coordinates (not to be confused with the space-time light cone coordinates of §7.7)

$$\sigma^{\pm} = \tau \pm \sigma \tag{7.121}$$

with conjugate derivatives

$$\partial_{\pm} = \tfrac{1}{2}(\partial_{\tau} \pm \partial_{\sigma}). \tag{7.122}$$

The components of the world sheet metric may be rewritten in terms of

$$\eta_{++} = \eta_{--} = 0 \qquad \eta_{+-} = \eta_{-+} = \tfrac{1}{2}. \tag{7.123}$$

Choosing the covariant gauge of (7.13) is then equivalent to imposing the gauge conditions

$$F_{++}(h_{\alpha\beta}) \equiv h_{++} = 0 \tag{7.124}$$

and

$$F_{--}(h_{\alpha\beta}) \equiv h_{--} = 0. \tag{7.125}$$

World sheet reparametrizations

$$\tau \to \tau + \xi^0 \qquad \sigma \to \sigma + \xi^1 \tag{7.126}$$

may be recast as

$$\sigma^{\pm} \to \sigma^{\pm} + \xi^{\pm}. \tag{7.127}$$

The effect of this reparametrization, as in (7.5), may be written as

$$\delta h_{\alpha\beta} = \nabla_{\alpha}\xi_{\beta} + \nabla_{\beta}\xi_{\alpha} \tag{7.128}$$

where

$$\nabla_\alpha \xi_\beta \equiv (\partial_\alpha - \Gamma^\gamma_{\alpha\beta})\xi_\gamma \tag{7.129}$$

with $\Gamma^\gamma_{\alpha\beta}$ the affine connection derived from the world sheet metric $h_{\alpha\beta}$. In terms of world sheet 'light cone' coordinates

$$\delta h_{++} = 2\,\nabla_+ \xi_+ \tag{7.130}$$

and

$$\delta h_{--} = 2\,\nabla_- \xi_-. \tag{7.131}$$

To apply the Faddeev–Popov quantization procedure to the string theory, we require the path integral

$$Z \propto \int \mathscr{D}X^\mu \int \mathscr{D}h_{++}\,\mathscr{D}h_{--}\,\mathscr{D}h_{+-}\,\delta[F_{++}]\delta[F_{--}]$$

$$\times \det\left(\frac{\delta F_{++}}{\delta \xi_+}\right) \det\left(\frac{\delta F_{--}}{\delta \xi_-}\right) \mathrm{e}^{\mathrm{i}S} \tag{7.132}$$

where S is the bosonic string action of (7.4). The effect of the functional δ-functions, with F_{++} and F_{--} as in (7.124) and (7.125), is to select the gauge $h_{++} = h_{--} = 0$ and to leave an integral over X^μ and h_{+-} (or equivalently, in this latter case, over φ of (7.13)). The determinants may be cast in terms of Faddeev–Popov ghosts (not to be confused with the negative-norm ghosts of §7.5) by using the identity[6]

$$\det B = \int \mathscr{D}c\,\mathscr{D}b \exp\frac{\mathrm{i}}{\pi}\int \mathrm{d}^2\sigma\,\mathrm{d}^2\sigma'\,c_i(\tau',\sigma')B_{ij}(\tau',\sigma';\tau,\sigma)b_j(\tau,\sigma) \tag{7.133}$$

where the ghosts c_i and anti-ghosts b_j are anti-commuting Grassmann variables, and

$$\int \mathrm{d}^2\sigma \equiv \int_{\tau_\mathrm{i}}^{\tau_\mathrm{f}} \mathrm{d}\tau \int_0^\pi \mathrm{d}\sigma \tag{7.134}$$

as in (7.4). With the aid of (7.124) and (7.130) we see that

$$\frac{\delta F_{++}(\tau',\sigma')}{\delta \xi_+(\tau,\sigma)} = 2\,\nabla_+{}'\delta(\tau'-\tau)\,\delta(\sigma'-\sigma). \tag{7.135}$$

Thus, (7.133) (with the ghosts given the conventional labelling c^- and b_{--}) gives

$$\det\frac{\delta F_{++}}{\delta \xi_+} = \int \mathscr{D}c^-\,\mathscr{D}b_{--} \exp\frac{\mathrm{i}}{\pi}\int \mathrm{d}^2\sigma\,c^-(\tau,\sigma)\,\nabla_+ b_{--}(\tau,\sigma). \tag{7.136}$$

Similarly,

$$\det \frac{\delta F_{--}}{\delta \xi_-} = \int \mathscr{D}c^+ \mathscr{D}b_{++} \exp \frac{i}{\pi} \int d^2\sigma \, c^+(\tau, \sigma) \nabla_- b_{++}(\tau, \sigma). \tag{7.137}$$

The path integral (7.132) may therefore be written as

$$Z \propto \int \mathscr{D}\varphi \int \mathscr{D}X^\mu \mathscr{D}c^- \mathscr{D}c^+ \mathscr{D}b_{--} \mathscr{D}b_{++} \, e^{i(S + S_{FP})} \tag{7.138}$$

where

$$S_{FP} = \frac{1}{\pi} \int d^2\sigma \, (c^- \nabla_+ b_{--} + c^+ \nabla_- b_{++}). \tag{7.139}$$

Although it is beyond our scope here, it may be shown[(10)] that in this formulation of the theory, with the Faddeev–Popov ghost contribution to the Virasoro generators included, there is no anomaly in the Virasoro algebra when

$$D = 26 \qquad a = 1. \tag{7.140}$$

Exercises

7.1 Show that the general solution of the string wave equation with closed-string boundary conditions is as in (7.22).

7.2 Derive the commutator (7.34) for string centre-of-mass coordinate x^μ and centre-of-mass momentum p^μ.

7.3 Derive the expression (7.42) for the string Hamiltonian in terms of oscillators.

7.4 Show that the general solution of the string wave equation with open-string boundary conditions is (7.43).

7.5 Show that in terms of oscillators the open-string Hamiltonian is (7.47).

7.6 Derive the Virasoro algebra commutator $[L_m, L_n]$ when $m + n \neq 0$.

7.7 Calculate the constant $b(m)$ in (7.73) by considering the expectation value of $[L_m, L_{-m}]$.

7.8 Derive the expression (7.108) for the squared mass for the closed string in terms of oscillators.

7.9 Repeat the calculation of Exercise 7.8 for the open string to obtain (7.117).

References

General references

Brink L 1984 Superstrings in *Supersymmetry* ed K Dietz, R Flume, G v Gehlen and V Rittenburg (New York: Plenum) p 89

Green M B, Schwarz J H and Witten E 1987 *Superstring Theory* (Cambridge: Cambridge University Press)

Schwarz J H 1982 *Phys. Rep.* **89** 224

West P 1989 An introduction to string theory *Preprint* TH 51565/88 (CERN); 1989 *Acta Phys. Pol.* B **20** 471

References in the text

1 Green M B, Schwarz J H and Witten E 1987 *Superstring Theory* (Cambridge: Cambridge University Press) Chapter 1

2 Scherk J and Schwarz J H 1974 *Nucl. Phys.* B **81** 118; 1974 *Phys. Lett.* **52B** 347

3 Green M B and Schwarz J H 1982 *Phys. Lett.* **109B** 444

4 Brower R C 1972 *Phys. Rev.* D **6** 1655
Goddard P and Thorn C B 1972 *Phys. Lett.* **40B** 235
Green M B, Schwarz J H and Witten E 1987 *Superstring Theory* (Cambridge: Cambridge University Press) §2.3.3

5 Goddard P, Goldstone J, Rebbi C and Thorn C B 1973 *Nucl. Phys.* B **56** 109

6 Ramond P 1981 *Field Theory: a Modern Primer* (New York: Benjamin/ Cummings)

7 Polyakov A M 1981 *Phys. Lett.* **103B** 207, 211

8 Faddeev L D and Popov V N 1967 *Phys. Lett.* **25B** 29

9 Bailin D and Love A 1986 *Introduction to Gauge Field Theory* (Bristol: Institute of Physics Publishing)

10 Green M B, Schwarz J H and Witten E 1987 *Superstring Theory* (Cambridge: Cambridge University Press) §3.1

8

THE SUPERSTRING

8.1 Introduction

The bosonic string theory of Chapter 7 is not entirely satisfactory as a complete theory, first because it does not possess any fermionic states, and secondly because the string ground state is tachyonic (as discussed in §7.8). Both these difficulties can be overcome by constructing a theory in which, associated with each bosonic degree of freedom $X^\mu(\tau, \sigma), \mu = 0, \ldots, D-1$, there is a world sheet spinor fermionic degree of freedom $\Psi^\mu(\tau, \sigma)$ (described by a two-component Majorana spinor). The action for this theory needs to be formulated in such a way as to avoid the occurrence of ghost states of negative norm.

For the bosonic string, such ghosts were removed in §7.5 by the physical state conditions (7.68)–(7.70) which had their origin in the constraint equations (7.17) and (7.18), which in turn derived from variation of the action with respect to the world sheet metric $h_{\alpha\beta}$. Thus, for the bosonic string, the absence of negative-norm ghosts depends crucially on the reparametrization-invariant form (7.4) of the action. This action is formally an action for the coupling of the D scalar fields X^μ to two-dimensional gravity. We may therefore suspect that an appropriate construction of the action in the case when fermionic degrees of freedom Ψ^μ are also present might be to treat X^μ and Ψ^μ as supersymmetric partners for a world sheet supersymmetry, and to couple them to two-dimensional supergravity. It turns out that such an action does indeed provide a theory in which all negative-norm ghosts are removed by constraint equations arising as Euler–Lagrange equations of the action. We shall see in §8.8 that consistency of the theory requires D to be 10 for the superstring (rather than 26 as for the bosonic string).

In the first instance, although the superstring possesses world sheet supersymmetry, it does not possess space-time supersymmetry. However, we shall see in §8.8 that a potential problem of tachyonic ground states is avoided by applying a projection to the states of the theory, and after this projection has been applied a ten-dimensional space-time supersymmetric theory is obtained. Whether space-time supersymmetry survives in the final four-dimensional theory depends on the way in which the ten-dimensional theory is compactified to four dimensions, as will be discussed in Chapter 10.

DOI: 10.1201/9780367805807-8

8.2 The superstring action

As discussed in §8.1, the required action is formally an action coupling supersymmetric partner bosonic and fermionic fields $X^\mu(\tau, \sigma)$ and $\Psi^\mu(\tau, \sigma)$ to two-dimensional supergravity[1]. Such an action may be derived using the Noether procedure in a way similar to that for the construction of the action for the chiral superfield coupled to four-dimensional supergravity in §5.2.

A suitable starting point for the Noether procedure is the action

$$S_0 = -\frac{1}{2\pi}\int d^2\sigma\,(-\det h)^{1/2}(h^{\alpha\beta}\,\partial_\alpha X^\mu\,\partial_\beta X_\mu + i\,\bar{\Psi}^\mu\rho^\alpha\,\partial_\alpha\Psi_\mu) \tag{8.1}$$

where

$$\rho^0 = \begin{pmatrix} 0 & -i \\ i & 0 \end{pmatrix} \qquad \rho^1 = \begin{pmatrix} 0 & i \\ i & 0 \end{pmatrix} \tag{8.2}$$

are two-dimensional γ-matrices satisfying

$$\{\rho^\alpha, \rho^\beta\} = 2\eta^{\alpha\beta}I \tag{8.3}$$

with $\eta_{\alpha\beta}$ as in (7.14).

The action S_0 possesses world sheet reparametrization invariance and an on-shell global world sheet supersymmetry under the supersymmetry transformation

$$\delta X^\mu = \bar{\xi}\Psi^\mu \tag{8.4}$$

and

$$\delta\Psi^\mu = -\,i\,\rho^\alpha\,\partial_\alpha X^\mu\,\xi \tag{8.5}$$

where ξ is a two-component Majorana spinor parameter. It may be checked (Exercise 8.1) that $\mathcal{S}_0$ is invariant under this transformation, and that the transformation is truly a supersymmetry in the sense that the commutator of two transformations δ_1 and δ_2 is a translation:

$$[\delta_1, \delta_2]X^\mu = -\,2i\,\bar{\xi}_2\rho^\alpha\xi_1\,\partial_\alpha X^\mu \tag{8.6}$$

and

$$[\delta_1, \delta_2]\Psi^\mu = -\,2i\,\bar{\xi}_2\rho^\alpha\xi_1\,\partial_\alpha\Psi^\mu. \tag{8.7}$$

In checking (8.6) and (8.7) some properties of two-component world sheet Majorana spinors are required. For instance,

$$\bar{\xi}_1\xi_2 = \bar{\xi}_2\xi_1 \tag{8.8}$$

and

$$\bar{\xi}_1\rho^\alpha\xi_2 = -\,\bar{\xi}_2\rho^\alpha\xi_1 \tag{8.9}$$

which follow more or less immediately from the fact that the components of the Majorana spinors are real and anti-commuting.

The action S_0 is no longer invariant when ξ is replaced by a local variable $\xi(\tau, \sigma)$. Then,

$$\delta S_0 = \frac{2}{\pi}\int d^2\sigma\,(-\det h)^{1/2}\,\partial_\alpha\bar{\xi}J^\alpha \tag{8.10}$$

where

$$J^\alpha = \tfrac{1}{2}\rho^\beta\rho^\alpha\Psi^\mu\,\partial_\beta X_\mu \tag{8.11}$$

is the world sheet supercurrent. To cancel this variation of S_0 we need to introduce the two-dimensional supergravity 'gravitino' χ_α with the transformation

$$\delta\chi_\alpha = \partial_\alpha\xi. \tag{8.12}$$

Then δS_0 is cancelled by adding to the action:

$$S_1 = -\frac{1}{\pi}\int d^2\sigma\,(-\det h)^{1/2}\bar{\chi}_\alpha\rho^\beta\rho^\alpha\Psi^\mu\,\partial_\beta X_\mu\,. \tag{8.13}$$

However, the transformation of X_μ gives an extra term in δS_1,

$$\begin{aligned}\delta S_1 &= -\frac{1}{\pi}\int d^2\sigma\,(-\det h)^{1/2}\bar{\chi}_\alpha\rho^\beta\rho^\alpha\Psi^\mu\bar{\Psi}_\mu\,\partial_\beta\xi + \cdots \\ &= \frac{1}{2\pi}\int d^2\sigma\,(-\det h)^{1/2}\bar{\Psi}_\mu\Psi^\mu\bar{\chi}_\alpha\rho^\beta\rho^\alpha\,\partial_\beta\xi + \cdots.\end{aligned} \tag{8.14}$$

This variation is cancelled in its turn by adding a further term to the action:

$$S_2 = -\frac{1}{4\pi}\int d^2\sigma\,(-\det h)^{1/2}\bar{\Psi}_\mu\Psi^\mu\bar{\chi}_\alpha\rho^\beta\rho^\alpha\chi_\beta\,. \tag{8.15}$$

Then with the addition of some terms to the local supersymmetry transformations the action

$$S = S_0 + S_1 + S_2 \tag{8.16}$$

is invariant under

$$\delta X^\mu = \bar{\xi}\Psi^\mu \tag{8.17}$$

$$\delta\Psi^\mu = -\mathrm{i}\,\rho^\alpha\xi(\partial_\alpha X^\mu - \bar{\Psi}^\mu\chi_\alpha) \tag{8.18}$$

$$\delta\chi_\alpha = \partial_\alpha\xi \tag{8.19}$$

and

$$\delta e^a_\alpha = -2\mathrm{i}\,\bar{\xi}\rho^a\chi_\alpha \tag{8.20}$$

where e^a_α is the zweibein, satisfying $h_{\alpha\beta} = e^a_\alpha e^b_\beta \eta_{ab}$. This is the locally world sheet supersymmetric action we were seeking, which we shall use as our action for the superstring.

At least at the classical level, the superstring action, like the bosonic string action, also possesses a local Weyl scaling or conformal invariance under the coordinate-dependent rescaling

$$\delta h_{\alpha\beta} = \Lambda(\tau, \sigma)h_{\alpha\beta} \tag{8.21}$$

$$\delta\chi_\alpha = \tfrac{1}{2}\Lambda(\tau, \sigma)\chi_\alpha \tag{8.22}$$

$$\delta\Psi^\mu = -\tfrac{1}{2}\Lambda(\tau, \sigma)\Psi^\mu \tag{8.23}$$

and

$$\delta X^\mu = 0 \tag{8.24}$$

which now scale χ_α and Ψ^μ as well as $h_{\alpha\beta}$. Moreover, there is a local (superconformal) symmetry that acts only on the 'gravitino' for two-dimensional world sheet supergravity,

$$\delta\chi_\alpha = \mathrm{i}\,\rho_\alpha\eta(\tau, \sigma) \tag{8.25}$$

and

$$\delta h_{\alpha\beta} = \delta\Psi^\mu = \delta X^\mu = 0 \tag{8.26}$$

where η is a two-dimensional world sheet Majorana spinor parameter.

8.3 Equations of motion and the covariant gauge

Variation of the action (8.16) with respect to X^μ, Ψ^μ, $h_{\alpha\beta}$ and χ_α, under the assumption that the surface terms vanish, leads to Euler–Lagrange equations. These equations take a particularly simple form if we choose a covariant gauge, much as we did for the bosonic string. For the bosonic string we were able to exploit the world sheet reparametrization invariance and the conformal invariance to write the world sheet metric in the form

$$h_{\alpha\beta} = \eta_{\alpha\beta} \tag{8.27}$$

with

$$\eta_{\alpha\beta} = \mathrm{diag}(1, -1) \tag{8.28}$$

and the same is true for the superstring. In addition, the two-dimensional world sheet supersymmetry together with the superconformal symmetry (8.25), (8.26) may be employed to choose

$$\chi_\alpha = 0. \tag{8.29}$$

In this covariant gauge, the Euler–Lagrange equations are

$$\partial_\alpha \partial^\alpha X^\mu = 0 \tag{8.30}$$

$$\mathrm{i}\, \rho^\alpha \partial_\alpha \Psi^\mu = 0 \tag{8.31}$$

$$T_{\alpha\beta} = -\,\partial_\alpha X^\mu \partial_\beta X_\mu - \frac{\mathrm{i}}{4} \bar{\Psi}^\mu (\rho_\alpha \partial_\beta + \rho_\beta \partial_\alpha) \Psi_\mu + \frac{\eta_{\alpha\beta}}{2} \left(\partial^\gamma X^\mu \partial_\gamma X_\mu + \frac{\mathrm{i}}{2} \bar{\Psi}^\mu \rho^\gamma \partial_\gamma \Psi_\mu \right) = 0 \tag{8.32}$$

and

$$J^\alpha = \tfrac{1}{2} \rho^\beta \rho^\alpha \Psi^\mu \partial_\beta X_\mu = 0 \tag{8.33}$$

where $T_{\alpha\beta}$ may be interpreted as the energy–momentum tensor for the two-dimensional supergravity of the free scalar and Majorana spinor fields X^μ and Ψ^μ, and J^α is the world sheet supercurrent. These last two equations amount to constraint equations.

For the derivation of the Euler–Lagrange equations (8.30) and (8.31) to be valid it is necessary for surface terms to vanish. In the case of the bosonic degrees of freedom this allows open- and closed-string boundary conditions as in (7.20), (7.21). The surface terms for the fermionic degrees of freedom Ψ^μ are a little more subtle and are best discussed by writing the world sheet spinor Ψ^μ in components as

$$\Psi^\mu = \begin{pmatrix} \Psi^\mu_{\mathrm{R}} \\ \Psi^\mu_{\mathrm{L}} \end{pmatrix}. \tag{8.34}$$

Then the equation of motion (8.31) reduces to the pair of equations

$$\left(\frac{\partial}{\partial \tau} + \frac{\partial}{\partial \sigma} \right) \Psi^\mu_{\mathrm{R}} = 0 \tag{8.35}$$

and

$$\left(\frac{\partial}{\partial \tau} - \frac{\partial}{\partial \sigma} \right) \Psi^\mu_{\mathrm{L}} = 0. \tag{8.36}$$

Consequently, Ψ^μ_{R} is a function of $\tau - \sigma$ only and describes right-moving degrees of freedom, and Ψ^μ_{L} is a function of $\tau + \sigma$ only, and describes left-moving degrees of freedom. Moreover, the term $S(\Psi^\mu)$ involving Ψ^μ in the action in covariant gauge may be written as

$$S(\Psi^\mu) = \frac{\mathrm{i}}{\pi} \int \mathrm{d}^2\sigma \, (\Psi^\mu_{\mathrm{L}} \partial_- \Psi_{\mu\mathrm{L}} + \Psi^\mu_{\mathrm{R}} \partial_+ \Psi_{\mu\mathrm{R}}) \tag{8.37}$$

where

$$\partial_{\pm} \equiv \tfrac{1}{2}(\partial_\tau \pm \partial_\sigma). \tag{8.38}$$

It may then be seen that, in the case of closed strings, the surface terms arising from the variation of the action (8.37) vanish for boundary conditions

$$\Psi^\mu_{\mathrm{R}}(\tau, \sigma + \pi) = \pm\, \Psi^\mu_{\mathrm{R}}(\tau, \sigma) \tag{8.39}$$

and

$$\Psi^\mu_{\mathrm{L}}(\tau, \sigma + \pi) = \pm\, \Psi^\mu_{\mathrm{L}}(\tau, \sigma). \tag{8.40}$$

Notice that periodic or anti-periodic boundary conditions may be chosen independently for the right movers Ψ^μ_{R} and the left movers Ψ^μ_{L} (periodic boundary conditions for fermionic degrees of freedom are usually referred to as Ramond[2] boundary conditions denoted by R, and anti-periodic boundary conditions as Neveu–Schwarz[3] boundary conditions, denoted by NS).

For open strings, the surface terms vanish for boundary conditions

$$\Psi^\mu_{\mathrm{L}}(\tau, \sigma) = \Psi^\mu_{\mathrm{R}}(\tau, \sigma) \tag{8.41}$$

and

$$\Psi^\mu_{\mathrm{L}}(\tau, \pi) = \pm\, \Psi^\mu_{\mathrm{R}}(\tau, \pi). \tag{8.42}$$

8.4 Mode expansions and quantization

For the closed superstring, the mode expansions for the bosonic degrees of freedom, and their quantization, are just as in §7.3 for the closed bosonic string. The mode expansions for the fermionic right and left movers depend on whether the choice of periodic (Ramond, R) or anti-periodic (Neveu–Schwarz, NS) boundary conditions is made, and this choice may be made independently for the right and left movers. Thus, for the right movers the mode expansion is

$$\Psi^\mu_{\mathrm{R}} = \sum_{n \in \mathbb{Z}} d^\mu_n\, \mathrm{e}^{-2\mathrm{i}\, n(\tau - \sigma)} \qquad \mathrm{R} \tag{8.43}$$

or

$$\Psi^\mu_{\mathrm{R}} = \sum_{r \in \mathbb{Z} + 1/2} b^\mu_r\, \mathrm{e}^{-2\mathrm{i}\, r(\tau - \sigma)} \qquad \mathrm{NS} \tag{8.44}$$

where n runs over all the integers, and r runs over all the half-integers. For the left movers, the mode expansion is

$$\Psi_L^\mu = \sum_{n \in \mathbb{Z}} \tilde{d}_n^\mu \, e^{-2i\,n(\tau + \sigma)} \qquad \text{R} \tag{8.45}$$

or

$$\Psi_L^\mu = \sum_{r \in \mathbb{Z} + 1/2} \tilde{b}_r^\mu \, e^{-2i\,r(\tau + \sigma)} \qquad \text{NS.} \tag{8.46}$$

Modular invariance of the theory (discussed in Chapter 11) requires us to consider all four possible pairings of boundary conditions (string sectors) in a consistent theory.

Quantization of the fermionic degrees of freedom is achieved by imposing the canonical anti-commutation relations (for the canonical momenta $(i/2\pi)\Psi^\mu_{\mathrm{R\,or\,L}}$)

$$\begin{aligned}\{\Psi_R^\mu(\tau, \sigma), \Psi_R^\nu(\tau, \sigma')\} &= \{\Psi_L^\mu(\tau, \sigma), \Psi_L^\nu(\tau, \sigma')\} \\ &= -2\pi\delta(\sigma - \sigma')\eta^{\mu\nu}\end{aligned} \tag{8.47}$$

and

$$\{\Psi_R^\mu(\tau, \sigma), \Psi_L^\nu(\tau, \sigma')\} = 0. \tag{8.48}$$

Substituting the mode expansions (8.43)–(8.46) in (8.47) and (8.48) it may be seen (Exercise 8.2) that the corresponding anti-commutators for the oscillators in the mode expansions are, for Ramond boundary conditions,

$$\{d_m^\mu, d_n^\nu\} = \{\tilde{d}_m^\mu, \tilde{d}_n^\nu\} = -\delta_{m+n,0}\eta^{\mu\nu} \tag{8.49}$$

and for Neveu–Schwarz boundary conditions

$$\{b_r^\mu\, b_s^\nu\} = \{\tilde{b}_r^\mu, \tilde{b}_s^\nu\} = -\delta_{r+s,0}\eta^{\mu\nu} \tag{8.50}$$

with left- and right-mover oscillators anti-commuting. (Notice again that Ramond or Neveu–Schwarz boundary conditions may be chosen independently for the right and left movers.)

The Hamiltonian for the closed superstring in covariant gauge may be calculated much as in §7.4 for the closed bosonic string with the result (Exercise 8.3) that

$$\begin{aligned}H = -\frac{1}{2\pi}\int_0^\pi d\sigma\, (&\partial_\tau X^\mu\, \partial_\tau X_\mu + \partial_\sigma X^\mu\, \partial_\sigma X_\mu \\ &+ i\,\Psi_L^\mu\, \partial_\sigma \Psi_{\mu L} - i\,\Psi_R^\mu\, \partial_\sigma \Psi_{\mu R})\end{aligned} \tag{8.51}$$

where we have set l of (7.23) equal to 1. In terms of the oscillators in the mode expansions (7.22), (8.43) and (8.45) for Ramond boundary conditions for both right and left movers,

$$H = -\tfrac{1}{2}p^\mu p_\mu - \sum_{n \neq 0} (\alpha^\mu_{-n}\alpha_{\mu n} + \tilde{\alpha}^\mu_{-n}\tilde{\alpha}_{\mu n})$$
$$- \sum_{n \in \mathbb{Z}} n(d^\mu_{-n}d_{\mu n} + \tilde{d}^\mu_{-n}\tilde{d}_{\mu n}). \tag{8.52}$$

An exactly similar expression applies when the right and/or left movers have Neveu–Schwarz boundary conditions with the replacement of d^μ_n by b^μ_r and/or $\tilde{d}^\mu_n$ or $\tilde{b}^\mu_r$, where $r \in \mathbb{Z} - \frac{1}{2}$.

For the open superstring, the mode expansions of the right and left movers are not independent of each other. For periodic (Ramond, R) boundary conditions in (8.42), the mode expansions are

$$\Psi^\mu_R = \frac{1}{\sqrt{2}} \sum_{n \in \mathbb{Z}} d^\mu_n \, e^{-in(\tau - \sigma)} \tag{8.53}$$

and

$$\Psi^\mu_L = \frac{1}{\sqrt{2}} \sum_{n \in \mathbb{Z}} d^\mu_n \, e^{-in(\tau + \sigma)}. \tag{8.54}$$

The factors of $1/\sqrt{2}$ are conventional, and $2n$ of (8.43) is replaced by n in the exponent because Ψ^μ_R and Ψ^μ_L are not now required to be separately periodic over the range 0 to π of σ. For anti-periodic (Neveu–Schwarz, NS) boundary conditions on (8.42), the mode expansions are instead

$$\Psi^\mu_R = \frac{1}{\sqrt{2}} \sum_{r \in \mathbb{Z} + 1/2} b^\mu_r \, e^{-ir(\tau - \sigma)} \tag{8.55}$$

and

$$\Psi^\mu_L = \frac{1}{\sqrt{2}} \sum_{r \in \mathbb{Z} + 1/2} b^\mu_r \, e^{-ir(\tau + \sigma)}. \tag{8.56}$$

Quantization of the fermionic degrees of freedom is again achieved by imposing (8.47) and (8.48), and the corresponding anti-commutators for the oscillators in the mode expansions are

$$\{d^\mu_m, d^\nu_n\} = -\delta_{m+n,0}\eta^{\mu\nu} \tag{8.57}$$

and

$$\{b^\mu_r, b^\nu_s\} = -\delta_{r+s,0}\eta^{\mu\nu}. \tag{8.58}$$

The Hamiltonian is again given by (8.51), and in terms of the oscillators in the mode expansions (7.43), (8.53) and (8.54) for Ramond boundary conditions

$$H = -\tfrac{1}{2}p^\mu p_\mu - \tfrac{1}{2}\sum_{n \neq 0} \alpha^\mu_{-n}\alpha_{\mu n} - \tfrac{1}{2}\sum_{n \in \mathbb{Z}} n d^\mu_{-n} d_{\mu n} \tag{8.59}$$

and similarly for Neveu–Schwarz boundary conditions with d_n^μ replaced by b_r^μ.

In subsequent sections, we shall concentrate our attention upon the closed superstring which, by providing the right movers for the heterotic string of Chapter 9, has been the basis for all phenomenologically promising superstring theories to date.

8.5 Super-Virasoro algebra for the closed string

The superstring degrees of freedom in the covariant gauge not only obey the wave equations (8.30) and (8.31) but are also subject to the constraint equations (8.32) and (8.33). In detail these constraints are

$$T_{00} = T_{11} = -\tfrac{1}{2}(\partial_\tau X^\mu \, \partial_\tau X_\mu + \partial_\sigma X^\mu \, \partial_\sigma X_\mu) - \frac{i}{2}(\Psi_L^\mu \, \partial_+ \Psi_{\mu L} + \Psi_R^\mu \, \partial_- \Psi_{\mu R}) = 0 \tag{8.60}$$

$$T_{01} = T_{10} = -\partial_\tau X^\mu \, \partial_\sigma X_\mu - \frac{i}{2}(\Psi_L^\mu \, \partial_+ \Psi_{\mu L} - \Psi_R^\mu \, \partial_- \Psi_{\mu R}) = 0 \tag{8.61}$$

$$J^0 = \begin{pmatrix} \Psi_R^\mu \, \partial_- X_\mu \\ \Psi_L^\mu \, \partial_+ X_\mu \end{pmatrix} = 0 \tag{8.62}$$

$$J^1 = \begin{pmatrix} \Psi_R^\mu \, \partial_- X_\mu \\ -\Psi_L^\mu \, \partial_+ X_\mu \end{pmatrix} = 0. \tag{8.63}$$

It is convenient to write the independent constraint equations as

$$T_{++} = T_{--} = J_+ = J_- = 0 \tag{8.64}$$

where

$$T_{++} \equiv \tfrac{1}{2}(T_{00} + T_{01}) = -\partial_+ X_L^\mu \, \partial_+ X_{\mu L} - \frac{i}{2}\Psi_L^\mu \, \partial_+ \Psi_{\mu L} \tag{8.65}$$

$$T_{--} \equiv \tfrac{1}{2}(T_{00} - T_{01}) = -\partial_- X_R^\mu \, \partial_- X_{\mu R} - \frac{i}{2}\Psi_R^\mu \, \partial_- \Psi_{\mu R} \tag{8.66}$$

$$J_+ \equiv \Psi_L^\mu \, \partial_+ X_{\mu L} \tag{8.67}$$

and

$$J_- \equiv \Psi_R^\mu \, \partial_- X_{\mu R} \tag{8.68}$$

with $\partial_\pm$ as in (8.38).

As for the bosonic string, it is not possible quantum mechanically to impose the full content of these constraints without inconsistency, and in these circumstances it is useful to work with the Fourier components of T_{--} and T_{++},

$$L_m = \frac{1}{2\pi}\int_0^\pi d\sigma\, e^{2i\, m(\tau - \sigma)} T_{--} \qquad m \neq 0 \tag{8.69}$$

and

$$\tilde{L}_m = \frac{1}{2\pi}\int_0^\pi d\sigma\, e^{2i\, m(\tau + \sigma)} T_{++} \qquad m \neq 0 \tag{8.70}$$

where we are using units where $l = 1$ in (7.23) and correspondingly $T = \pi^{-1}$. It is also useful to introduce Fourier components of J_- and J_+. For Neveu–Schwarz boundary conditions for the right movers we define

$$G_r = \frac{1}{2\pi}\int_0^\pi d\sigma\, e^{2i\, r(\tau - \sigma)} J_- \qquad r \in \mathbb{Z} + \tfrac{1}{2} \tag{8.71}$$

and for Ramond boundary conditions for the right movers we define

$$F_m = \frac{1}{2\pi}\int_0^\pi d\sigma\, e^{2i\, m(\tau - \sigma)} J_- \qquad m \in \mathbb{Z}, m \neq 0. \tag{8.72}$$

Similarly, for Neveu–Schwarz boundary conditions for the left movers we define

$$\tilde{G}_r = \frac{1}{2\pi}\int_0^\pi d\sigma\, e^{2i\, r(\tau + \sigma)} J_+ \qquad r \in \mathbb{Z} + \tfrac{1}{2} \tag{8.73}$$

and for Ramond boundary conditions

$$\tilde{F}_m = \frac{1}{2\pi}\int_0^\pi d\sigma\, e^{2i\, m(\tau + \sigma)} J_+ \qquad m \in \mathbb{Z}. \tag{8.74}$$

As for the bosonic string in §7.5, we shall handle the case $m = 0$ in (8.69) and (8.70) separately.

These super-Virasoro operators may be expressed (Exercise 8.4) in terms of the oscillators in the expansions (7.26) and (7.27) with the definitions (7.58) and (8.43)–(8.46). For Neveu–Schwarz boundary conditions for the right-mover fermionic degrees of freedom,

$$L_m = -\tfrac{1}{2}\sum_{n \in \mathbb{Z}} \alpha^\mu_{m-n}\alpha_{\mu n} + \tfrac{1}{2}\sum_{r \in \mathbb{Z}+1/2}\left(\frac{m}{2} - r\right) b^\mu_{m-r} b_{\mu r} \qquad m \neq 0, \text{NS} \tag{8.75}$$

and

$$G_r = -\tfrac{1}{2} \sum_{n \in \mathbb{Z}} b^{\mu}_{m-n} \alpha_{\mu n}. \tag{8.76}$$

For Ramond boundary conditions for the right movers we have instead

$$L_m = -\tfrac{1}{2} \sum_{n \in \mathbb{Z}} \alpha^{\mu}_{m-n} \alpha_{\mu n} + \tfrac{1}{2} \sum_{n \in \mathbb{Z}} \left(\frac{m}{2} - n\right) d^{\mu}_{m-n} d_{\mu n} \qquad m \neq 0, \text{R} \tag{8.77}$$

and

$$F_m = -\tfrac{1}{2} \sum_{n \in \mathbb{Z}} d^{\mu}_{m-n} \alpha_{\mu n}. \tag{8.78}$$

Similarly, for Neveu–Schwarz boundary conditions for the left-mover fermionic degrees of freedom

$$\tilde{L}_m = -\tfrac{1}{2} \sum_{n \in \mathbb{Z}} \tilde{\alpha}^{\mu}_{m-n} \tilde{\alpha}^{\mu}_{n} + \tfrac{1}{2} \sum_{r \in \mathbb{Z}+1/2} \left(\frac{m}{2} - r\right) \tilde{b}^{\mu}_{m-r} \tilde{b}_{\mu r} \qquad m \neq 0, \text{NS} \tag{8.79}$$

and

$$\tilde{G}_r = -\tfrac{1}{2} \sum_{n \in \mathbb{Z}} \tilde{b}^{\mu}_{r-n} \tilde{\alpha}_{\mu n} \tag{8.80}$$

and for Ramond boundary conditions for the left movers

$$\tilde{L}_m = -\tfrac{1}{2} \sum_{n \in \mathbb{Z}} \tilde{\alpha}^{\mu}_{m-n} \tilde{\alpha}_{\mu n} + \tfrac{1}{2} \sum_{n \in \mathbb{Z}} \left(\frac{m}{2} - n\right) \tilde{d}^{\mu}_{m-n} \tilde{d}_{\mu n} \qquad m \neq 0, \text{R} \tag{8.81}$$

and

$$\tilde{F}_m = -\tfrac{1}{2} \sum_{n \in \mathbb{Z}} \tilde{d}^{\mu}_{m-n} \tilde{\alpha}_{\mu n}. \tag{8.82}$$

In all cases, m denotes an integer, and r denotes a half-integer.

As for the bosonic string, if we were to use (8.69) and (8.70) with $m = 0$ to define L_0 and $\tilde{L}_0$ then (8.75), (8.77), (8.79) and (8.81) would also apply for $m = 0$. However, as in §7.5, we shall denote the expressions obtained in this way by L_0' and $\tilde{L}_0'$, and reserve L_0 and $\tilde{L}_0$ for certain normal-ordered quantities to be introduced shortly. Thus,

$$L_0' \equiv \frac{1}{2\pi} \int_0^{\pi} d\sigma \, T_{--} \tag{8.83}$$

and

$$\tilde{L}_0' \equiv \frac{1}{2\pi} \int_0^{\pi} d\sigma \, T_{++}. \tag{8.84}$$

For Neveu–Schwarz boundary conditions for the right movers,

$$L_0' = -\tfrac{1}{2}\sum_{n\in\mathbb{Z}} \alpha^{\mu}_{-n}\alpha_{\mu n} - \tfrac{1}{2}\sum_{r\in\mathbb{Z}+1/2} r b^{\mu}_{-r} b_{\mu r} \qquad \text{NS} \tag{8.85}$$

and for Ramond boundary conditions

$$\tilde{L}_0' = -\tfrac{1}{2}\sum_{n\in\mathbb{Z}} \alpha^{\mu}_{-n}\alpha_{\mu n} - \tfrac{1}{2}\sum_{n\in\mathbb{Z}} n d^{\mu}_{-n} d_{\mu n} \qquad \text{R.} \tag{8.86}$$

Similarly, for Neveu–Schwarz boundary conditions for the left movers

$$\tilde{L}_0' = -\tfrac{1}{2}\sum_{n\in\mathbb{Z}} \tilde{\alpha}^{\mu}_{-n}\tilde{\alpha}_{\mu n} - \tfrac{1}{2}\sum_{r\in\mathbb{Z}+1/2} r \tilde{b}^{\mu}_{-r} \tilde{b}_{\mu r} \qquad \text{NS} \tag{8.87}$$

and for Ramond boundary conditions

$$\tilde{L}_0' = -\tfrac{1}{2}\sum_{n\in\mathbb{Z}} \tilde{\alpha}^{\mu}_{-n}\tilde{\alpha}_{\mu n} - \tfrac{1}{2}\sum_{n\in\mathbb{Z}} n \tilde{d}^{\mu}_{-n} \tilde{d}_{\mu n} \qquad \text{R.} \tag{8.88}$$

Arranging the bosonic and fermionic oscillators in L_0' and $\tilde{L}_0'$ in normal-ordered form with all annihilation operators α^{μ}_{n} and d^{μ}_{n}, and b^{μ}_{r}, n and r positive, to the right of all creation operators α^{μ}_{-n} and d^{μ}_{-n}, and b^{μ}_{-r}, n and r positive, we have for right movers in the Neveu–Schwarz sector

$$L_0' = -\tfrac{1}{2}\alpha^{\mu}_{0}\alpha_{\mu 0} - \sum_{n>0} \alpha^{\mu}_{-n}\alpha_{\mu n} - \sum_{r>0} r b^{\mu}_{-r} b_{\mu r} - a_{\text{NS}} \qquad \text{NS} \tag{8.89}$$

and for right movers in the Ramond sector

$$L_0' = -\tfrac{1}{2}\alpha^{\mu}_{0}\alpha_{\mu 0} - \sum_{n>0} \alpha^{\mu}_{-n}\alpha_{\mu n} - \sum_{n>0} n d^{\mu}_{-n} d_{\mu n} - a_{\text{R}} \qquad \text{R.} \tag{8.90}$$

In (8.89) and (8.90), a_{NS} and a_{R} are formally infinite constants arising from the commutators and anti-commutators. We shall discuss the values of the constants upon regularization in §8.7, and we shall find that a_{NS} and a_{R} have the values $\frac{1}{2}$ and 0, respectively. Similarly, for left movers in the Neveu–Schwarz sector

$$\tilde{L}_0' = -\tfrac{1}{2}\tilde{\alpha}^{\mu}_{0}\tilde{\alpha}_{\mu 0} - \sum_{n>0} \tilde{\alpha}^{\mu}_{-n}\tilde{\alpha}_{\mu n} - \sum_{r>0} r \tilde{b}^{\mu}_{-r} \tilde{b}_{\mu r} - a_{\text{NS}} \qquad \text{NS} \tag{8.91}$$

and for left movers in the Ramond sector

$$\tilde{L}_0' = -\tfrac{1}{2}\tilde{\alpha}^{\mu}_{0}\tilde{\alpha}_{\mu 0} - \sum_{n>0} \tilde{\alpha}^{\mu}_{-n}\tilde{\alpha}_{\mu n} - \sum_{n>0} n \tilde{d}^{\mu}_{-n} \tilde{d}_{\mu n} - a_{\text{R}} \qquad \text{R.} \tag{8.92}$$

(The same constants a_{NS} and a_{R} must occur for both left and right movers because the same numbers of oscillators with the same commutation and anti-commutation relations are involved.)

On the other hand, a common convention is to define L_0 and $\tilde{L}_0$ to be just the normal-ordered products. Thus

$$L_0 = -\tfrac{1}{2} \sum_{n > \mathbb{Z}} :\alpha^{\mu}_{-n} \alpha_{\mu n}: - \tfrac{1}{2} \sum_{r \in \mathbb{Z} + 1/2} r : b^{\mu}_{-r} b_{\mu r} : \qquad \text{NS} \quad (8.93)$$

and

$$L_0 = -\tfrac{1}{2} \sum_{n \in \mathbb{Z}} \alpha^{\mu}_{-n} \alpha_{\mu n}: - \tfrac{1}{2} \sum_{n \in \mathbb{Z}} n : d^{\mu}_{-n} d_{\mu n} : \qquad \text{R} \quad (8.94)$$

with similar expressions for the left movers. With these definitions

$$L_0 = L_0' + a_{\text{NS}} \qquad \text{NS} \quad (8.95)$$

$$L_0 = L_0' + a_{\text{R}} \qquad \text{R} \quad (8.96)$$

$$\tilde{L}_0 = \tilde{L}_0' + a_{\text{NS}} \qquad \text{NS} \quad (8.97)$$

and

$$\tilde{L}_0 = \tilde{L}_0' + a_{\text{R}} \qquad \text{R.} \quad (8.98)$$

The constraint equations (8.64) may now be formulated in terms of the super-Virasoro operators, which are the Fourier components of T_{++}, T_{--}, J_+ and J_-. For the right movers, there are conditions on physical states $|\varphi\rangle$:

$$L_m|\varphi\rangle = 0 \qquad m > 0 \quad (8.99)$$

and

$$L_0'|\varphi\rangle = 0 \quad (8.100)$$

or equivalently, depending on whether the boundary conditions for the fermionic right-mover degrees of freedom are Neveu–Schwarz or Ramond,

$$L_0|\varphi\rangle = a_{\text{NS}}|\varphi\rangle \qquad \text{NS} \quad (8.101)$$

or

$$L_0|\varphi\rangle = a_{\text{R}}|\varphi\rangle \qquad \text{R.} \quad (8.102)$$

Also, in the right-mover Neveu–Schwarz sector,

$$G_r|\varphi\rangle = 0 \qquad r > 0, \text{NS} \quad (8.103)$$

and in the right-mover Ramond sector,

$$F_m|\varphi\rangle = 0 \qquad m > 0, \text{R.} \quad (8.104)$$

Exactly similar conditions apply for the left movers. These physical state conditions are not applied for m or r negative for reasons to do with the super-Virasoro algebra that we shall discuss later.

The Hamiltonian (8.52), for right and left movers both in the Ramond sector is related to the Virasoro operators by

$$H = 2(L_0' + \tilde{L}_0'). \tag{8.105}$$

The same expression applies for the other possible choices of boundary conditions for the right- and left-mover fermionic degrees of freedom.

The super-Virasoro algebra for the right movers with Neveu–Schwarz boundary conditions for the right-mover fermionic degrees of freedom can be derived from the explicit expressions (8.75), (8.76) and (8.93) using the commutators (7.35) of the bosonic oscillators α_n^μ and the anti-commutators (8.50) of the fermionic oscillators b_r^μ, with the result

$$[L_m, L_n] = (m - n)L_{m+n} + A(m)\delta_{m+n,0} \tag{8.106}$$

$$[L_m, G_r] = \left(\frac{m}{2} - r\right) G_{m+r} \tag{8.107}$$

and

$$\{G_r, G_s\} = 2L_{r+s} + B(r)\delta_{r+s,0}. \tag{8.108}$$

Similarly, for Ramond boundary conditions for the right-mover fermionic degrees of freedom

$$[L_m, L_n] = (m - n)L_{m+n} + C(m)\delta_{m+n,0} \tag{8.109}$$

$$[L_m, F_n] = \left(\frac{m}{2} - n\right) F_{m+n} \tag{8.110}$$

and

$$\{F_m, F_n\} = 2L_{m+n} + E(m)\delta_{m+n,0}. \tag{8.111}$$

Exactly similar algebras apply for the left movers.

In (8.106), (8.108), (8.109) and (8.111), $A(m)$, $B(r)$, $C(m)$ and $E(m)$ are c-number anomaly terms deriving from normal-ordering problems. They may be evaluated much as in §7.5, by studying expectation values of commutators in the ground state, and this will be the subject of the next section. For convenience, we record the conclusions here, namely

$$A(m) = \frac{D}{8} m(m^2 - 1) \tag{8.112}$$

$$B(r) = \frac{D}{8} (r^2 - \tfrac{1}{4}) \tag{8.113}$$

$$C(m) = \frac{D}{8} m^3 \tag{8.114}$$

and

$$E(m) = \frac{D}{8} m^2 . \tag{8.115}$$

8.6 Closed superstring ground states and superconformal anomalies

The mass-squared operator for the closed superstring may be obtained from (8.101) and (8.102), using the expansion (8.93) and (8.94) (and the analogous expressions for the left movers), and recalling that the centre-of-mass momentum p^μ enters as in (7.58) (with $l = 1$ in the units being employed). By a similar line of argument to that pursued at the end of §7.5, we may conveniently write the mass-squared operator in the form

$$M^2 = M_R^2 + M_L^2 \tag{8.116}$$

where

$$\tfrac{1}{4} M_R^2 = -\sum_{n=1}^{\infty} \alpha_{-n}^{\mu} \alpha_{\mu n} - \sum_{r=1/2}^{\infty} r b_{-r}^{\mu} b_{\mu r} - a_{NS} \tag{8.117}$$

for NS-sector right movers,

$$\tfrac{1}{4} M_R^2 = -\sum_{n=1}^{\infty} \alpha_{-n}^{\mu} \alpha_{\mu n} - \sum_{n=1}^{\infty} n d_{-n}^{\mu} d_{\mu n} - a_R \tag{8.118}$$

for R-sector right movers, and, analogously,

$$\tfrac{1}{4} M_L^2 = -\sum_{n=1}^{\infty} \tilde{\alpha}_{-n}^{\mu} \tilde{\alpha}_{\mu n} - \sum_{r=1/2}^{\infty} r \tilde{b}_{-r}^{\mu} \tilde{b}_{\mu r} - a_{NS} \tag{8.119}$$

for NS-sector left movers, and

$$\tfrac{1}{4} M_L^2 = -\sum_{n=1}^{\infty} \tilde{\alpha}_{-n}^{\mu} \tilde{\alpha}_{\mu n} - \sum_{n=1}^{\infty} n \tilde{d}_{-n}^{\mu} \tilde{d}_{\mu n} - a_R \tag{8.120}$$

for R-sector left movers, with

$$M_L^2 = M_R^2 \tag{8.121}$$

as a consequence of (8.101), (8.102) and the corresponding expansions for left movers.

The oscillators in the mode expansions act as step-up and step-down operators for the eigenvalues of M_R^2 and M_L^2 because of the commutators (7.35) and (7.36), and the anti-commutators (8.49) and (8.50). Acting with the oscillator α^{μ}_{-m} or d^{μ}_{-m} $(m > 0)$ on a state increases the value of $\frac{1}{4}M_{\mathrm{R}}^2$ by m, and acting with the oscillator b^{μ}_{-r} $(r > 0)$ increases the value of $\frac{1}{4}M_{\mathrm{R}}^2$ by r. The oscillators $\tilde{\alpha}^{\mu}_{-m}$, $\tilde{d}^{\mu}_{-m}$ and $\tilde{b}^{\mu}_{-r}$ have a similar effect on $\frac{1}{4}M_{\mathrm{L}}^2$. The ground state is the product of the ground state for the right movers and the ground state for the left movers. Let us concentrate on the right-mover ground state. If the fermionic right movers are in the NS sector then the ground state $|0\rangle_{\mathrm{R}}$ is obtained by requiring

$$\alpha^{\mu}_{m}|0\rangle_{\mathrm{R}} = b^{\mu}_{r}|0\rangle_{\mathrm{R}} = 0 \qquad m, r > 0. \tag{8.122}$$

When the fermionic right movers are in the R sector, a collection of states forming a degenerate ground state occurs rather than a unique ground state as in the NS case. Analogously to (8.122), we require the right-mover ground state $|0\rangle_{\mathrm{R}}$ to satisfy

$$\alpha^{\mu}_{m}|0\rangle_{\mathrm{R}} = d^{\mu}_{m}|0\rangle_{\mathrm{R}} = 0 \qquad m > 0. \tag{8.123}$$

However, a subtlety results from the presence of the oscillators d^{μ}_{0}, which may act on a state without changing the value of M_{R}^2. The nature of this degenerate ground state may be elucidated by considering

$$\gamma^{\mu} \equiv \mathrm{i}\sqrt{2}d^{\mu}_{0} \qquad \mu = 0, 1, \ldots, D-1. \tag{8.124}$$

As a consequence of (8.49), the γ^{μ} obey the (Clifford) algebra

$$\{\gamma^{\mu}, \gamma^{\nu}\} = 2\eta^{\mu\nu} \tag{8.125}$$

with $\eta_{\mu\nu}$ as in (7.3), i.e. the γ^{μ} behave like D-dimensional γ-matrices. Since the various ground states transform amongst themselves under the action of the d^{μ}_{0} and so of the γ^{μ}, the degenerate ground state is an irreducible representation of the Clifford algebra. In turn, this means that the ground state is a spinor representation of SO$(1, D-1)$ because the generators $M^{\mu\nu}$ of this representation may be constructed from the above γ-matrices as

$$M^{\mu\nu} = \frac{\mathrm{i}}{4}[\gamma^{\mu}, \gamma^{\nu}] \tag{8.126}$$

which satisfy the Lie algebra of SO$(1, D-1)$,

$$[M^{\mu\nu}, M^{\rho\sigma}] = \mathrm{i}(\eta^{\mu\rho}M^{\nu\sigma} + \eta^{\nu\sigma}M^{\mu\rho} - \eta^{\mu\sigma}M^{\nu\rho} - \eta^{\nu\rho}M^{\mu\sigma}). \tag{8.127}$$

For the NS sector, the superconformal anomalies $A(m)$ and $B(r)$ may now be evaluated by calculating the expectation values of the commutator (8.106) and the anti-commutator (8.108) in a zero-momentum Fock space ground state $|0; 0\rangle$, with the results

$$A(m) = \frac{D}{8}m(m^2 - 1) \tag{8.128}$$

and

$$B(r) = \frac{D}{8}(r^2 - \tfrac{1}{4}). \tag{8.129}$$

(A ground state of the oscillator algebra Fock space with string centre-of-mass momentum is labelled as $|0; p\rangle$.) In (8.128), a fermionic contribution has been added to the bosonic contribution obtained in (7.74).

Similarly, for the R sector, the superconformal anomalies $C(m)$ and $E(m)$ may be calculated from the expectation values of (8.109) and (8.111) in any one of the degenerate zero-momentum Fock space ground states to yield

$$C(m) = \frac{D}{8}m^3 \tag{8.130}$$

and

$$E(m) = \frac{D}{8}m^2. \tag{8.131}$$

As for the bosonic string, ghost states of negative norm are created by the time components of the oscillators α^{μ}_{-m}. In addition, ghosts states are created by the time components of the oscillators d^{μ}_{-m} and b^{μ}_{-r} for the R and NS sectors respectively, because, for m positive,

$$\langle 0; 0|d^0_m d^0_{-m}|0; 0\rangle = \langle 0; 0|\{d^0_m, d^0_{-m}\}|0; 0\rangle = -1 \tag{8.132}$$

employing (8.123), and similarly for the NS sector. The physical state conditions (8.99) and (8.104) for the Ramond sector, and (8.99) and (8.103) for the Neveu–Schwarz sector, provide enough conditions to allow one to forbid the occurrence of all ghost states. For $D = 10$, $a_{\mathrm{R}} = 0$ and $a_{\mathrm{NS}} = \frac{1}{2}$, which we shall see in §8.8 are the appropriate choices for the superstring, all ghosts are indeed removed from the theory by these conditions[(4)].

8.7 The light cone gauge

For the bosonic string in §7.7, we saw that even after the covariant gauge of (7.15) had been selected, there remained further gauge freedom. This freedom could be removed by imposing additional gauge conditions (light cone gauge) in such a way as to reduce the number of non-trivial components of X^{μ} and leave only physical dynamical degrees of freedom. For the superstring, a covariant gauge was chosen in §8.3 in which the world sheet 'gravitino' χ_{α} was gauged away, and the world sheet metric was given by (8.27). Just as for the bosonic string, a combination of a world sheet reparametrization and a local Weyl scaling may be used to remove some residual gauge freedom and select a gauge in which X^+ is given by (see (7.101))

$$X^+(\tau, \sigma) = x^+ + p^+\tau \tag{8.133}$$

where x^+ and p^+ are constants (in units where l is 1). In addition, a local world sheet supersymmetry transformation may be used to choose

$$\Psi^+ = 0 \tag{8.134}$$

where

$$\Psi^\pm \equiv \frac{1}{\sqrt{2}}(\Psi^0 \pm \Psi^{D-1}) \tag{8.135}$$

without modifying (8.133), because of (8.17).

The coordinates X^- and Ψ^- may now be calculated in terms of the transverse degrees of freedom X^i and Ψ^i, $i = 1, \ldots, D-2$, as follows. Written in terms of light cone variables, the constraint equations (8.64)–(8.68) become

$$T_{++} = -2\,\partial_+X_L^+\,\partial_+X_L^- - \frac{i}{2}(\Psi_L^+\,\partial_+\Psi_L^- + \Psi_L^-\,\partial_+\Psi_L^+)$$

$$+\,\partial_+X_L^i\,\partial_+X_L^i - \frac{i}{2}\Psi_L^i\,\partial_+\Psi_L^i \tag{8.136}$$

$$T_{--} = -2\,\partial_-X_R^+\,\partial_-X_R^- - \frac{i}{2}(\Psi_R^+\,\partial_-\Psi_R^- + \Psi_R^-\,\partial_-\Psi_R^+)$$

$$+\,\partial_-X_R^i\,\partial_-X_R^i + \frac{i}{2}\Psi_R^i\,\partial_-\Psi_R^i \tag{8.137}$$

$$J_+ = \Psi_L^+\,\partial_+X_L^- + \Psi_L^-\,\partial_+X_L^+ - \Psi_L^i\,\partial_+X_L^i \tag{8.138}$$

and

$$J_- = \Psi_R^+\,\partial_-X_R^- + \Psi_R^-\,\partial_-X_R^+ - \Psi_R^i\,\partial_-X_R^i \tag{8.139}$$

where the world sheet derivatives $\partial_\pm$ are given by (8.38). Applying the light cone gauge conditions (8.133) and (8.134) leads to

$$p^+\,\partial_+X_L^- = \partial_+X_L^i\,\partial_+X_L^i + \frac{i}{2}\Psi_L^i\,\partial_+\Psi_L^i \tag{8.140}$$

and

$$p^+\,\partial_-X_R^- = \partial_-X_R^i\,\partial_-X_R^i + \frac{i}{2}\Psi_R^i\,\partial_-\Psi_R^i \tag{8.141}$$

which may be solved for X_L^- and X_R^- in terms of transverse degrees of freedom, and

$$\tfrac{1}{2}p^+\Psi_L^- = \Psi_L^i\,\partial_+X_L^i \tag{8.142}$$

and

$$\tfrac{1}{2}p^+\Psi_R^- = \Psi_R^i\,\partial_-X_R^i \tag{8.143}$$

which express Ψ_L^- and Ψ_R^- in terms of transverse degrees of freedom. Thus, only the transverse degrees of freedom X^i and Ψ^i, $i = 1, \ldots, D-2$, remain as dynamical degrees of freedom. These will have the usual mode expansions for the closed superstring (7.26), (7.27) and (8.43)–(8.46), with the space-time index μ restricted to transverse degrees of freedom, $i = 1, \ldots, D-2$.

The mass-squared operator in terms of left-mover oscillators can be derived from (8.140) by substituting the expansions (7.27) for X_L^μ and (8.45) or (8.46) for Ψ_L^i and integrating from 0 to π with respect to σ. All terms may be written in normal-ordered form with positive values of n or r to the right of negative values using the commutators (7.36) and anti-commutators (8.49) or (8.50). In similar fashion, an alternative expression for the mass-squared operator in terms of right-mover oscillators may be derived from (8.141). In the Ramond sector, the result is

$$\begin{aligned}\tfrac{1}{8}M^2 = \tfrac{1}{8}p^2 &= \sum_{n=1}^{\infty} \tilde{\alpha}_{-n}^i\tilde{\alpha}_n^i + \sum_{n=1}^{\infty} n\tilde{d}_{-n}^i\tilde{d}_n^i \\ &= \sum_{n=1}^{\infty} \alpha_{-n}^i\alpha_n^i + \sum_{n=1}^{\infty} d_{-n}^i d_n^i \end{aligned}\tag{8.144}$$

which it is convenient to cast in the form

$$M^2 = M_R^2 + M_L^2 \tag{8.145}$$

with

$$\tfrac{1}{4}M_R^2 = \sum_{n=1}^{\infty} \alpha_{-n}^i\alpha_n^i + \sum_{n=1}^{\infty} n d_{-n}^i d_n^i \tag{8.146}$$

$$\tfrac{1}{4}M_L^2 = \sum_{n=1}^{\infty} \tilde{\alpha}_{-n}^i\tilde{\alpha}_n^i + \sum_{n=1}^{\infty} n\tilde{d}_{-n}^i\tilde{d}_n^i \tag{8.147}$$

and

$$M_R^2 = M_L^2. \tag{8.148}$$

No normal-ordering constant occurs in (8.146) and (8.147) because of an exact cancellation between the bosonic and fermionic contributions in the Ramond sector. Thus, a_R in (8.118) or (8.120) is now identified to be

$$a_R = 0. \tag{8.149}$$

(Using ζ-function regularization, as in §7.7, each bosonic degree of freedom contributes $\frac{1}{24}$ to a_R and each fermionic degree of freedom, with periodic boundary conditions as appropriate to the Ramond sector, contributes $-\frac{1}{24}$ to a_R.)

In the Neveu–Schwarz sector, the corresponding result is

$$\tfrac{1}{4}M_R^2 = \sum_{n=1}^{\infty} \alpha_{-n}^i \alpha_n^i + \sum_{r=1/2}^{\infty} r b_{-r}^i b_r^i - \frac{(D-2)}{16} \tag{8.150}$$

and

$$\tfrac{1}{4}M_L^2 = \sum_{n=1}^{\infty} \tilde{\alpha}_{-n}^i \tilde{\alpha}_n^i + \sum_{r=1/2}^{\infty} r \tilde{b}_{-r}^i \tilde{b}_r^i - \frac{(D-2)}{16} \tag{8.151}$$

which determines a_{NS} in (8.117) or (8.119) to be

$$a_{NS} = \frac{(D-2)}{16}. \tag{8.152}$$

The normal-ordering constant has again been calculated using ζ-function regularization, as in §7.7, so each bosonic degree of freedom contributes $\frac{1}{24}$ to a_{NS} and each fermionic degree of freedom, with anti-periodic boundary conditions as appropriate to the Neveu–Schwarz sector, contributes $\frac{1}{48}$. We shall see in §8.8 that the appropriate choice of D is 10.

8.8 Superstring states, GSO projections and space-time supersymmetry

In the light cone gauge, superstring states may be constructed by acting with products of oscillators α_{-n}^i and d_{-n}^i or b_{-r}^i on the ground state $|0\rangle_R$ for the right movers, and with products of oscillators $\tilde{\alpha}_{-n}^i$ and $\tilde{d}_{-n}^i$ or $\tilde{b}_{-r}^i$ on the left-mover ground state $|0\rangle_L$, and forming the product of the resulting right- and left-mover states. A calculation of the masses of these states requires a determination of the space-time dimensionality D of the superstring theory to fix the normal-ordering constant $a_{NS} = (D-2)/16$ which occurs in (8.150) and (8.151). (The normal-ordering constant for the Ramond sector has already been obtained in (8.149).) As for the bosonic string, normal-ordering constants may be determined by requiring that the commutators of the Lorentz generators in the light cone gauge are free from quantum anomalies. For the superstring, this requires

$$D = 10 \tag{8.153}$$

and

$$a_{NS} = \tfrac{1}{2}. \tag{8.154}$$

It then follows from (8.150) and (8.151) that the NS sector ground state for right or left movers is tachyonic with M_R^2 or M_L^2 being $-\frac{1}{2}$.

The problem of tachyonic ground states may be avoided, and a (ten-dimensional) space-time supersymmetric theory obtained at the same time, by applying a projection to the states of the theory due to Gliozzi, Scherk and Olive[5] (the GSO projection). This projection is performed *separately* on left and right movers. For the Neveu–Schwarz sector, the GSO projection P is

$$P = \frac{(1 + (-1)^{F + 1})}{2} \qquad \text{NS sector} \qquad (8.155)$$

where the number operator F is defined for left movers by

$$F = \sum_{r = 1/2}^{\infty} ((\tilde{f}_r^{\alpha})^{\dagger} \tilde{f}_r^{\alpha} - (\tilde{g}_r^{\alpha})^{\dagger} \tilde{g}_r^{\alpha}) \qquad \text{NS sector} \qquad (8.156)$$

where $\tilde{f}_r^{\alpha}$ and $\tilde{g}_r^{\alpha}$, as defined in (9.101), are oscillators for complex fermionic degrees of freedom $\hat{\Psi}^{\alpha}$, $\alpha = 1, \ldots, 4$, replacing the real fermions Ψ^i, $i = 1, \ldots, 8$, and similarly for right movers with f_r^{α} and g_r^{α} replacing $\tilde{f}_r^{\alpha}$ and $\tilde{g}_r^{\alpha}$. Thus, only states with $(-1)^F = -1$ survive the projection in the NS sector. Among other things, this has the effect of deleting the tachyonic ground states with $F = 0$.

On the other hand, for the Ramond sector, the GSO projection is defined to be

$$P = \frac{1}{2}(1 + \eta(-1)^{F + 1}) \qquad \text{R sector} \qquad (8.157)$$

where η is either $+1$ or -1 and may be chosen independently for the left and right movers. In this case, F is defined for left movers by

$$F = \sum_{m = 1}^{\infty} ((\tilde{c}_{m-1}^{\alpha})^{\dagger} \tilde{c}_{m-1}^{\alpha} - (\tilde{e}_m^{\alpha})^{\dagger} \tilde{e}_m^{\alpha}) \qquad \text{R sector} \qquad (8.158)$$

where $\tilde{c}_{m-1}^{\alpha}$ and $\tilde{e}_m^{\alpha}$ are defined in (9.109), and similarly for right movers with c_{m-1}^{α} and e_m^{α} replacing $\tilde{c}_{m-1}^{\alpha}$ and $\tilde{e}_m^{\alpha}$. In this case, F contains a zero mode part $(\tilde{c}_0^{\alpha})^{\dagger} \tilde{c}_0^{\alpha}$ or $(c_0^{\alpha})^{\dagger} c_0^{\alpha}$. Theories where η has the same value for both left and right movers are, for historical reasons, referred to as type-IIB theories and those where η has opposite values for left and right movers as type-IIA theories.

The basic distinction between the states of type-IIA and type-IIB theories may be clarified by considering the zero-mode contribution to $(-1)^F$. As discussed in §8.6, the ground state for the Ramond sector in covariant gauge is degenerate, forming a spinor of $SO(1, D - 1)$. Correspondingly, in the light cone gauge and with $D = 10$, the ground state is an $SO(8)$ spinor. For definiteness, let us focus on the right-mover Ramond ground state. An exactly similar discussion applies for the left movers. The 16 independent components of this SO(8) spinor ground state may be chosen to be

$$|n_\alpha\rangle = \prod_{\alpha=1}^{4} (a_\alpha^\dagger)^{n_\alpha}|0\rangle_R \qquad n_\alpha = 0 \text{ or } 1 \tag{8.159}$$

where

$$a_\alpha \equiv c_0^\alpha = \frac{1}{\sqrt{2}}(d_0^{2\alpha-1} + \mathrm{i}\, d_0^{2\alpha}) \qquad \alpha = 1, \ldots, 4. \tag{8.160}$$

The spinor splits into two spinors of opposite chirality $\chi = \pm 1$ when we introduce the chirality operator

$$\chi = \prod_\alpha \chi_\alpha \tag{8.161}$$

where

$$\chi_\alpha = \gamma^{2\alpha-1}\gamma^{2\alpha} \qquad \alpha = 1, \ldots, 4 \tag{8.162}$$

and γ^μ is as in (8.124). It may be checked that $a_\alpha^\dagger$ anti-commutes with χ_α, so that $|0\rangle_R$ and $a_\alpha^\dagger|0\rangle_R$ have opposite chirality. Moreover,

$$[\chi_\alpha, a_\alpha^\dagger] = 2\mathrm{i}\, a_\alpha^\dagger \tag{8.163}$$

from which it follows that we may take the chiralities of $|0\rangle_R$ and $a_\alpha^\dagger|0\rangle_R$ to be $(-\mathrm{i})^4 = 1$ and $\mathrm{i}(-\mathrm{i})^3 = -1$ respectively. Consequently, $|n_\alpha\rangle$ of (8.159) has chirality

$$\chi = (-1)^{\Sigma_\alpha n_\alpha}. \tag{8.164}$$

This means that if F_0 is the zero-mode contribution to F, then the chirality χ of the SO(8) spinor ground state is

$$\chi = (-1)^{F_0}. \tag{8.165}$$

Thus, type-IIB theories have Ramond-sector spinor ground states for right and left movers of the same chirality, whereas type-IIA theories impose opposite chirality. Since there is no absolute definition of chirality, we may if we wish take $\eta = 1$ for both right and left movers for the type-IIB theory, and $\eta = 1$ for right movers and $\eta = -1$ for left movers for the type-IIA theory.

Before GSO projection, the massless states in the various Ramond and Neveu–Schwarz sectors for right and left movers are

$$|0\rangle_R \tilde{b}^i_{-1/2}|0\rangle_L \qquad \text{R–NS sector} \tag{8.166}$$

$$b^i_{-1/2}|0\rangle_R|0\rangle_L \qquad \text{NS–R sector} \tag{8.167}$$

$$b^i_{-1/2}|0\rangle_R \tilde{b}^j_{-1/2}|0\rangle_L \qquad \text{NS–NS sector} \tag{8.168}$$

and

$$|0\rangle_R|0\rangle_L \qquad \text{R–R sector} \tag{8.169}$$

in terms of the oscillators for real fermionic degrees of freedom. These oscillators may be written as linear combinations of the oscillators for complex fermionic degrees of freedom using (9.102), (9.103), (9.110) and (9.111). As a result they increase the eigenvalue of F by 1 modulo 2. Consequently, the effect of the GSO projection on the massless states is to preserve the states (8.168), to restrict the right movers in (8.166) and the left movers in (8.167) to a single SO(8) chirality, and the right and left movers in (8.169) each to a single chirality. For the type-IIB and type-IIA theories discussed above, the Ramond-sector SO(8) spinor ground states for right and left movers have the same and opposite chirality, respectively.

After GSO projection, the NS-sector vector $b^i_{-1/2}|0\rangle_R$ and the R-sector spinor ground state $|0\rangle_R$ have the same number of degrees of freedom, namely 8, and similarly for the left movers. Thus, the massless states of the theory given by (8.166)–(8.169) contain equal numbers of bosonic and fermionic degrees of freedom. This matching of the numbers of degrees of freedom is a necessary condition for supersymmetry and strongly suggests that, after GSO projection, we have a supersymmetric theory (in ten dimensions). This is confirmed by comparing the massless states with those of the two possible ten-dimensional $N = 2$ supergravity theories. For the type-IIA theory, the massless states of (8.166)–(8.169) surviving GSO projection are the products that can be formed by using an SO(8) vector or an SO(8) spinor of definite chirality for the right movers, and an SO(8) vector or an SO(8) spinor of *opposite* chirality for the left movers. These are precisely the massless states of the ten-dimensional $N = 2$ supergravity theory that was constructed[(6)] by dimensional reduction of eleven-dimensional supergravity. For the type-IIB theory, the massless states of (8.166)–(8.169) surviving GSO projection are the same sorts of products of right and left movers but with the *same* chirality of SO(8) spinor for right and left movers. These are the massless states of the chiral version[(7)] of ten-dimensional $N = 2$ supergravity. The supersymmetry of these theories can be made manifest in another formulation of superstring theory that we shall discuss briefly in the next section.

8.9 Other formulations of the superstring

A path integral quantization for the superstring[(8)] may be developed in much the same way as for the bosonic string in §7.9. In the case of the bosonic string, the Fadeev–Popov ghosts associated with the world sheet reparametrization-invariance gauge symmetries use anti-commuting Grassmann variables. For the superstring, these Fadeev–Popov ghosts are required, but so also are ghosts associated with local world sheet supersymmetry. Because world sheet supersymmetry transformations are parametrized by world sheet Majorana spinors, the Fadeev–Popov ghosts in this case are commuting degrees of freedom rather than anti-commuting Grassmann variables.

It is also possible to construct an alternative formulation of the superstring in the light cone gauge in which space-time supersymmetry is manifest[9]. In this approach, the theory is formulated in terms of 8 bosonic transverse degrees of freedom X^i, $i = 1, \ldots, 8$, providing 8 right movers and 8 left movers, and 16 fermionic transverse degrees of freedom which form a world sheet spinor, much as in (8.34), but with 8 right-mover degrees of freedom constituting an SO(8) spinor of definite chirality, and 8 left-mover degrees of freedom an SO(8) spinor of either the same or opposite chirality (corresponding to the type-IIB and type-IIA cases discussed in §8.8). Thus, right from the outset there are equal numbers of space-time bosonic and space-time fermionic degrees of freedom in the string degrees of freedom themselves. This allows the light cone gauge superstring action to be written in a form that displays space-time supersymmetry (as well as world sheet supersymmetry). This formulation is equivalent to the formulation of §8.8 after the GSO projection has been applied to render it space-time supersymmetric. Detailed expositions of these alternative formulations of the superstring may be found elsewhere[10].

Exercises

8.1 Check that the transformation given by (8.4) and (8.5) is a world sheet supersymmetry, and that the action (8.1) is invariant under this transformation.

8.2 Derive the anti-commutators (8.49) and (8.50) for oscillators in the superstring mode expansions.

8.3 Derive the Hamiltonian (8.51) for the closed superstring.

8.4 Express the super-Virasoro operators in terms of oscillators in the superstring mode expansions.

References

General references

Brink L 1984 Superstrings in *Supersymmetry* ed K Dietz, R Flume, G v Gehlen and V Rittenberg (New York: Plenum) p 89

Green M B, Schwarz J H and Witten E 1987 *Superstring Theory* (Cambridge: Cambridge University Press)

Schwarz J H 1982 *Phys. Rep.* **89** 224

References in the text

1 Brink L, Di Vecchia P and Howe P 1976 *Phys. Lett.* **65B** 471
Deser S and Zumino B 1976 *Phys. Lett.* **65B** 369
2 Ramond P 1971 *Phys. Rev.* D **3** 2415
3 Neveu A and Schwarz J H 1971 *Nucl. Phys.* B **31** 86
4 Goddard P and Thorn C B 1972 *Phys. Lett.* **40B** 235
Schwarz J H 1972 *Nucl. Phys.* B **46** 61
Brower R C and Friedman K A 1973 *Phys. Rev.* D **7** 535
5 Gliozzi F, Scherk J and Olive D 1976 *Phys. Lett.* **65B** 282; 1977 *Nucl. Phys.* B **122** 253
6 Cremmer E and Julia B 1979 *Nucl. Phys.* B **159** 141
Scherk J and Schwarz J H 1979 *Nucl. Phys.* B **153** 61
7 Schwarz J H 1982 *Phys. Rep.* **89** §2.3
8 Friedan D, Martinec E and Shenker S 1986 *Nucl. Phys.* B **271** 93 and references therein
9 Green M B and Schwarz J H 1981 *Nucl. Phys.* B **181** 502; 1982 *Nucl. Phys.* B **198** 252; 1982 *Phys. Lett.* **109B** 444
10 Schwarz J H 1982 *Phys. Rep.* **89** 224
Brink L 1984 Superstrings in *Supersymmetry* ed K Dietz, R Flume, G v Gehlen and V Rittenberg (New York: Plenum) p 89
Green M B, Schwarz J H and Witten E 1987 *Superstring Theory* (Cambridge: Cambridge University Press) Chapter 5

9

THE HETEROTIC STRING

9.1 Introduction

A realistic superstring theory will have to contain gauge fields for the electroweak and strong interactions and perhaps for a grand unified theory. The simplest way for this to happen is for gauge fields to be present in the ten-dimensional theory, though, *a priori*, there is the possibility that the gauge fields arise from the construction of a four-dimensional theory from the ten-dimensional theory, as will be discussed in Chapter 10. For open superstrings, this may be achieved(1) by associating with one end of the string an index labelling a state of a representation R of some group, and with the other end an index labelling a state of the conjugate representation $\bar{R}$. Such theories are only consistent(2) when the gauge group is SO(32) or Sp(32).

The type-IIA closed superstring is unsuitable to describe the real world because, as discussed in §8.8, it is non-chiral. The type-IIB closed superstring is also unsuitable, in the first instance, because the only massless states it contains are in the supergravity multiplet of $N = 2$ supergravity, so there are no non-abelian gauge fields present in the ten-dimensional theory.

Nonetheless, theories of only closed strings may be obtained with a non-abelian gauge group occurring in the ten-dimensional theory by the oblique method of using the right movers of a type-II superstring and the left movers of a bosonic string(3). The GSO projection (8.155) and (8.157) is performed on the right movers to ensure equal numbers of bosonic and fermionic degrees of freedom, as required for space-time supersymmetry. The freedom to choose the right and left movers of different kinds of string derives from the facts that the states of a bosonic string or of type-II superstring are direct products of the Fock space states for the right and left movers, and that interactions respect this structure. Such a string theory is referred to as a heterotic string. (To quote Gross *et al*(3), heterosis is 'increased vigour displayed by crossbred plants or animals'.) At first sight, such a theory makes little sense, because the right movers are in 10 space-time dimensions whereas the left movers are in 26 space-time dimensions. However, in this approach, 16 of the left-mover dimensions X^μ are 'compactified' by associating them with a 16-dimensional torus with radii on the assumedly extremely small scale set by the string fundamental length l of (7.23). As we shall see in §9.4, some gauge fields then arise in the Kaluza–Klein manner, as is to be expected when a higher-dimensional theory containing gravity is reduced to

DOI: 10.1201/9780367805807-9

a lower dimension by compactification[4]. There are also a large number of additional gauge fields of a stringy origin whose existence depends on the possibility of having winding numbers for the string on the torus. In this way, the extra 16 left-mover dimensions provide the gauge group of the resulting 10-dimensional theory.

For the heterotic string, it is found that the possible gauge groups consistent with gauge and gravitational anomaly cancellation[3] are SO(32) or $E_8 \times E_8$. The latter possibility has led to phenomenologically promising models because one E_8 factor can contain E_6 which in turn contains SO(10) with useful subgroups such as flipped SU(5) × U(1) or SO(6) × SO(4), while the other E_8 factor can be treated as a hidden-sector gauge group, much as in the discussion of supergravity in §5.7 and §5.8. We therefore focus in what follows on the heterotic string with the $E_8 \times E_8$ gauge group in ten dimensions.

9.2 Mode expansions and quantization

For the right movers, the mode expansions for the heterotic string are those of §8.4 for the closed superstring. For the bosonic degrees of freedom,

$$X_R^\mu(\tau - \sigma) = \tfrac{1}{2}x^\mu + \tfrac{1}{2}p^\mu(\tau - \sigma) + \frac{i}{2}\sum_{n \neq 0} \frac{\alpha_n^\mu}{n} e^{-2in(\tau - \sigma)} \tag{9.1}$$

and for the fermionic degrees of freedom

$$\Psi_R^\mu(\tau - \sigma) = \sum_{n \in \mathbb{Z}} d_n^\mu e^{-2in(\tau - \sigma)} \qquad \text{R sector} \tag{9.2}$$

and

$$\Psi_R^\mu(\tau - \sigma) = \sum_{r \in \mathbb{Z} + 1/2} b_r^\mu e^{-2ir(\tau - \sigma)} \qquad \text{NS sector} \tag{9.3}$$

where in each case $\mu = 0, 1, \ldots, 9$.

For the left movers, it is convenient to introduce a notation that distinguishes the first ten degrees of freedom, which we denote by X_L^μ, $\mu = 0, 1, \ldots, 9$, from the last 16 degrees of freedom. We denote these 16 'internal' degrees of freedom by X_L^I, $I = 1, \ldots, 16$. For the first ten degrees of freedom the mode expansion is just as in §7.4 for the closed-bosonic-string left movers

$$X_L^\mu(\tau + \sigma) = \tfrac{1}{2}x^\mu + \tfrac{1}{2}p^\mu(\tau + \sigma) + \frac{i}{2}\sum_{n \neq 0} \frac{\tilde{\alpha}_n^\mu}{n} e^{-2in(\tau + \sigma)} \tag{9.4}$$

where $\mu = 0, 1, \ldots, 9$.

For the 16 internal degrees of freedom we write

$$X_L^I(\tau+\sigma) = x_L^I + p_L^I(\tau+\sigma) + \frac{i}{2}\sum_{n\neq 0}\frac{\tilde{\alpha}_n^I}{n}e^{-2in(\tau+\sigma)} \tag{9.5}$$

where $I = 1, \ldots, 16$.

The quantization for $\mu = 0, 1, \ldots, 9$ is that of §7.4 and §8.4, so

$$[x^\mu, p^\nu] = -i\eta^{\mu\nu} \tag{9.6}$$

$$[\alpha_m^\mu, \alpha_n^\nu] = [\tilde{\alpha}_m^\mu, \tilde{\alpha}_n^\nu] = -m\delta_{m+n,0}\eta^{\mu\nu} \tag{9.7}$$

$$\{d_m^\mu, d_n^\nu\} = -\delta_{m+n,0}\eta^{\mu\nu} \qquad \text{R sector} \tag{9.8}$$

and

$$\{b_r^\mu, b_s^\nu\} = -\delta_{r+s,0}\eta^{\mu\nu} \qquad \text{NS sector} \tag{9.9}$$

where in each case $\mu, \nu = 0, 1, \ldots, 9$. More care is required for X_L^I, $I = 1, \ldots, 16$, which are not paired with a right mover. Proper account may be taken of the absence of corresponding right movers by first replacing X_L^I by X^I with both right- and left-mover degrees of freedom and then eliminating the right movers by imposing the constraint

$$(\partial_\tau - \partial_\sigma)X^I = 0. \tag{9.10}$$

In §7.4, we could have arrived at the quantization condition (7.33) by writing down the Poisson bracket of the classical theory, and then making the transition to the quantum theory by replacing the Poisson bracket by the corresponding commutator multiplied by $-i$. Here, a similar procedure may be followed but using the Dirac bracket in the classical theory, to incorporate the (second-class) constraint (9.10), before passing to the quantum theory by replacing this bracket by a commutator multiplied by $-i$. The result of this calculation is to replace (7.33) (with I, J substituted for μ, ν) by

$$[P^I(\tau,\sigma), X^J(\tau,\sigma')] = \frac{i}{2}\delta(\sigma-\sigma')\eta^{IJ} = -\frac{i}{2}\delta(\sigma-\sigma')\delta^{IJ} \tag{9.11}$$

which differs by a factor of $\frac{1}{2}$ from the naive expression. Consistency of the mode expansion (9.5) with (9.11) then requires the factor of $\frac{1}{2}$ to feed through into the commutator

$$[x_L^I, p_L^J] = -\frac{i}{2}\eta^{IJ} = \frac{i}{2}\delta^{IJ} \tag{9.12}$$

while the oscillators $\tilde{\alpha}_n^I$ have the commutators

$$[\tilde{\alpha}_m^I, \tilde{\alpha}_n^J] = -m\delta_{m+n,0}\eta^{IJ} = m\delta_{m+n,0}\delta^{IJ} \tag{9.13}$$

(without $\frac{1}{2}$) because there is no contribution to (9.11) from right-mover oscillators. (No similar modification is required for the right-mover fermionic degrees of freedom Ψ^{μ}_{R}, which have no corresponding left movers, because the anti-commutation relations (8.47) are already separated into right and left movers.)

9.3 Compactification of the bosonic string on a circle

As discussed in §9.1, the 16 extra left-mover dimensions of the heterotic string are to be compactified on a torus to obtain a ten-dimensional theory with a gauge group arising from the compactified dimensions. In this section, we warm up to this task by considering the simpler example of a bosonic string with one dimension (for both right and left movers) compactified on a circle, say X^{25}, Then, mode expansions for X^{μ}, $\mu = 0, 1, \ldots, 24$ are as in §7.4:

$$X^{\mu}_{\text{R}}(\tau - \sigma) = \tfrac{1}{2}x^{\mu} + \tfrac{1}{2}p^{\mu}(\tau - \sigma) + \frac{\mathrm{i}}{2}\sum_{n \neq 0}\frac{1}{n}\alpha^{\mu}_{n}\,\mathrm{e}^{-2\mathrm{i}\,n(\tau - \sigma)} \qquad (9.14)$$

and

$$X^{\mu}_{\text{L}}(\tau + \sigma) = \tfrac{1}{2}x^{\mu} + \tfrac{1}{2}p^{\mu}(\tau + \sigma) + \frac{\mathrm{i}}{2}\sum_{n \neq 0}\frac{1}{n}\tilde{\alpha}^{\mu}_{n}\,\mathrm{e}^{-2\mathrm{i}\,n(\tau + \sigma)} \qquad (9.15)$$

for $\mu = 0, 1, \ldots, 24$. However, in the case of X^{25}, which is compactified on a circle of radius R we must make the identification

$$x^{25} \equiv x^{25} + 2\pi Rn \qquad (9.16)$$

for any integer n. There are then extra ways of satisfying the closed-string boundary condition (7.20) corresponding to winding the string n times round the circle, so

$$X^{25}(\tau, \sigma + \pi) = X^{25}(\tau, \sigma) + 2\pi Rn. \qquad (9.17)$$

Thus, the general mode expansion is

$$X^{25}(\tau, \sigma) = x^{25} + p^{25}\tau + 2L\sigma + \frac{\mathrm{i}}{2}\sum_{n \neq 0}\frac{1}{n}\left(\alpha^{25}_{n}\,\mathrm{e}^{-2\mathrm{i}\,n(\tau - \sigma)} + \tilde{\alpha}^{25}_{n}\,\mathrm{e}^{-2\mathrm{i}\,n(\tau + \sigma)}\right) \qquad (9.18)$$

where

$$L = nR. \qquad (9.19)$$

The integer n in (9.19) is the winding number for the string configuration. The momentum p^{25} is then constrained by the requirement that $\exp(\mathrm{i}\,p^{25}x^{25})$ should be single valued when we replace x^{25} by the equivalent coordinate $x^{25} + 2\pi Rn$. Consequently, we must take

$$p^{25} = \frac{m}{R} \tag{9.20}$$

where m is any integer. Decomposing into right and left movers by writing τ and σ in terms of $\tau - \sigma$ and $\tau + \sigma$,

$$X^{25} = X_{\mathrm{R}}^{25} + X_{\mathrm{L}}^{25} \tag{9.21}$$

where

$$X_{\mathrm{R}}^{25}(\tau - \sigma) = x_{\mathrm{R}}^{25} + p_{\mathrm{R}}^{25}(\tau - \sigma) + \frac{\mathrm{i}}{2} \sum_{n \neq 0} \frac{1}{n} \alpha_n^{25} \, \mathrm{e}^{-2\mathrm{i}\, n(\tau - \sigma)} \tag{9.22}$$

and

$$X_{\mathrm{L}}^{25}(\tau + \sigma) = x_{\mathrm{L}}^{25} + p_{\mathrm{L}}^{25}(\tau + \sigma) + \frac{\mathrm{i}}{2} \sum_{n \neq 0} \frac{1}{n} \tilde{\alpha}_n^{25} \, \mathrm{e}^{-2\mathrm{i}\, n(\tau + \sigma)} \tag{9.23}$$

with

$$p_{\mathrm{R}}^{25} = \tfrac{1}{2}(p^{25} - 2L) \tag{9.24}$$

$$p_{\mathrm{L}}^{25} = \tfrac{1}{2}(p^{25} + 2L) \tag{9.25}$$

and

$$x^{25} = x_{\mathrm{R}}^{25} + x_{\mathrm{L}}^{25}. \tag{9.26}$$

To obtain the 25-dimensional mass-squared operator for the physical states we need to amend L_0 and $\tilde{L}_0$ of (7.65) and (7.66) to take account of the effect of one string dimension being compactified on a circle. This amounts to replacing $\Sigma_{\mu=0}^{25} p^\mu p_\mu$ by $\Sigma_{\mu=0}^{24} p^\mu p_\mu - (2p_{\mathrm{R}}^{25})^2$ for L_0, and by $\Sigma_{\mu=0}^{24} p^\mu p_\mu - (2p_{\mathrm{L}}^{25})^2$ for $\tilde{L}_0$. It is convenient in the following exposition to use $p^\mu p_\mu$ to denote $\Sigma_{\mu=0}^{24} p^\mu p_\mu$, which is the 25-dimensional mass-squared operator M^2 (rather than the 26-dimensional mass-squared operator as in (7.82) where the sum over μ is from 0 to 25). Then, we find

$$\tfrac{1}{8}M^2 = \tfrac{1}{8}p^\mu p_\mu = \tfrac{1}{2}(p_{\mathrm{R}}^{25})^2 - \sum_{n=1}^{\infty} (\alpha_{-n}^{\mu} \alpha_{\mu n} - \alpha_{-n}^{25} \alpha_{25,\, n}) - 1$$

$$= \tfrac{1}{2}(p_{\mathrm{L}}^{25})^2 - \sum_{n=1}^{\infty} (\tilde{\alpha}_{-n}^{\mu} \tilde{\alpha}_{\mu n} - \tilde{\alpha}_{-n}^{25} \tilde{\alpha}_{25,\, n}) - 1 \tag{9.27}$$

where the normal-ordering constant a of (7.82) has been set to 1 as required by (7.118), and the sum over μ for the oscillator terms is also from 0 to 24. The mass-squared operator may now be cast in the form (analogous to (7.83) to (7.86))

$$M^2 = M_{\mathrm{R}}^2 + M_{\mathrm{L}}^2 \tag{9.28}$$

where

$$\tfrac{1}{4}M_{\mathrm{R}}^2 = \tfrac{1}{2}(p_{\mathrm{R}}^{25})^2 - \sum_{n=1}^{\infty} (\alpha^{\mu}_{-n}\alpha_{\mu n} - \alpha^{25}_{-n}\alpha_{25,n}) - 1 \tag{9.29}$$

$$\tfrac{1}{4}M_{\mathrm{L}}^2 = \tfrac{1}{2}(p_{\mathrm{L}}^{25})^2 - \sum_{n=1}^{\infty} (\tilde{\alpha}^{\mu}_{-n}\tilde{\alpha}_{\mu n} - \tilde{\alpha}^{25}_{-n}\tilde{\alpha}_{25,n}) - 1 \tag{9.30}$$

and

$$M_{\mathrm{R}}^2 = M_{\mathrm{L}}^2 \tag{9.31}$$

which allows us to think of the squared mass as having equal contributions from right and left movers.

Alternatively, substituting the explicit expressions for p_{R}^{25} and p_{L}^{25} of (9.24) and (9.25), we have

$$\begin{aligned}\tfrac{1}{8}M^2 &= \frac{(p^{25})^2}{8} + \frac{L^2}{2} - \frac{1}{2}\sum_{n=1}^{\infty} (\alpha^{\mu}_{-n}\alpha_{\mu n} - \alpha^{25}_{-n}\alpha_{25,n}) \\ &\quad - \frac{1}{2}\sum_{n=1}^{\infty} (\tilde{\alpha}^{\mu}_{-n}\tilde{\alpha}_{\mu n} - \tilde{\alpha}^{25}_{-n}\tilde{\alpha}_{25,n}) - 1 \\ &= \frac{n^2R^2}{2} + \frac{m^2}{8R^2} - \frac{1}{2}\sum_{n=1}^{\infty} (\alpha^{\mu}_{-n}\alpha_{\mu n} - \alpha^{25}_{-n}\alpha_{25,n}) \\ &\quad - \frac{1}{2}\sum_{n=1}^{\infty} (\tilde{\alpha}^{\mu}_{-n}\tilde{\alpha}_{\mu n} - \tilde{\alpha}^{25}_{-n}\tilde{\alpha}_{25,n}) - 1\end{aligned} \tag{9.32}$$

with

$$\begin{aligned}&\sum_{n=1}^{\infty} (\tilde{\alpha}^{\mu}_{-n}\tilde{\alpha}_{\mu n} - \tilde{\alpha}^{25}_{-n}\tilde{\alpha}_{25,n}) - \sum_{n=1}^{\infty} (\alpha^{\mu}_{-n}\alpha_{\mu n} - \alpha^{25}_{-n}\alpha_{25,n}) \\ &\qquad = p^{25}L = mn.\end{aligned} \tag{9.33}$$

Equation (9.32) displays the contributions to the 25-dimensional mass squared from the winding number and momentum in the compactified dimension.

The mass-squared operator may be written entirely in terms of physical dynamical degrees of freedom by adopting the light cone gauge. It is convenient for our present purposes to define the light cone coordinates to be

$$X^{\pm} = \frac{1}{\sqrt{2}}(X^0 \pm X^{24}) \tag{9.34}$$

so that the transverse degrees of freedom are X^i, $i = 1, \ldots, 23$, and X^{25}, which corresponds to the compactified dimension. (This differs from the choice (7.89) which involves X^{25} instead of X^{24}.) Following the argument of §7.7, the mass-squared operator then takes the form

$$M^2 = M_{\mathrm{R}}^2 + M_{\mathrm{L}}^2 \tag{9.35}$$

where

$$\tfrac{1}{4}M_{\mathrm{R}}^2 = \tfrac{1}{2}(p_{\mathrm{R}}^{25})^2 + N - 1 \tag{9.36}$$

$$\tfrac{1}{4}M_{\mathrm{L}}^2 = \tfrac{1}{2}(p_{\mathrm{L}}^{25})^2 + \tilde{N} - 1 \tag{9.37}$$

with

$$N = \sum_{n=1}^{\infty} (\alpha_{-n}^i \alpha_n^i + \alpha_{-n}^{25}\alpha_{25,n}) \tag{9.38}$$

$$\tilde{N} = \sum_{n=1}^{\infty} (\tilde{\alpha}_{-n}^i \tilde{\alpha}_n^i + \tilde{\alpha}_{-n}^{25}\tilde{\alpha}_{25,n}) \tag{9.39}$$

and

$$M_{\mathrm{R}}^2 = M_{\mathrm{L}}^2. \tag{9.40}$$

Alternatively,

$$\tfrac{1}{8}M^2 = \frac{n^2R^2}{2} + \frac{m^2}{8R^2} + \frac{N}{2} + \frac{\tilde{N}}{2} - 1 \tag{9.41}$$

with

$$N - \tilde{N} = mn. \tag{9.42}$$

After compactification of one dimension on a circle, the massless states of the resulting 25-dimensional theory include (in analogy with (7.120)) the graviton, the dilaton, and the antisymmetric tensor for 25 dimensions, as the traceless symmetric, trace and anti-symmetric parts of $\alpha_{-1}^i|0\rangle_{\mathrm{R}}\tilde{\alpha}_{-1}^j|0\rangle_{\mathrm{L}}$. However, there are also massless vector particles V^i and $\tilde{V}^i$ obtained by taking one index to be associated with 25-dimensional space-time and the other with the compactified dimension:

$$V^i = \alpha_{-1}^i|0\rangle_{\mathrm{R}}\tilde{\alpha}_{-1}^{25}|0\rangle_{\mathrm{L}} \tag{9.43}$$

and

$$\tilde{V}^i = \alpha_{-1}^{25}|0\rangle_{\mathrm{R}}\tilde{\alpha}_{-1}^i|0\rangle_{\mathrm{L}}. \tag{9.44}$$

One such massless vector could have been expected to arise in Kaluza–Klein fashion when a dimension was compactified on a circle, with components of the 26-dimensional metric tensor with one index associated with the compactified dimension becoming the components of a U(1) gauge field for 25 dimensions. Here there are two such gauge fields, deriving from the 26-dimensional metric tensor and the anti-symmetric tensor field, characteristic of string theory, as $(1/\sqrt{2})(V^i \pm \tilde{V}^i)$.

The gauge group arising from the compactification on a circle can be larger than $U(1) \times U(1)$ for a particular choice of the radius R. This enhancement of the gauge group is a stringy phenomenon depending on the existence of winding number. Let us use $|m, n\rangle$ to denote a state obtained from the ground state by taking momentum $p^{25} = m/R$ in the compactified dimension, and winding number $L = nR$, but without applying any oscillators to the ground state. Then, for a special choice of R, four extra massless vector fields may be constructed, namely

$$W^i_+ = \alpha^i_{-1}|1, 1\rangle \tag{9.45}$$

$$W^i_- = \alpha^i_{-1}|-1, -1\rangle \tag{9.46}$$

$$\tilde{W}^i_+ = \tilde{\alpha}^i_{-1}|1, -1\rangle \tag{9.47}$$

and

$$\tilde{W}^i_- = \tilde{\alpha}^i_{-1}|-1, 1\rangle. \tag{9.48}$$

This may be seen as follows. As a consequence of (9.35) and (9.40), a massless state requires

$$M^2_{\mathrm{R}} = M^2_{\mathrm{L}} = 0. \tag{9.49}$$

One way to arrange this in (9.36) and (9.37) is to take N and $\tilde{N}$ both to be one, and p^{25}_{R} and p^{25}_{L} both to be zero (which implies that m and n are both zero). This yields the massless vectors of (9.43) and (9.44). An alternative way of obtaining massless vectors is to take $N = 1$ and $p^{25}_{\mathrm{R}} = 0$ to arrange $M^2_{\mathrm{R}} = 0$, and to take $\tilde{N} = 0$ and $(p^{25}_{\mathrm{L}})^2 = 2$ to arrange $M^2_{\mathrm{L}} = 0$. Using the explicit expressions (9.24) and (9.25) for p^{25}_{R} and p^{25}_{L}, and noticing that in this case

$$mn = N - \tilde{N} = 1 \tag{9.50}$$

it follows that

$$m = n = \pm 1 \tag{9.51}$$

and

$$R = \frac{1}{\sqrt{2}}. \tag{9.52}$$

These are therefore massless vectors W^i_+ and W^i_- provided that the radius of the circle on which a dimension is compactified is given by (9.52). Similarly, massless vectors $\tilde{W}^i_+$ and $\tilde{W}^i_-$ also occur for this value of R. To sum up, for the special choice (9.52) for the radius of the compactified dimension, there are present in the theory the six massless vector fields required for the adjoint representation of the gauge group SU(2) × SU(2). (That these massless vector fields are indeed the gauge fields of an SU(2) × SU(2) gauge group may be confirmed by explicitly constructing the generators of the gauge group and checking that they commute with the mass-squared operator.)

9.4 Compactification of the heterotic string on a torus

A ten-dimensional theory, with the 'internal' degrees of freedom $X^I_{\rm L}$, $I = 1, \ldots, 16$, of §9.2 providing a gauge group, may be constructed by compactifying these degrees of freedom on a 16-dimensional torus. Much as for the quantization, it is necessary to impose boundary conditions on X^I, with both right and left movers, and then to eliminate the right movers. A 16-dimensional torus may be defined by introducing a lattice Γ with basis vectors e^I_a, $a = 1, \ldots, 16$, chosen to have length $\sqrt{2}$, and by making the identification

$$x^I \equiv x^I + \sqrt{2}\pi \sum_{a=1}^{16} n_a R_a e^I_a \tag{9.53}$$

where the R_a are radii and the n_a are arbitrary integers. There are then extra ways of satisfying the closed-string boundary conditions (7.20) by winding the string round the torus so that

$$X^I(\tau, \sigma + \pi) = X^I(\tau, \sigma) + \sqrt{2}\pi \sum_{a=1}^{16} n_a R_a e^I_a = X^I(\tau, \sigma) + 2\pi L^I \tag{9.54}$$

where

$$L^I = \frac{1}{\sqrt{2}} \sum_{a=1}^{16} n_a R_a e^I_a. \tag{9.55}$$

The L^I are usually referred to as the winding numbers. Then, the mode expansions are

$$X^I(\tau, \sigma) = x^I + p^I \tau + 2L^I \sigma + \frac{\mathrm{i}}{2} \sum_{n \neq 0} \frac{1}{n} (\alpha^I_n \,\mathrm{e}^{-2\mathrm{i}n(\tau - \sigma)} + \tilde{\alpha}^I_n \,\mathrm{e}^{-2\mathrm{i}n(\tau + \sigma)}). \tag{9.56}$$

Decomposing into right and left movers by writing τ and σ in terms of $\tau - \sigma$ and $\tau + \sigma$,

$$X^I = X^I_{\mathrm{R}} + X^I_{\mathrm{L}} \tag{9.57}$$

where

$$X^I_{\mathrm{R}}(\tau - \sigma) = x^I_{\mathrm{R}} + p^I_{\mathrm{R}}(\tau - \sigma) + \frac{\mathrm{i}}{2}\sum_{n \neq 0} \frac{1}{n} \alpha^I_n \, \mathrm{e}^{-2\mathrm{i}\, n(\tau - \sigma)} \tag{9.58}$$

and

$$X^I_{\mathrm{L}}(\tau + \sigma) = x^I_{\mathrm{L}} + p^I_{\mathrm{L}}(\tau + \sigma) + \frac{\mathrm{i}}{2}\sum_{n \neq 0} \frac{1}{n} \tilde{\alpha}^I_n \, \mathrm{e}^{-2\mathrm{i}\, n(\tau + \sigma)} \tag{9.59}$$

with

$$p^I_{\mathrm{R}} = \tfrac{1}{2}(p^I - 2L^I) \tag{9.60}$$

$$p^I_{\mathrm{L}} = \tfrac{1}{2}(p^I + 2L^I) \tag{9.61}$$

and

$$x^I = x^I_{\mathrm{R}} + x^I_{\mathrm{L}}. \tag{9.62}$$

We now wish to eliminate the right movers, which in particular means that we should take

$$p^I_{\mathrm{R}} = 0 \tag{9.63}$$

with the consequence that

$$p^I_{\mathrm{L}} = 2L^I. \tag{9.64}$$

(Notice that if we were not to compactify the internal dimensions, the L^I would all be zero and the internal momenta p^I_{L} would have to be zero.) Moreover, we should take

$$x^I_{\mathrm{R}} = 0 \tag{9.65}$$

which together with (9.62) and (9.53) means that

$$x^I_{\mathrm{L}} \equiv x^I_{\mathrm{L}} + \sqrt{2}\pi \sum_{a=1}^{16} n_a R_a e^I_a. \tag{9.66}$$

The commutation relation (9.12) implies that it is $2p^I_{\mathrm{L}}$ rather than p^I_{L} that generates translations of x^I_{L}, and therefore we should require $\exp(2\mathrm{i}\,\Sigma^{16}_{I=1}\, p^I_{\mathrm{L}} x^I_{\mathrm{L}})$ to be singled valued when x^I_{L} is replaced by the equivalent coordinates of (9.66). If the lattice $\tilde{\Gamma}$, with basis vectors denoted by e^{*I}_a, dual to Γ, is defined by

$$\sum_{I=1}^{16} e_a^I e_b^{*I} = \delta_{ab} \tag{9.67}$$

then p_L^I must be given by

$$p_L^I = \frac{1}{\sqrt{2}} \sum_{a=1}^{16} \frac{m_a}{R_a} e_a^{*I} \tag{9.68}$$

where m_a are arbitrary integers.

In the light cone gauge, the ten-dimensional mass-squared operator for the physical states is given by

$$M^2 = M_R^2 + M_L^2 \tag{9.69}$$

with M_R^2 and M_L^2 as follows. For the superstring right movers, using (8.146) and (8.150) with $D = 10$,

$$\tfrac{1}{4} M_R^2 = N \tag{9.70}$$

where

$$N = \sum_{n=1}^{\infty} (\alpha_{-n}^i \alpha_n^i + n d_{-n}^i d_n^i) \qquad \text{R sector} \tag{9.71}$$

and

$$N = \sum_{n=1}^{\infty} \alpha_{-n}^i \alpha_n^i + \sum_{r=1/2}^{\infty} r b_{-r}^i b_r^i - \tfrac{1}{2} \qquad \text{NS sector.} \tag{9.72}$$

For the bosonic string left movers with 16 dimensions compactified on a torus, the analogue of (9.37) is

$$\tfrac{1}{4} M_L^2 = \frac{1}{2} \sum_{I=1}^{16} (p_L^I)^2 + \tilde{N} - 1 \tag{9.73}$$

with

$$\tilde{N} = \sum_{n=1}^{\infty} (\tilde{\alpha}_{-n}^i \tilde{\alpha}_{in} + \tilde{\alpha}_{-n}^I \tilde{\alpha}_{In}) \tag{9.74}$$

where a sum over i from 1 to 8, and over I from 1 to 16 is understood. Also, for physical states,

$$M_R^2 = M_L^2 . \tag{9.75}$$

(For the bosonic string, as well as the superstring, the light cone coordinates have been chosen to be $(1/\sqrt{2})(X^0 \pm X^9)$, so that the transverse degrees of freedom for ten-dimensional space-time are $i = 1, \ldots, 8$.)

Massless states require

$$M_R^2 = M_L^2 = 0. \tag{9.76}$$

For massless vector bosons, the form (9.72) of M_R^2 appropriate to the NS sector has to be employed. There are 16 massless vectors $\tilde{V}_I^i$, $I = 1, \ldots, 16$, of Kaluza–Klein type (analogous to (9.44)):

$$\tilde{V}_I^i = b_{-1/2}^i|0\rangle_R \tilde{\alpha}_{-1}^I|0\rangle_L \qquad I = 1, \ldots, 16 \tag{9.77}$$

providing a $U^{16}(1)$ gauge group. As in §9.3, the gauge group can be enhanced for special choices of the lattice Γ and the radii R_a. Let us use the notation $|p_L^I\rangle$ to denote a state obtained from the left-mover ground state by taking momentum p_L^I in the compactified dimensions (without acting with any oscillators). In the present case, the extra gauge fields $W^i(p_L^I)$ arising in a stringy way are

$$W^i(p_L^I) = b_{-1/2}^i|0\rangle_R|p_L^I\rangle \tag{9.78}$$

with

$$\sum_I (p_L^I)^2 = 2 \tag{9.79}$$

which, as can be seen from (9.73), ensures that M_L^2 is zero. Whether there are internal momenta for which (9.79) is satisfied depends on the lattice Γ, its dual $\tilde{\Gamma}$, and the choice of radii R_a.

As we shall now discuss, there are very few choices of the lattice Γ consistent with an acceptable string theory. In Chapter 10, we shall see that there is a fundamental constraint on string theories, referred to as modular invariance, that is needed to ensure absence of gauge and gravitational anomalies and finiteness of string loop contributions to scattering amplitudes. Demanding a modular invariant theory restricts the radii of the torus to be

$$R_a = 1/\sqrt{2} \qquad a = 1, \ldots, 16 \tag{9.80}$$

and the lattice Γ to be an even self-dual lattice, i.e. a lattice for which

$$\tilde{\Gamma} = \Gamma \tag{9.81}$$

and

$$g_{aa} \equiv \sum_{I=1}^{16} e_a^I e_a^I = \text{even integer}. \tag{9.82}$$

There are only two such lattices in 16 dimensions, denoted by Γ_{16} and $\Gamma_8 \times \Gamma_8$. The first of these (Γ_{16}) contains the root lattice of SO(32) as a sublattice (and leads to an SO(32) gauge group). We shall focus on the second possibility ($\Gamma_8 \times \Gamma_8$), which is the direct product of two E_8 root lattices. The momenta p_L^I are then on the root lattice of $E_8 \times E_8$, and the momenta of length two required to satisfy (9.79) are the weight vectors of the adjoint representation of $E_8 \times E_8$. It would therefore by unsurprising if the extra massless vector fields (9.78), taken together with the Kaluza–Klein gauge fields (9.77), were the gauge fields of an $E_8 \times E_8$ gauge group. This can be demonstrated by an explicit construction of the generators of the gauge group, and a check that they commute with the mass-squared operator. In this way, an $E_8 \times E_8$ gauge group arises from a toroidal compactification of the left-mover internal degrees of freedom of the heterotic string.

Other massless states may be constructed by using the superstring right movers $b^i_{-1/2}|0\rangle_R$, $i = 1, \ldots, 8$, for the NS sector, and $|0\rangle_R$ for the R sector, with $M_R^2 = 0$, and the bosonic string left movers $\tilde{\alpha}^j_{-1}|0\rangle_L$, $j = 1, \ldots, 8$, with $M_L^2 = 0$. In this way, we obtain in the NS sector the massless states

$$b^i_{-1/2}|0\rangle_R \tilde{\alpha}^j_{-1}|0\rangle_L \qquad i, j = 1, \ldots, 8 \tag{9.83}$$

which decompose into a traceless symmetric ten-dimensional graviton, a scalar (dilaton), and an antisymmetric tensor. In the R sector there occur the states

$$|0\rangle_R \tilde{\alpha}^j_{-1}|0\rangle_L \qquad j = 1, \ldots, 8. \tag{9.84}$$

The decomposition of the product of the ten-dimensional spinor right mover and the ten-dimensional vector left mover provides a ten-dimensional gravitino together with an eight-component ten-dimensional spinor. In this way, the complete supergravity multiplet for ten-dimensional $N = 1$ supergravity[5] is generated.

The theory contains *no* tachyons because the only right-mover state with negative M_R^2 is the NS-sector ground state $|0\rangle_R$ with $M_R^2 = -2$, and the only left-mover state with negative M_L^2 is the bosonic string ground state $|0\rangle_L$ with $M_L^2 = -4$, and we cannot then satisfy $M_R^2 = M_L^2$ as required by (9.75). (Notice that, unlike in the case of the superstring, the absence of tachyons is *not* enforced by a GSO projection.)

9.5 Fermionization and bosonization

In the next section, we shall present an alternative formulation of the heterotic string which replaces the internal bosonic degrees of freedom compactified on a 16-dimensional torus by 32 world sheet fermionic degrees of freedom, a process referred to as fermionization. That such a replacement should be possible *a priori* depends on the fact that spin is not defined in two dimensions and the two-dimensional nature of the world sheet. In this section, we shall illustrate fermionization (and the inverse process of bosonization) by considering the simplest case of a single bosonic degree of freedom compactified on a circle. This bosonic model has been discussed in §9.3, where the compactified dimension was denoted by X^{25}. It will be convenient here to display only the part of the squared mass due to X^{25}, including the contribution $-\frac{1}{24}$ to the normal-ordering constant. It will also be convenient to suppress the index and denote X^{25} by X. Thus, following (9.22) and (9.23) we write the mode expansions

$$X_R(\tau - \sigma) = x_R + p_R(\tau - \sigma) + \frac{i}{2} \sum_{n \neq 0} \frac{1}{n} \alpha_n \, e^{-2i\, n(\tau - \sigma)} \tag{9.85}$$

and

$$X_L(\tau + \sigma) = x_L + p_L(\tau + \sigma) + \frac{i}{2} \sum_{n \neq 0} \frac{1}{n} \tilde{\alpha}_n \, e^{-2i\, n(\tau + \sigma)} \tag{9.86}$$

with

$$p_R = \tfrac{1}{2}(p - 2L) \tag{9.87}$$

$$p_L = \tfrac{1}{2}(p + 2L) \tag{9.88}$$

and

$$x = x_R + x_L \tag{9.89}$$

where

$$p = m/R \tag{9.90}$$

and

$$L = nR \tag{9.91}$$

where m and n are integers. The contribution of the single compactified dimension to the squared mass

$$M^2 = M_R^2 + M_L^2 \tag{9.92}$$

is given by

$$\tfrac{1}{4}M_R^2 = \tfrac{1}{2}p_R^2 + N - \tfrac{1}{24} \tag{9.93}$$

and

$$\tfrac{1}{4}M_L^2 = \tfrac{1}{2}p_L^2 + \tilde{N} - \tfrac{1}{24} \tag{9.94}$$

where

$$N = \sum_{n=1}^{\infty} \alpha_{-n}\alpha_n \tag{9.95}$$

and

$$\tilde{N} = \sum_{n=1}^{\infty} \tilde{\alpha}_{-n} \tilde{\alpha}_n . \tag{9.96}$$

We wish to show that for the choice of radius

$$R = 1 \tag{9.97}$$

there is an alternative formulation of the single compactified bosonic degree of freedom X_L as single complex fermionic degree of freedom. (An exactly similar discussion can of course be given for X_R.) From §8.4 and §8.7, a pair of real fermionic degrees of freedom has left-mover mode expansions for Neveu–Schwarz boundary conditions

$$\Psi_L^k = \sum_{r \in \mathbb{Z} + 1/2} \tilde{b}_r^k \, e^{-2i\, r(\tau + \sigma)} \qquad k = 1, 2 \tag{9.98}$$

and the contribution of this pair of real fermions to the squared mass M_L^2 is given by

$$\tfrac{1}{4} M_L^2 = \sum_{r=1/2}^{\infty} r \tilde{b}_{-r}^k \tilde{b}_r^k - \tfrac{1}{24} \tag{9.99}$$

recalling that each NS fermion contributes $-\frac{1}{48}$ to the normal-ordering constant. A reformulation in terms of a single complex fermionic degree of freedom $\hat{\Psi}_L$ may be made by writing

$$\hat{\Psi}_L = \frac{1}{\sqrt{2}} (\Psi_L^1 + i\, \Psi_L^2) \tag{9.100}$$

Then,

$$\hat{\Psi}_L = \sum_{r=1/2}^{\infty} (\tilde{f}_r \, e^{-2i\, r(\tau + \sigma)} + \tilde{g}_r^{\dagger} \, e^{2i\, r(\tau + \sigma)}) \tag{9.101}$$

with

$$\tilde{f}_r = \frac{1}{\sqrt{2}} (\tilde{b}_r^1 + i\, \tilde{b}_r^2) \tag{9.102}$$

and

$$\tilde{g}_r^{\dagger} = \frac{1}{\sqrt{2}} (\tilde{b}_{-r}^1 + i\, \tilde{b}_{-r}^2). \tag{9.103}$$

The contribution of $\hat{\Psi}_L$ to $\frac{1}{4} M_L^2$ may then be written as

$$\tfrac{1}{4}M_{\rm L}^2 = \sum_{r=1/2}^{\infty} r(\bar{f}_r^\dagger \bar{f}_r + \bar{g}_r^\dagger \bar{g}_r) - \tfrac{1}{24} \tag{9.104}$$

where we have used the fact that

$$(\tilde{b}_r^k)^\dagger = \tilde{b}_{-r}^k \qquad k = 1, 2 \tag{9.105}$$

which follows from the mode expansion (8.46) for real fermionic degrees of freedom. As a consequence of the anti-commutation relations (8.50), the oscillators for the complex fermion obey

$$\{\bar{f}_r^\dagger, \bar{f}_s\} = \{\bar{g}_r^\dagger, \bar{g}_s\} = \delta_{rs} \tag{9.106}$$

from which it follows that $\bar{f}_r^\dagger$ and $\bar{g}_r^\dagger$ increase the eigenvalue of $\frac{1}{4}M_{\rm L}^2$ by r.

On the other hand, for Ramond boundary conditions the pair of real fermionic degrees of freedom has left-mover mode expansion

$$\Psi_{\rm L}^k = \sum_{n \in \mathbb{Z}} \tilde{d}_n^k\, {\rm e}^{-2{\rm i}n(\tau + \sigma)} \qquad k = 1, 2 \tag{9.107}$$

and contributes to $M_{\rm L}^2$ according to

$$\tfrac{1}{4}M_{\rm L}^2 = \sum_n n\tilde{d}_{-n}^k \tilde{d}_n^k + \tfrac{1}{12} \tag{9.108}$$

recalling that each Ramond fermion contributes $\frac{1}{24}$ to the normal-ordering constant. Reformulating in terms of a single complex fermionic degree of freedom $\hat{\Psi}_{\rm L}$ defined by (9.100) we obtain the mode expansion

$$\hat{\Psi}_{\rm L} = \sum_{m=1}^{\infty} (\bar{c}_{m-1}\, {\rm e}^{-2{\rm i}(m-1)(\tau+\sigma)} + \bar{e}_m^\dagger\, {\rm e}^{2{\rm i}m(\tau+\sigma)}) \tag{9.109}$$

with

$$\bar{c}_{m-1} = \frac{1}{\sqrt{2}}(\tilde{d}_{m-1}^1 + {\rm i}\,\tilde{d}_{m-1}^2) \qquad m = 1, \ldots, \infty \tag{9.110}$$

and

$$\bar{e}_m^\dagger = \frac{1}{\sqrt{2}}(\tilde{d}_{-m}^1 + {\rm i}\,\tilde{d}_{-m}^2) \qquad m = 1, \ldots, \infty. \tag{9.111}$$

The contribution of $\hat{\Psi}_{\rm L}$ to $\frac{1}{4}M_{\rm L}^2$ in this case is

$$\tfrac{1}{4}M_{\rm L}^2 = \sum_{n=1}^{\infty} n(\bar{c}_n^\dagger \bar{c}_n + \bar{e}_n^\dagger \bar{e}_n) + \tfrac{1}{12} \tag{9.112}$$

where we have used the fact following from the mode expansion (8.45) that

$$(\tilde{d}_n^k)^\dagger = \tilde{d}_{-n}^k \qquad k = 1, 2. \tag{9.113}$$

As a consequence of the anticommutation relations (8.49) the oscillators for the complex fermions obey

$$\{\bar{c}_m^{\dagger}, \bar{c}_m\} = \{\bar{e}_m^{\dagger}, \bar{e}_n\} = \delta_{mn} \tag{9.114}$$

from which it follows that $\bar{c}_n^{\dagger}$ and $\bar{e}_n^{\dagger}$ increase the eigenvalue of $\frac{1}{4}M_{\mathrm{L}}^2$ by n.

A comparison of the values of $\frac{1}{2}M_{\mathrm{L}}^2$ in the bosonic and fermionic formulations may now be made and such a comparison for some low-lying states is given in tables 9.1 and 9.2 for the NS sector and for the R sector of the fermionic formulation, respectively. It can be seen that the number of states of the fermionic theory at each value of $\frac{1}{4}M_{\mathrm{L}}^2$ is the same as the number of states of the bosonic theory. A more detailed correspondence of states in the two formulations may be achieved by defining a charge lattice momentum q_{L} for the fermionic theory by

$$q_{\mathrm{L}} = N_{\mathrm{L}}^{\mathrm{F}} + v_{\mathrm{L}} - \tfrac{1}{2} \tag{9.115}$$

where

$$v_{\mathrm{L}} = \begin{cases} \frac{1}{2} & \text{NS sector} \\ 0 & \text{R sector} \end{cases} \tag{9.116}$$

and $N_{\mathrm{L}}^{\mathrm{F}}$ is the eigenvalue of the fermionic 'number' operator. For the NS sector, the ground state is defined to have $N_{\mathrm{L}}^{\mathrm{F}} = 0$ and f_r^{+} creates 1 unit and g_r^{+} creates -1 unit of $N_{\mathrm{L}}^{\mathrm{F}}$. For the R sector, $N_{\mathrm{F}}^{\mathrm{L}}$ is 0 or 1 for the two components of the spinor ground state, c_n^{+} creates 1 unit and e_n^{+} creates -1 unit of $N_{\mathrm{L}}^{\mathrm{F}}$. Then for each value of $\frac{1}{4}M_{\mathrm{L}}^2$, the number of fermionic states with $q_{\mathrm{L}} = p_{\mathrm{L}}$ matches the number of bosonic states with this value of p_{L}, as also displayed in tables 9.1 and 9.2. In the case of table 9.2, $|0\rangle_{\mathrm{L}}$ denotes the degenerate spinor ground state and so denotes two states with different values of q_{L}.

The equivalence of the theory of a single left-moving complex fermionic degree of freedom and the theory of a single left-moving bosonic degree of freedom compactified on a circle of radius $R = 1$ may be made more explicit by writing

$$\hat{\Psi}_{\mathrm{L}} = :\mathrm{e}^{2\mathrm{i}X_{\mathrm{L}}}: \tag{9.117}$$

where the normal ordering is taken to mean

$$:\mathrm{e}^{2\mathrm{i}X_{\mathrm{L}}}: = \exp\left(-\sum_{n<0}\frac{1}{n}\bar{\alpha}_n\,\mathrm{e}^{-2\mathrm{i}n(\tau+\sigma)}\right)\exp\left(-\sum_{n>0}\frac{1}{n}\bar{\alpha}_n\,\mathrm{e}^{-2\mathrm{i}n(\tau+\sigma)}\right)$$
$$\times\exp[2\mathrm{i}(x_{\mathrm{L}} + (p_{\mathrm{L}} + \tfrac{1}{2})(\tau+\sigma))]. \tag{9.118}$$

To show that the two theories are equivalent means checking that all correlation functions are the same, which can be done by showing that the replacement of $\hat{\Psi}_{\mathrm{L}}$ by $:\mathrm{e}^{2\mathrm{i}X_{\mathrm{L}}}:$ gives the correct operator product expansions

Table 9.1 Values of $\frac{1}{4}M_L^2$ for fermionic states in the NS sector compared with bosonic states for $R = 1$.

$\frac{1}{4}M_L^2$	Charge lattice momentum q_L	NS sector	Bosonic theory
$-\frac{1}{24}$	$q_L = p_L = 0$	$\|0\rangle_L$	$\|p_L = 0\rangle$
$-\frac{1}{24}+\frac{1}{2}$	$q_L = p_L = 1$	$\tilde{f}^\dagger_{1/2}\|0\rangle_L$	$\|p_L = 1\rangle$
	$q_L = p_L = -1$	$\tilde{g}^\dagger_{1/2}\|0\rangle_L$	$\|p_L = -1\rangle$
$-\frac{1}{24}+1$	$q_L = p_L = 0$	$\tilde{f}^\dagger_{1/2}\tilde{g}^\dagger_{1/2}\|0\rangle_L$	$\tilde{\alpha}_{-1}\|p_L = 0\rangle$
$-\frac{1}{24}+\frac{3}{2}$	$q_L = p_L = 1$	$\tilde{f}^\dagger_{3/2}\|0\rangle_L$	$\tilde{\alpha}_{-1}\|p_L = 1\rangle$
	$q_L = p_L = -1$	$\tilde{g}^\dagger_{3/2}\|0\rangle_L$	$\tilde{\alpha}_{-1}\|p_L = -1\rangle$
$-\frac{1}{24}+2$	$q_L = p_L = 0$	$\tilde{f}^\dagger_{3/2}\tilde{g}^\dagger_{1/2}\|0\rangle_L$, $\tilde{g}^\dagger_{3/2}\tilde{f}^\dagger_{1/2}\|0\rangle_L$	$\tilde{\alpha}_{-2}\|p_L = 0\rangle$, $\tilde{\alpha}_{-1}\tilde{\alpha}_{-1}\|p_L = 0\rangle$
	$q_L = p_L = 2$	$\tilde{f}^\dagger_{3/2}\tilde{f}^\dagger_{1/2}\|0\rangle_L$	$\|p_L = 2\rangle$
	$q_L = p_L = -2$	$\tilde{g}^\dagger_{3/2}\tilde{g}^\dagger_{1/2}\|0\rangle_L$	$\|p_L = -2\rangle$
$-\frac{1}{24}+\frac{5}{2}$	$q_L = p_L = 1$	$\tilde{f}^\dagger_{5/2}\|0\rangle_L$, $\tilde{f}^\dagger_{3/2}\tilde{f}^\dagger_{1/2}\tilde{g}^\dagger_{1/2}\|0\rangle_L$	$\tilde{\alpha}_{-2}\|p_L = 1\rangle$, $\tilde{\alpha}_{-1}\tilde{\alpha}_{-1}\|p_L = 1\rangle$
	$q_L = p_L = -1$	$\tilde{g}^\dagger_{5/2}\|0\rangle_L$, $\tilde{g}^\dagger_{3/2}\tilde{f}^\dagger_{1/2}\tilde{g}^\dagger_{1/2}\|0\rangle_L$	$\tilde{\alpha}_{-2}\|p_L = -1\rangle$, $\tilde{\alpha}_{-1}\tilde{\alpha}_{-1}\|p_L = -1\rangle$
$-\frac{1}{24}+3$	$q_L = p_L = 0$	$\tilde{f}^\dagger_{5/2}\tilde{g}^\dagger_{1/2}\|0\rangle_L$, $\tilde{g}^\dagger_{5/2}\tilde{f}^\dagger_{1/2}\|0\rangle_L$, $\tilde{f}^\dagger_{3/2}\tilde{g}^\dagger_{3/2}\|0\rangle_L$	$\tilde{\alpha}_{-3}\|p_L = 0\rangle$, $\tilde{\alpha}_{-2}\tilde{\alpha}_{-1}\|p_L = 0\rangle$, $\tilde{\alpha}_{-1}\tilde{\alpha}_{-1}\tilde{\alpha}_{-1}\|p_L = 0\rangle$
	$q_L = p_L = 2$	$\tilde{f}^\dagger_{5/2}\tilde{f}^\dagger_{1/2}\|0\rangle_L$	$\tilde{\alpha}_{-1}\|p_L = 2\rangle$
	$q_L = p_L = -2$	$\tilde{g}^\dagger_{5/2}\tilde{g}^\dagger_{1/2}\|0\rangle_L$	$\tilde{\alpha}_{-1}\|p_L = -2\rangle$

for $\hat{\Psi}_L$ and currents constructed from it[6]. The only unexpected feature of the definition of normal ordering in (9.117) is the occurrence of $p_L + \frac{1}{2}$ rather than p_L; otherwise (9.117) is just normal ordered in the usual sense. However, from the discussion above of corresponding states in the two formulations, anti-periodic (NS) boundary conditions for the complex fermionic degree of freedom should correspond to integral bosonic momentum p_L, and periodic (R) boundary conditions should correspond to half-integral bosonic momentum p_L. The presence of $p_L + \frac{1}{2}$ in (9.117) ensures that this is the case.

9.6 Fermionic formulation of the compactified heterotic string

The discussion of §9.5 suggests that it should be possible to find an alternative formulation of the heterotic string in which the toroidally compactified degree of freedom X^I_L, $I = 1, \ldots, 16$, are replaced by 16 complex left-moving fermionic degrees of freedom or 32 real fermionic degrees of freedom. In this section, we shall show that by a suitable choice of

Table 9.2 Values of $\frac{1}{4}M_L^2$ for fermionic states in the Ramond sector compared with bosonic states for $R = 1$.

$\frac{1}{4}M_L^2$	Charge lattice momentum q_L	R sector	Bosonic theory
$\frac{1}{12}$	$q_L = p_L = \pm\frac{1}{2}$	$\vert 0\rangle_L$	$\vert p_L = \pm\frac{1}{2}\rangle$
$\frac{1}{12} + 1$	$q_L = p_L = \frac{3}{2}, \frac{1}{2}$	$\tilde{c}_1^\dagger\vert 0\rangle_L$	$\vert p_L = \frac{3}{2}\rangle$, $\tilde{\alpha}_{-1}\vert p_L = \frac{1}{2}\rangle$
	$q_L = p_L = -\frac{3}{2}, -\frac{1}{2}$	$\tilde{e}_1^\dagger\vert 0\rangle_L$	$\vert p_L = -\frac{3}{2}\rangle$, $\tilde{\alpha}_{-1}\vert p_L = -\frac{1}{2}\rangle$
$\frac{1}{12} + 2$	$q_L = \frac{3}{2}, \frac{1}{2}$	$\tilde{c}_2^\dagger\vert 0\rangle_L$	$\tilde{\alpha}_{-1}\vert p_L = \frac{3}{2}\rangle$, $\tilde{\alpha}_{-2}\vert p_L = \frac{1}{2}\rangle$, $\tilde{\alpha}_{-1}\tilde{\alpha}_{-1}\vert p_L = \frac{1}{2}\rangle$
	$q_L = \frac{1}{2}, -\frac{1}{2}$	$\tilde{c}_1^\dagger\tilde{e}_1^\dagger\vert 0\rangle_L$	
			$\tilde{\alpha}_{-2}\vert p_L = -\frac{1}{2}\rangle$, $\tilde{\alpha}_{-1}\tilde{\alpha}_{-1}\vert p_L = -\frac{1}{2}\rangle$
	$q_L = -\frac{1}{2}, -\frac{3}{2}$	$\tilde{e}_2^\dagger\vert 0\rangle_L$	$\tilde{\alpha}_{-1}\vert p_L = -\frac{3}{2}\rangle$
$\frac{1}{12} + 3$	$q_L = \frac{5}{2}, \frac{3}{2}$	$\tilde{c}_2^\dagger\tilde{c}_1^\dagger\vert 0\rangle_L$	$\vert p_L = \frac{5}{2}\rangle$, $\tilde{\alpha}_{-2}\vert p_L = \frac{3}{2}\rangle$, $\tilde{\alpha}_{-1}\tilde{\alpha}_{-1}\vert p_L = \frac{3}{2}\rangle$
	$q_L = \frac{3}{2}, \frac{1}{2}$	$\tilde{c}_3^\dagger\vert 0\rangle_L$	
			$\tilde{\alpha}_{-3}\vert p_L = \frac{1}{2}\rangle$, $\tilde{\alpha}_{-2}\tilde{\alpha}_{-1}\vert p_2 = \frac{1}{2}\rangle$, $\tilde{\alpha}_{-1}\tilde{\alpha}_{-1}\tilde{\alpha}_{-1}\vert p_L = \frac{1}{2}\rangle$
	$q_L = \frac{1}{2}, -\frac{1}{2}$	$\tilde{c}_2^\dagger\tilde{e}_1^\dagger\vert 0\rangle_L$, $\tilde{e}_2^\dagger\tilde{c}_1^\dagger\vert 0\rangle_L$	
			$\tilde{\alpha}_{-3}\vert p_L = -\frac{1}{2}\rangle$, $\tilde{\alpha}_{-2}\tilde{\alpha}_{-1}\vert p_L = -\frac{1}{2}\rangle$, $\tilde{\alpha}_{-1}\tilde{\alpha}_{-1}\tilde{\alpha}_{-1}\vert p_L = -\frac{1}{2}\rangle$
	$q_L = -\frac{1}{2}, -\frac{3}{2}$	$\tilde{e}_3^\dagger\vert 0\rangle_L$	
			$\tilde{\alpha}_{-2}\vert p_L = -\frac{3}{2}\rangle$, $\tilde{\alpha}_{-1}\tilde{\alpha}_{-1}\vert p_L = -\frac{3}{2}\rangle$
	$q_L = -\frac{3}{2}, -\frac{5}{2}$	$\tilde{e}_2^\dagger\tilde{e}_1^\dagger\vert 0\rangle_L$	$\vert p_L = -\frac{5}{2}\rangle$

boundary conditions for these 32 real fermionic degrees of freedom, and by applying a suitable GSO projection, we can indeed reproduce the $E_8 \times E_8$ gauge fields of §9.4.

Our desire to construct the version of the theory with the $E_8 \times E_8$ gauge group rather than the SO(32) gauge group suggests that we should separate the 32 internal real fermionic left-mover degrees of freedom into two sets of 16 which we denote by $\lambda^A, A = 1, \ldots, 16$, and $\bar{\lambda}^A, A = 1, \ldots, 16$. Because of the possibility of assigning Ramond (R) or Neveu–Schwarz (NS) boundary conditions independently to the λ^A and $\bar{\lambda}^A$, the internal theory then possesses four sectors (R, R), (NS, NS), (R, NS) and (NS, R), where the first boundary condition of the pair refers to the λ^A and the second to the $\bar{\lambda}^A$. The mode expansions for the λ^A are

$$\lambda^A = \sum_{n \in \mathbb{Z}} \lambda_n^A \, \mathrm{e}^{-2\mathrm{i}\,n(\tau + \sigma)} \qquad \text{R sector} \tag{9.119}$$

and

$$\lambda^A = \sum_{r \in \mathbb{Z} + 1/2} \lambda_r^A \, e^{-2i\,r(\tau + \sigma)} \qquad \text{NS sector} \tag{9.120}$$

and similarly for the $\bar{\lambda}^A$.

To determine the masses of the states, the normal-ordering constants for the four sectors are required. As discussed in §8.7, each real bosonic degree of freedom contributes $\frac{1}{24}$ to the normal-ordering constant, each real fermionic degree of freedom with periodic (R) boundary conditions contributes $-\frac{1}{24}$, and each real fermionic degree of freedom with anti-periodic (NS) boundary conditions contributes $\frac{1}{48}$. Thus the normal-ordering constants $\tilde{a}$ for the left movers for the above four sectors are

$$\tilde{a}(\text{R}, \text{R}) = -1 \tag{9.121}$$

$$\tilde{a}(\text{NS}, \text{NS}) = +1 \tag{9.122}$$

and

$$\tilde{a}(\text{R}, \text{NS}) = \tilde{a}(\text{NS}, \text{R}) = 0 \tag{9.123}$$

including the contribution of the eight transverse left-mover bosonic degrees of freedom associated with ten-dimensional space-time.

The mass-squared operator for the left movers is then

$$\tfrac{1}{4}M_{\text{L}}^2 = \tilde{N}(\alpha, \beta) - \tilde{a}(\alpha, \beta) \tag{9.124}$$

where $\alpha, \beta = \text{R}$ or NS depending on the boundary conditions for the two sets of real fermions λ^A and $\bar{\lambda}^A$. The normal-ordering constants $\tilde{a}(\alpha, \beta)$ are given by (9.121)–(9.123). The oscillator term $\tilde{N}(\alpha, \beta)$ is given by

$$\tilde{N}(\text{R}, \text{R}) = \sum_{n=1}^{\infty} \tilde{\alpha}_{-n}^i \tilde{\alpha}_n^i + \sum_{n=1}^{\infty} n(\lambda_{-n}^A \lambda_n^A + \bar{\lambda}_{-n}^A \bar{\lambda}_n^A) \tag{9.125}$$

$$\tilde{N}(\text{NS}, \text{NS}) = \sum_{n=1}^{\infty} \tilde{\alpha}_{-n}^i \tilde{\alpha}_n^i + \sum_{r=1/2}^{\infty} r(\lambda_{-r}^A \lambda_r^A + \bar{\lambda}_{-r}^A \bar{\lambda}_r^A) \tag{9.126}$$

$$\tilde{N}(\text{R}, \text{NS}) = \sum_{n=1}^{\infty} \tilde{\alpha}_{-n}^i \tilde{\alpha}_n^i + \sum_{n=1}^{\infty} n\lambda_{-n}^A \lambda_n^A + \sum_{r=1/2}^{\infty} r\bar{\lambda}_{-r}^A \bar{\lambda}_r^A \tag{9.127}$$

and

$$\tilde{N}(\text{NS}, \text{R}) = \sum_{n=1}^{\infty} \tilde{\alpha}_{-n}^i \tilde{\alpha}_n^i + \sum_{r=1/2}^{\infty} r\lambda_{-r}^A \lambda_r^A + \sum_{n=1}^{\infty} n\bar{\lambda}_{-n}^A \bar{\lambda}_n^A \tag{9.128}$$

where $i = 1, \ldots, 8$, runs over the transverse ten-dimensional space-time degrees of freedom. The mass-squared operator for the right movers is as in (9.70)–(9.72) with the usual requirement of equality of M_R^2 and M_L^2 for physical states.

The $E_8 \times E_8$ gauge fields now arise from the four sectors in the following way. In the (NS, NS) sector, there are the massless vector states

$$b^i_{-1/2}|0\rangle_R \lambda^A_{-1/2}\lambda^B_{-1/2}|0\rangle_L \tag{9.129}$$

$$b^i_{-1/2}|0\rangle_R \bar{\lambda}^A_{-1/2}\bar{\lambda}^B_{-1/2}|0\rangle_L \tag{9.130}$$

and

$$b^i_{-1/2}|0\rangle_R \lambda^A_{-1/2}\bar{\lambda}^B_{-1/2}|0\rangle_L \tag{9.131}$$

which provide the representation **(120, 1)**, **(1, 120)** and **(16, 16)** of SO(16) × SO(16), respectively. In the (R, NS) sector, there are the massless vector fields

$$b^i_{-1/2}|0\rangle_R|0\rangle_L. \tag{9.132}$$

The left-mover ground state for this sector is the spinor representation of the first SO(16) factor of SO(16) × SO(16) and so is the representation **(128, + 128′, 1)** where **128** and **128′** are the two chiralities of the SO(16) spinor constructed by applying an even or odd number of zero-mode oscillators analogously to the two chiralities of the SO(10) spinor discussed in §8.8. Similarly, in the (NS, R) sector there are massless vector fields in **(1, 128 + 128′)** of SO(16) × SO(16). The (R, R) sector contains no massless states.

The 248-dimensional adjoint representation of E_8 has the decomposition under SO(16)

$$\mathbf{248} = \mathbf{120} + \mathbf{128}. \tag{9.133}$$

Thus, to obtain precisely the gauge fields of $E_8 \times E_8$ it is necessary to delete from the theory the massless vector states in **(16, 16)**, **(128′,1)** and **(1, 128′)** of SO(16) × SO(16). This may be achieved by introducing a pair of GSO projections, one for the fermionic degrees of freedom λ^A and the other for the $\bar{\lambda}^A$, as follows. For the λ^A, we make the projection

$$P = \frac{1 + (-1)^F}{2} \tag{9.134}$$

where F is defined by

$$F = \sum_{r=1/2}^{\infty} \lambda^A_{-r}\lambda^A_r \qquad \text{NS sector} \tag{9.135}$$

and

$$F = \sum_{n=0}^{\infty} \lambda^A_{-n} \lambda^A_n \qquad \text{R sector.} \tag{9.136}$$

Then, surviving states have an even number of λ^A oscillators acting on the ground state. In the Ramond sector, this includes the zero-mode oscillators and so a single chirality of the SO(16) spinor is selected. An exactly similar GSO projection is made for the $\bar{\lambda}^A$. As a consequence of this pair of GSO projections, the surviving massless vector states are, in the adjoint representation of $E_8 \times E_8$,

$$(\mathbf{248}, \mathbf{1}) + (\mathbf{1}, \mathbf{248}) = (\mathbf{120}, \mathbf{1}) + (\mathbf{128}, \mathbf{1}) + (\mathbf{1}, \mathbf{120}) + (\mathbf{1}, \mathbf{128}) \tag{9.137}$$

The massless supergravity multiplet is constructed in exactly the same way as in §9.4. In this way, we obtain exactly the same spectrum of massless states as in the bosonic formulation of §9.4, and it can be shown that this extends to the spectrum of massive states (Exercise 9.1).

Exercises

9.1 Construct the massive states of the $E_8 \times E_8$ heterotic string at the first excited level in both the fermionic and bosonic formulations.

9.2 By treating the real fermionic left-mover degrees of freedom as a single set of 32 with the same boundary conditions, construct the SO(32) heterotic string in the fermionic formulation.

References

General references

Green M B, Schwarz J H and Witten E 1987 *Superstring Theory* (Cambridge: Cambridge University Press)
Gross D J, Harvey J A, Martinec E and Rohm R 1985 *Nucl. Phys.* B **256** 253

References in the text

1 Paton J E and Chan H M 1969 *Nucl. Phys.* B **10** 516
Neveu A and Scherk J 1972 *Nucl. Phys.* B **36** 155
2 Marcus N and Sagnotti A 1982 *Phys. Lett.* **119B** 97
3 Gross D J, Harvey J A, Martinec E and Rohm R 1985 *Phys. Rev. Lett.* **54** 502; 1985 *Nucl. Phys.* B **256** 253
4 Bailin D and Love A 1987 *Rep. Prog. Phys.* **50** 1087
Duff M J, Nilson B E W and Pope C N 1986 *Phys. Rep.* **130** 1

5 Gliozzi F, Scherk J and Olive D 1977 *Nucl. Phys.* B **122** 253
Chamseddine A H 1981 *Nucl. Phys.* B **185** 403
Bergshoeff M, de Roo B, de Witt B and Van Nieuwenhuizen P 1983 *Nucl. Phys.* B **195** 97
Chapline G and Manton N S 1983 *Phys. Lett.* **120B** 105
6 For example,
Lerche W and Schellekens A S 1987 *Preprint* TH 4925 (CERN)

10

COMPACTIFICATION OF THE TEN-DIMENSIONAL HETEROTIC STRING TO FOUR DIMENSIONS

10.1 Introduction

Any string theory that is to be a candidate theory of the world we live in will have to possess just four observable space-time dimensions, or, if there are extra spatial dimensions, they will have to be compactified on a sufficiently small scale as to be unobservable with the energies that are currently available to us. In Chapter 9 a heterotic string theory with ten space-time dimensions was constructed by either fermionizing 16 of the left-mover bosonic degrees of freedom, or equivalently by compactifying these degrees of freedom on a torus in such a way that an $E_8 \times E_8'$ gauge group arose from these 16 extra left-mover dimensions. To complete the construction of a four-dimensional theory it is necessary next to compactify six of these ten dimensions in some way for both right and left movers. The simplest possibility is a toroidal compactification. However, we shall see that such a compactification produces a theory with $N = 4$ space-time supersymmetry rather than the $N = 1$ space-time supersymmetry that we saw in Chapter 1 was required to obtain a chiral theory. Fortunately, theories with $N = 1$ space-time supersymmetry can be obtained by simple modifications of toroidal compactifications, referred to as orbifolds, in which points on the torus are identified by a symmetry of the lattice of the torus. Alternatively, compactification on a special class of manifolds, called Calabi–Yau manifolds, may be employed.

A somewhat different approach to the construction of four-dimensional heterotic string theories, which we shall discuss in the next chapter, is to return to the original heterotic string with the right movers of a superstring in ten dimensions and the uncompactified left movers of a bosonic string in sixteen dimensions, and to reduce the number of space-time dimensions to four directly by fermionizing all other string degrees of freedom. In this approach, there is no intermediate theory with ten space-time dimensions that is compactified to four dimensions.

10.2 Toroidal compactifications

The simplest way of producing a four-dimensional theory from the ten-dimensional heterotic string constructed in Chapter 9 is by compactifying six

DOI: 10.1201/9780367805807-10

of the remaining spatial dimensions on a torus. A great attraction of such a possibility is that the simple linear string equations of motion of previous chapters are unmodified, as a consequence of the fact that a torus is locally flat and so may be taken to have the same X^μ-independent metric as flat space, differing from flat space only in the imposition of spatial periodicity.

In the light cone gauge, the mode expansions for the transverse space-time degrees of freedom X^i_{R}, Ψ^i_{R} and X^i_{L}, $i = 1, 2$ associated with four-dimensional space-time are as before in (9.1)–(9.4), the mode expansions for the 16 left-mover internal degrees of freedom X^I_{L}, $I = 1, \ldots, 16$ are as in (9.58) and (9.59), and the mode expansions for the fermionic degrees of freedom associated with the remaining six spatial degrees of freedom Ψ^k_{R}, $k = 3, \ldots, 8$ are also unmodified and given by (9.2) and (9.3). On the other hand, the mode expansions for the bosonic degrees of freedom X^k_{R} and X^k_{L}, $k = 3, \ldots, 8$ need to be amended to take account of the toroidal compactification of these six spatial dimensions. If the lattice defining the six-dimensional torus has basis vectors e^k_t, $k, t = 3, \ldots, 8$ chosen to have length $\sqrt{2}$, then for the centre-of-mass string coordinates x^k there is the identification

$$x^k \equiv x^k + \sqrt{2}\pi \sum_{t=3}^{8} n_t R_t e^k_t \tag{10.1}$$

where the R_t are radii, and the n_t are arbitrary integers. This results in extra ways of satisfying the closed-string boundary conditions (7.20) by winding the string round the torus, so

$$X^k(\tau, \sigma + \pi) = X^k(\tau, \sigma) + 2\pi L^k \tag{10.2}$$

where the winding numbers L^k are given by

$$L^k = \frac{1}{\sqrt{2}} \sum_{t=3}^{8} n_t R_t e^k_t . \tag{10.3}$$

By analogy with §9.4, we can then write

$$X^k_{\mathrm{R}}(\tau - \sigma) = x^k_{\mathrm{R}} + p^k_{\mathrm{R}}(\tau - \sigma) + \frac{\mathrm{i}}{2} \sum_{n \neq 0} \frac{1}{n} \alpha^k_n \,\mathrm{e}^{-2\mathrm{i}\, n(\tau - \sigma)} \tag{10.4}$$

and

$$X^k_{\mathrm{L}}(\tau + \sigma) = x^k_{\mathrm{L}} + p^k_{\mathrm{L}}(\tau + \sigma) + \frac{\mathrm{i}}{2} \sum_{n \neq 0} \frac{1}{n} \tilde{\alpha}^k_n \,\mathrm{e}^{-2\mathrm{i}\, n(\tau + \sigma)} \tag{10.5}$$

with

$$p^k_{\mathrm{R}} = \tfrac{1}{2}(p^k - 2L^k) \tag{10.6}$$

$$p_{\mathrm{L}}^{k} = \tfrac{1}{2}(p^k + 2L^k) \tag{10.7}$$

and

$$x^k = x_{\mathrm{R}}^k + x_{\mathrm{L}}^k. \tag{10.8}$$

However, unlike in §9.4, the six dimensions being compactified here have both right movers X_{R}^k and left movers X_{L}^k, and we do not wish to eliminate the right movers. We must require $\exp(\mathrm{i}\sum_{k=3}^{8} p^k x^k)$ to be single valued when x^k is replaced by the equivalent coordinates of (10.1). If the lattice with basis vectors e_t^k has a dual lattice with basis vectors e_t^{*k} where

$$\sum_{k=3}^{8} e_t^k e_u^{*k} = \delta_{tu} \tag{10.9}$$

then p^k must be given by

$$p^k = \sqrt{2} \sum_{k=3}^{8} \frac{m_t}{R_t} e_t^{*k} \tag{10.10}$$

where the m_t are arbitrary integers.

The four-dimensional mass-squared operator for the physical states is given by

$$M^2 = M_{\mathrm{R}}^2 + M_{\mathrm{L}}^2 \tag{10.11}$$

with

$$M_{\mathrm{R}}^2 = M_{\mathrm{L}}^2 \tag{10.12}$$

and M_{R}^2 and M_{L}^2 as follows. For the superstring right movers,

$$\tfrac{1}{4}M_{\mathrm{R}}^2 = N + \frac{1}{2}\sum_{k=3}^{8} (p_{\mathrm{R}}^k)^2 \tag{10.13}$$

with p_{R}^k given by (10.6), and N by (9.71) or (9.72). (Notice that $2p_{\mathrm{R}}^k$ is the coefficient of $\frac{1}{2}(\tau - \sigma)$ and substitutes for p^k of (7.26) in the uncompactified case.) For the left movers in the bosonic formulation of the heterotic string,

$$\tfrac{1}{4}M_{\mathrm{L}}^2 = \tilde{N} - 1 + \frac{1}{2}\sum_{I=1}^{16} (p_{\mathrm{L}}^I)^2 + \frac{1}{2}\sum_{k=3}^{8} (p_{\mathrm{L}}^k)^2 \tag{10.14}$$

with $\tilde{N}$ given by (9.74), p_{L}^I by (9.61) and p_{L}^k by (10.7). Alternatively, in the fermionic formulation of the heterotic string,

$$\tfrac{1}{4}M_{\mathrm{L}}^2 = \tilde{N}(\alpha, \beta) - \tilde{a}(\alpha, \beta) + \frac{1}{2}\sum_{k=3}^{8} (p_{\mathrm{L}}^k)^2 \tag{10.15}$$

with $\tilde{N}(\alpha, \beta)$ and $\tilde{a}(\alpha, \beta)$, where $\alpha, \beta =$ R or NS, as in (9.125)–(9.128) and (9.121)–(9.123). It is not too difficult to show (Exercise 10.1) that massless states generically have *no* momentum or winding number on the compact manifold. This is analogous to the situation for the heterotic string left movers in §9.4 where massless states with non-zero winding number only occur for special choices of the lattice and radii.

The graviton for four-dimensional gravity for the toroidally compactified theory is the Neveu–Schwarz sector state

$$b^i_{-1/2}|0\rangle_R \tilde{\alpha}^j_{-1}|0\rangle_L \qquad i, j = 1, 2. \tag{10.16}$$

The four-dimensional gravity supermultiplet also contains four gravitini,

$$|0\rangle_R \tilde{\alpha}^j_{-1}|0\rangle_L \qquad j = 1, 2. \tag{10.17}$$

Here, $|0\rangle_R$ is the SO(8) spinor Ramond sector ground state, which can, in the first instance, be written as the direct product of a transverse space-time spinor and an SO(6) spinor with eight components in $\mathbf{4} + \bar{\mathbf{4}}$ of SO(6). However, since the SO(8) spinor is of definite chirality because of the GSO projection, each space-time chirality is associated with only $\mathbf{4}$ or only $\bar{\mathbf{4}}$, so we get a total of eight spinor states, as required for four gravitini. Moreover, there are six graviphotons (massless vector fields in the supergravity multiplet) given by the Neveu–Schwarz sector states

$$b^k_{-1/2}|0\rangle_R \tilde{\alpha}^j_{-1}|0\rangle_L \qquad k = 3, \ldots, 8, \ j = 1, 2. \tag{10.18}$$

Thus, we have the spin-$\frac{3}{2}$ and spin-1 content of the gravity supermultiplet for $N = 4$ supergravity, as in table 1.3. As discussed in §1.6, theories based on $N \geqslant 2$ supersymmetry do not provide a suitable description of the world in which we live because they are always non-chiral. This means that toroidal compactification of the extra six spatial dimensions is not appropriate. Fortunately, we shall see in the next section that there is a simple modification of toroidal compactification, compactification on an orbifold, which overcomes this difficulty.

10.3 Orbifold compactifications

A simple modification of toroidal compactification is compactification on an orbifold[1], a six-dimensional space obtained by identifying points on the torus that are mapped into one another by certain discrete symmetries of the lattice of the torus, referred to as the point group. This approach to construction of a four-dimensional theory retains the advantage of toroidal compactification that the linear string equations of motion of previous chapters are still unmodified. At the same time, we shall see that it is possible for orbifold compactification to produce a four-dimensional theory with

$N = 1$ supersymmetry rather than the undesired $N = 4$ supersymmetry of toroidal compactification.

We shall focus attention here on one particular example of an orbifold, namely the Z_3 orbifold. It will be convenient to use a complex basis for the six spatial degrees of freedom X^k, $k = 3, \ldots, 8$ associated with the compact manifold, and to employ complex coordinates Z^α, $\alpha = 1, 2, 3$, where

$$Z^1 = \frac{1}{\sqrt{2}}(X^3 + \mathrm{i}\, X^4) \tag{10.19}$$

$$Z^2 = \frac{1}{\sqrt{2}}(X^5 + \mathrm{i}\, X^6) \tag{10.20}$$

and

$$Z^3 = \frac{1}{\sqrt{2}}(X^7 + \mathrm{i}\, X^8). \tag{10.21}$$

The lattice for the underlying torus of the Z_3 orbifold is defined by making the identifications, for $\alpha = 1, 2, 3$,

$$Z^\alpha \equiv Z^\alpha + 1 \tag{10.22}$$

and

$$Z^\alpha \equiv Z^\alpha + \mathrm{e}^{2\pi \mathrm{i}/3}. \tag{10.23}$$

Thus, if we assemble Z^1, Z^2 and Z^3 into a vector $\boldsymbol{Z}$,

$$\boldsymbol{Z} \equiv \boldsymbol{Z} + \sum_{p=1}^{3} (m_p \boldsymbol{e}_p + n_p \boldsymbol{f}_p) \tag{10.24}$$

where the basis vectors $\boldsymbol{e}_p$ and $\boldsymbol{f}_p$ for the lattice are defined by

$$\boldsymbol{e}_1 = (100) \qquad \boldsymbol{e}_2 = (010) \qquad \boldsymbol{e}_3 = (001) \tag{10.25}$$

and

$$\boldsymbol{f}_p = \mathrm{e}^{2\pi \mathrm{i}/3} \boldsymbol{e}_p \qquad p = 1, 2, 3 \tag{10.26}$$

and m_p and n_p are integers. The point group for the Z_3 orbifold is the Z_3 discrete group generated by the element

$$\omega = \mathrm{diag}(\mathrm{e}^{2\pi \mathrm{i}/3}, \mathrm{e}^{2\pi \mathrm{i}/3}, \mathrm{e}^{2\pi \mathrm{i}/3}) \tag{10.27}$$

acting on $\boldsymbol{Z}$. The complex coordinates $\boldsymbol{Z}$ provide a basis for the three-dimensional representation of the SU(3) subgroup of the SO(6) rotation group for the six real coordinates X^k, $k = 3, \ldots, 8$ and ω is a (finite) element of this SU(3). It is not difficult to show (Exercise 10.2) that ω has the action on the basis vectors of the lattice of the torus

$$\omega e_\rho = f_\rho \tag{10.28}$$

and

$$\omega f_\rho = -e_\rho - f_\rho. \tag{10.29}$$

Thus, the discrete group generated by ω maps torus lattice vectors to torus lattice vectors, i.e. is a symmetry of the torus. The construction of the Z_3 orbifold is completed by identifying points on the torus that are mapped into one another by elements of the Z_3 group generated by ω, referred to as the point group of the orbifold.

The orbifold is not quite a manifold because of the existence of a finite number of fixed points on the torus that are mapped to themselves, up to a lattice vector, by a point group element. The characteristic property of a fixed point may be restated using the notion of a space group element $(\theta, \boldsymbol{l})$, which is a point group rotation θ followed by a displacement by an amount $\boldsymbol{l}$ on the lattice,

$$(\theta, \boldsymbol{l})\boldsymbol{Z} = \theta\boldsymbol{Z} + \boldsymbol{l} \tag{10.30}$$

where in general $\boldsymbol{l}$ is of the form

$$\boldsymbol{l} = \sum_{\rho=1}^{3} (m_\rho e_\rho + n_\rho f_\rho) \tag{10.31}$$

where m_ρ and n_ρ are integers. A fixed point is then a point $\boldsymbol{Z}$ that is strictly mapped to itself by a space group element, not just up to a lattice vector. We shall see in §10.5 that the fixed points of an orbifold are of great importance, the various twisted sectors of the orbifold discussed there having centre-of-mass coordinates, which are fixed points. To find the fixed points we have to solve

$$(\omega, \boldsymbol{l})\boldsymbol{Z} = \boldsymbol{Z} \tag{10.32}$$

with ω as in (10.27) and $\boldsymbol{l}$ as in (10.31). (Fixed points of ω are also fixed points of $\omega^2 = \omega^{-1}$.) It is not difficult to show (Exercise 10.3) that the fixed points may be written in the form

$$\boldsymbol{Z} = \frac{e^{i\pi/6}}{\sqrt{3}}(m_1 + n_1, m_2 + n_2, m_3 + n_3) - (n_1, n_2, n_3). \tag{10.33}$$

Thus, there are 27 inequivalent fixed points

$$\boldsymbol{Z} = \frac{e^{i\pi/6}}{\sqrt{3}}(p_1, p_2, p_3) \tag{10.34}$$

with $p_\rho = 0, \pm 1$ for $\rho = 1, 2, 3$, all other fixed points differing from these by a lattice vector, and so being the same points on the torus. The fixed points (10.34) satisfy (10.32) with

$$l = \sum_{\rho=1}^{3} p_\rho e_\rho. \tag{10.35}$$

However, equation (10.34) also gives a fixed point of (ω, l) with

$$l = \sum_{\rho=1}^{3} p_\rho e_\rho + (I - \omega)k \tag{10.36}$$

for *any* lattice vector k, because $Z - k$ is equivalent to Z on the torus.

The toroidal compactification of the heterotic string theory possesses an $E_8 \times E_8'$ gauge group. A first step may be taken towards obtaining a realistic gauge group by embedding the point group in the gauge group, i.e. by associating with ω an action Ω on the gauge degrees of freedom that is a finite global element of the gauge group. The simplest possibility is to take Ω to be the element

$$\Omega = \omega \tag{10.37}$$

of the SU(3) subgroup of an $E_6 \times SU(3)$ contained in the first E_8 factor of the $E_8 \times E_8'$ gauge group (the one to be associated with the observable sector). We shall see in the next section that this breaks the observable sector E_8 to $E_6 \times SU(3)$. Further breaking of the gauge symmetry may be achieved by the Wilson line mechanism discussed in §10.6.

The states of the toroidally compactified theory are not all *bona fide* states of the orbifold theory. To define consistent states on the orbifold, such that equivalent points on the orbifold are on the same footing, we must restrict ourselves to states that are invariant under the action of the point group (including the action of the embedding of the point group in the gauge group). States of the orbifold theory derived as point-group-invariant states of the corresponding toroidally compactified theory constitute the so-called untwisted sector of the orbifold. In addition, the orbifold theory possesses extra states not to be found in the toroidally compactified theory, referred to as twisted-sector states. The twisted sectors of the orbifold are obtained by observing that it is no longer necessary for the closed heterotic string boundary conditions to be strictly satisfied. It is sufficient for them to be satisfied up to the action of a point group element (which links equivalent points on the orbifold). We shall discuss these twisted sectors in §10.5.

10.4 The untwisted sector of the Z_3 orbifold

As discussed in §10.3, the untwisted sector of the Z_3 orbifold consists of those states that can be constructed as a subset of the states of the underlying toroidal compactification by demanding point group invariance. The massless states, from which the light states we observe originate, are of particular interest. To construct these states we require the mode expansions of the string degrees of freedom which are exactly those of the toroidal compactification of §10.2, although it will usually be convenient here to cast the bosonic and fermionic degrees of freedom associated with the compact manifold in the complex basis of (10.19)–(10.21).

It will now be shown that the undesired $N = 4$ supergravity of the toroidal compactification has been reduced to $N = 1$ supergravity by the requirement of point group invariance by checking that the number of gravitini has been reduced from 4 to 1. In (10.17) we saw that the gravitini were the states

$$|0\rangle_R \tilde{\alpha}^j_{-1}|0\rangle_L \qquad j = 1, 2 \tag{10.38}$$

where $|0\rangle_R$ was the SO(8) spinor Ramond sector ground state of definite chirality. This spinor decomposed into a **4** of SO(6) with (say) right chiral space-time chirality, and a $\bar{\mathbf{4}}$ of SO(6) with left chiral space-time chirality. Under the SU(3) subgroup of SO(6) we have the decomposition

$$\mathbf{4} = \mathbf{3} + \mathbf{1} \tag{10.39}$$

and, under the action of the point group element ω of (10.27), the **3** transforms with a phase factor $e^{2\pi i/3}$ and the **1** is invariant. On the other hand, the left mover (10.38), which is associated entirely with four-dimensional space-time is invariant under the action of ω. Thus, to make a right chiral point-group-invariant state we must retain only the singlet in (10.39). Similarly, to make a left-chiral point-group-invariant state we must retain only the singlet in

$$\bar{\mathbf{4}} = \bar{\mathbf{3}} + \mathbf{1}. \tag{10.40}$$

These are then just the right and left chiral states of a single gravitino, as required for $N = 1$ supergravity.

The $E_8 \times E_8'$ gauge group of the toroidal compactification is also modified by the requirement of point group invariance when ω has the embedding Ω in the gauge group of (10.37). This cannot affect the (hidden sector) E_8' but reduces the (observable sector) E_8 to $E_6 \times SU(3)$ as follows. Under $E_6 \times SU(3)$, the 248-dimensional adjoint of E_8 decomposes as

$$\mathbf{248} = (\mathbf{78}, \mathbf{1}) + (\mathbf{1}, \mathbf{8}) + (\mathbf{27}, \mathbf{3}) + (\overline{\mathbf{27}}, \bar{\mathbf{3}}). \tag{10.41}$$

The gauge fields of E_8 arising from (9.129)–(9.132) have right mover $b^i_{-1/2}|0\rangle_R$, $i = 1, 2$, which is invariant under the action of ω. To obtain a point-group-invariant state, it is therefore necessary for the left mover to be invariant under the action of Ω. This eliminates $(\mathbf{27}, \mathbf{3})$ and $(\overline{\mathbf{27}}, \bar{\mathbf{3}})$ in (10.41) leaving only the states $(\mathbf{78}, \mathbf{1})$ of the adjoint of E_6 and $(\mathbf{1}, \mathbf{8})$ of the adjoint of SU(3), so the gauge group is reduced to $E_6 \times SU(3)$. (Notice that the **8** of

SU(3) transforms trivially under Ω because it is contained in $\mathbf{3} \times \bar{\mathbf{3}}$ and Ω acts with opposite phase factors on $\mathbf{3}$ and $\bar{\mathbf{3}}$.)

Apart from providing the $E_6 \times SU(3) \times E_8'$ gauge fields, the states (9.129)–(9.132) of the uncompactified theory also provide for the orbifold compactified-theory matter fields in $(\mathbf{27}, \mathbf{3})$ of $E_6 \times SU(3)$. These are the point-group-invariant states with right movers of the type $b^{\alpha}_{-1/2}|0\rangle_R$, $\alpha = 1, 2, 3$, in the complex basis of (10.19)–(10.21), and left movers in the $(\mathbf{27}, \mathbf{3})$ component of the adjoint of E_8. (Recall that in (9.129)–(9.132) the index i runs over space-time and the compact manifold to form a ten-dimensional vector state which in the four-dimensional sense is a vector and six scalars.) In more detail, the right movers $b^{*\alpha}_{-1/2}|0\rangle_R$ transform as $\mathbf{3}$ of the SU(3) subgroup of SO(6) associated with the six spatial degrees of freedom of the compact manifold and so transform with a phase factor $e^{2\pi i/3}$ under the action of ω. We can therefore obtain point-group-invariant states by linking these right movers with left movers in $(\overline{\mathbf{27}}, \bar{\mathbf{3}})$ of $E_6 \times SU(3)$, which transforms with phase factor $e^{-2\pi i/3}$ under the action of Ω. Similarly, linking the right movers $b^{\alpha}_{-1/2}|0\rangle_R$ with left movers in $(\mathbf{27}, \mathbf{3})$ of $E_6 \times SU(3)$ yields point-group-invariant states. In this way, we obtain scalar matter fields in three copies of $(\mathbf{27}, \mathbf{3})$ and three copies of $(\overline{\mathbf{27}}, \bar{\mathbf{3}})$, one for each value of α. If instead we consider Ramond-sector right movers, then the space-time right chiral component of the SO(6) spinor in $\mathbf{3}$ of SU(3) in (10.39) links to $(\overline{\mathbf{27}}, \bar{\mathbf{3}})$ left movers, and the space-time left chiral component of the SO(6) spinor in $\bar{\mathbf{3}}$ of SU(3) in (10.40) links to $(\mathbf{27}, \mathbf{3})$ left movers, to provide point-group-invariant states. Taken together with the above scalar states, this completes three left chiral supermultiplets in $(\mathbf{27}, \mathbf{3})$ of $E_6 \times SU(3)$ together with their anti-particles.

The untwisted sector also contains certain E_6 singlet scalars, referred to as moduli, which are point-group-invariant states constructed using the left-mover bosonic oscillators for the compact manifold $\tilde{\alpha}^k_{-1}$. In the complex basis of (10.19)–(10.21), these are the nine states

$$b^{\alpha}_{-1/2}|0\rangle_R \tilde{\alpha}^{*\beta}_{-1}|0\rangle_L \qquad \alpha, \beta = 1, 2, 3. \tag{10.42}$$

The possibility of giving expectation values to these moduli scalars is related to the possibility of modifying the scale or shape of the orbifold by adjusting various radii and angles characterizing the underlying torus.

10.5 The twisted sector of the Z_3 orbifold

As observed in §10.3, the orbifold theory possesses additional states that cannot occur in the toroidally compactified theory, the so-called twisted-sector states, whose existence depends on the fact that it is sufficient in an orbifold theory for the boundary conditions to be satisfied up to the action of a point group element. Indeed, the fundamental modular invariance of string theory (see Chapter 11) requires the twisted-sector states to be included in a consistent theory. For the Z_3 orbifold there are two twisted sectors, the ω and $\omega^2 = \omega^{-1}$ twisted sectors, in which the boundary conditions are satisfied up to a 'twist' by ω or ω^2, respectively. Thus, in the ω twisted sector the boundary conditions for the bosonic degrees of freedom for the compactified dimensions in the complex basis of (10.19)–(10.21) are

$$\mathbf{Z}(\tau, \sigma + \pi) = \omega \mathbf{Z}(\tau, \sigma) \tag{10.43}$$

and for the fermionic degrees of freedom for the compactified dimensions, again in the complex basis, the boundary conditions are

$$\boldsymbol{\Psi}_{\mathrm{R}}(\tau, \sigma + \pi) = \omega \boldsymbol{\Psi}_{\mathrm{R}}(\tau, \sigma) \qquad \text{R sector} \tag{10.44}$$

and

$$\boldsymbol{\Psi}_{\mathrm{R}}(\tau, \sigma + \pi) = -\omega \boldsymbol{\Psi}_{\mathrm{R}}(\tau, \sigma) \qquad \text{NS sector} \tag{10.45}$$

with similar expressions for the ω^2 twisted sector with ω^2 replacing ω. In addition, in the fermionic formulation of the heterotic string, there are three complex left-moving fermions with their boundary conditions twisted by $\Omega = \omega$, analogously to (10.44) and (10.45).

The mode expansions in twisted sectors must differ from those for the untwisted sector. Proceeding a little more generally than is required for the Z_3 orbifold, when a bosonic degree of freedom (in the complex basis) Z^α has its boundary conditions twisted by $\mathrm{e}^{-2\pi \mathrm{i}\eta}$ with $0 < \eta < 1$, and so

$$Z^\alpha(\tau, \sigma + \pi) = \mathrm{e}^{-2\pi\mathrm{i}\,\eta} Z^\alpha(\tau, \sigma) \tag{10.46}$$

then the appropriate mode expansion for the right mover Z^α_{R} is

$$Z^\alpha_{\mathrm{R}}(\tau - \sigma) = z^\alpha_{\mathrm{R}} + \frac{\mathrm{i}}{2} \sum_{n=1}^{\infty} \frac{1}{(n-\eta)} \beta^\alpha_{n-\eta}\, \mathrm{e}^{-2\mathrm{i}(n-\eta)(\tau-\sigma)}$$
$$- \frac{\mathrm{i}}{2} \sum_{n=0}^{\infty} \frac{1}{(n+\eta)} (\gamma^\alpha_{n+\eta})^\dagger\, \mathrm{e}^{2\mathrm{i}(n+\eta)(\tau-\sigma)} \tag{10.47}$$

where the oscillators $\beta^\alpha_{n-\eta}$ and $\gamma^\alpha_{n+\eta}$ have commutation relations

$$[\beta^\alpha_{m-\eta}, (\beta^\gamma_{n-\eta})^\dagger] = \delta_{\alpha\gamma}(n-\eta)\delta_{mn} \tag{10.48}$$

and

$$[\gamma^\alpha_{m+\eta}, (\gamma^\beta_{n+\eta})^\dagger] = \delta_{\alpha\beta}(n+\eta)\delta_{mn}. \tag{10.49}$$

The appropriate mode expansion for the left mover Z^α_{L} is

$$Z_{\mathrm{L}}^{\alpha}(\tau+\sigma) = z_{\mathrm{L}}^{\alpha} + \frac{\mathrm{i}}{2}\sum_{n=0}^{\infty}\frac{1}{(n+\eta)}\tilde{\beta}_{n+\eta}^{\alpha}\,\mathrm{e}^{-2\mathrm{i}(n+\eta)(\tau+\sigma)}$$

$$-\frac{\mathrm{i}}{2}\sum_{n=1}^{\infty}\frac{1}{(n-\eta)}(\tilde{\gamma}_{n-\eta}^{\alpha})^{\dagger}\,\mathrm{e}^{2\mathrm{i}(n-\eta)(\tau+\sigma)} \tag{10.50}$$

where the oscillators $\tilde{\beta}_{n+\eta}^{\alpha}$ and $\tilde{\gamma}_{n-\eta}^{\alpha}$ have commutation relations

$$[\tilde{\beta}_{m+\eta}^{\alpha}, (\tilde{\beta}_{n+\eta}^{\alpha})^{\dagger}] = \delta_{\alpha\gamma}(n+\eta)\delta_{mn} \tag{10.51}$$

and

$$[\tilde{\gamma}_{m-\eta}^{\alpha}, (\tilde{\gamma}_{n-\eta}^{\beta})^{\dagger}] = \delta_{\alpha\beta}(n-\eta)\delta_{mn}. \tag{10.52}$$

It is important to notice that consistency with the boundary condition (10.46) implies that there can be no momentum p^{α} for degrees of freedom Z^{α} for which the boundary conditions are twisted. Notice also that the boundary condition (10.46) demands that the centre-of-mass coordinate

$$z^{\alpha} = z_{\mathrm{R}}^{\alpha} + z_{\mathrm{L}}^{\alpha} \tag{10.53}$$

of the string satisfies

$$z^{\alpha} = \mathrm{e}^{-2\pi\mathrm{i}\,\eta}z^{\alpha}. \tag{10.54}$$

For the ω twisted sector of the Z_3 orbifold, η is $\frac{2}{3}$ for each value of α, and consistency with the boundary conditions requires that

$$z = \omega z \tag{10.55}$$

up to a lattice vector (since points that differ by a lattice vector are the same point on the torus or orbifold). Thus, we must require that

$$(\omega, \boldsymbol{l})z = z \tag{10.56}$$

for some lattice vector $\boldsymbol{l}$. This is just the fixed point condition (10.32). Thus, in twisted sectors of the string theory the centre-of-mass coordinate is required to be at a fixed point of the corresponding point group element.

When a right-moving fermionic degree of freedom (in the complex basis) $\Psi_{\mathrm{R}}^{\alpha}$ has its boundary conditions twisted by $\mathrm{e}^{-2\pi\mathrm{i}\,\eta}$ with $0<\eta<1$, then, in the Ramond sector,

$$\Psi_{\mathrm{R}}^{\alpha}(\tau, \sigma+\pi) = \mathrm{e}^{-2\pi\mathrm{i}\,\eta}\Psi_{\mathrm{R}}^{\alpha}(\tau, \sigma) \qquad \text{R sector} \tag{10.57}$$

and, in the Neveu–Schwarz sector, where the twist is superimposed upon the underlying Neveu–Schwarz sector boundary condition,

$$\Psi_{\mathrm{R}}^{\alpha}(\tau, \sigma+\pi) = -\,\mathrm{e}^{2\pi\mathrm{i}\,\eta}\Psi_{\mathrm{R}}^{\alpha}(\tau, \sigma) \qquad \text{NS sector}. \tag{10.58}$$

Then, the appropriate mode expansion for the Ramond sector is

$$\Psi_{\mathrm{R}}^{\alpha}(\tau-\sigma)=\sum_{n=1}^{\infty} c_{n-\eta}^{\alpha}\,\mathrm{e}^{-2\mathrm{i}(n-\eta)(\tau-\sigma)}$$
$$+\sum_{n=0}^{\infty}(e_{n+\eta}^{\alpha})^{\dagger}\,\mathrm{e}^{2\mathrm{i}(n+\eta)(\tau-\sigma)} \qquad \text{R sector} \tag{10.59}$$

where the oscillators have anti-commutation relations

$$\{c_{m-\eta}^{\alpha},(c_{n-\eta}^{\beta})^{\dagger}\}=\{e_{m+\eta}^{\alpha},(e_{n+\eta}^{\beta})^{\dagger}\}=\delta_{mn}\delta_{\alpha\beta}. \tag{10.60}$$

For the Neveu–Schwarz sector, the form of mode expansion depends on whether the twist η is less than or greater than $\frac{1}{2}$. The corresponding expansions are

$$\Psi_{\mathrm{R}}^{\alpha}(\tau-\sigma)=\sum_{n=1}^{\infty}(c_{n-\eta-1/2}^{\alpha}\,\mathrm{e}^{-2\mathrm{i}(n-\eta-1/2)(\tau-\sigma)}$$
$$+(e_{n+\eta-1/2}^{\alpha})^{\dagger}\,\mathrm{e}^{2\mathrm{i}(n+\eta-1/2)(\tau-\sigma)})$$
$$\text{NS sector, } 0<\eta<\tfrac{1}{2} \tag{10.61}$$

and

$$\Psi_{\mathrm{R}}^{\alpha}(\tau-\sigma)=\sum_{n=1}^{\infty}(c_{n-\eta+1/2}^{\alpha}\,\mathrm{e}^{-2\mathrm{i}(n-\eta+1/2)(\tau-\sigma)}$$
$$+(e_{n+\eta-3/2}^{\alpha})^{\dagger}\,\mathrm{e}^{2\mathrm{i}(n+\eta-3/2)(\tau-\sigma)})$$
$$\text{NS sector, } \tfrac{1}{2}<\eta<1 \tag{10.62}$$

where

$$\{c_{m-\eta\pm1/2}^{\alpha},(c_{n-\eta\pm1/2}^{\beta})^{\dagger}\}=\{e_{m+\eta-1/2}^{\alpha},(e_{n+\eta-1/2}^{\beta})^{\dagger}\}$$
$$=\{e_{m+\eta-3/2}^{\alpha},(e_{n+\eta-3/2}^{\beta})^{\dagger}\}=\delta_{mn}\delta_{\alpha\beta}. \tag{10.63}$$

In the fermionic formulation of the heterotic string of §9.6, there are also left-moving real fermionic degrees of freedom λ^A, $A=1,\ldots,16$, some of which have twisted boundary conditions when the point group is embedded in the gauge group. (To realize the point group embedding $\Omega=\omega$ on the fermionic degrees of freedom involves rotating these degrees of freedom by Ω. This can be seen from (9.129) where the adjoint representation of the SO(16) subgroup of E_8 for the gauge fields arises from the action of a pair of fermionic oscillators on the vacuum.) It is convenient to assemble these 16 real fermionic degrees of freedom into 8 complex fermionic degrees of freedom, which we shall denote by λ^p, $p=1,\ldots,8$. For a left-moving fermionic degree of freedom λ^p with it boundary conditions twisted by $e^{-2\pi i\eta}$, with $0<\eta<1$, in the Ramond sector,

$$\lambda^p(\tau, \sigma+\pi) = e^{-2\pi i\eta}\lambda^p(\tau, \sigma) \qquad \text{R sector} \tag{10.64}$$

and in the Neveu–Schwarz sector,

$$\lambda^p(\tau, \sigma+\pi) = -e^{-2\pi i\eta}\lambda^p(\tau, \sigma) \qquad \text{NS sector.} \tag{10.65}$$

The corresponding mode expansion for the Ramond sector is

$$\lambda^p(\tau+\sigma) = \sum_{n=0}^{\infty} \rho^p_{n+\eta}\, e^{-2i(n+\eta)(\tau+\sigma)} + \sum_{n=1}^{\infty} (\mu^p_{n-\eta})^\dagger\, e^{2i(n-\eta)(\tau+\sigma)}$$

$$\text{R sector} \tag{10.66}$$

where the oscillators have anti-commutation relations

$$\{\rho^p_{m+\eta}, (\rho^q_{n+\eta})^\dagger\} = \{\mu^p_{m-\eta}, (\mu^q_{n-\eta})^\dagger\} = \delta_{mn}\delta^{pq}. \tag{10.67}$$

For the Neveu-Schwarz sector, the form of the mode expansion depends on whether the twist η is less than or greater than $\frac{1}{2}$. The corresponding expansions are

$$\lambda^p(\tau+\sigma) = \sum_{n=1}^{\infty} \Big(\rho^p_{n+\eta-1/2}\, e^{-2i(n+\eta-1/2)(\tau+\sigma)}$$

$$+ (\mu^p_{n-\eta-1/2})^\dagger\, e^{2i(n-\eta-1/2)(\tau+\sigma)}\Big)$$

$$\text{NS sector, } 0<\eta<\tfrac{1}{2} \tag{10.68}$$

and

$$\lambda^p(\tau+\sigma) = \sum_{n=1}^{\infty} \Big(\rho^p_{n+\eta-3/2}\, e^{-2i(n+\eta-3/2)(\tau+\sigma)}$$

$$+ (\mu^p_{n-\eta+1/2})^\dagger\, e^{2i(n-\eta+1/2)(\tau+\sigma)}\Big)$$

$$\text{NS sector, } \tfrac{1}{2}<\eta<1 \tag{10.69}$$

where the oscillators have anti-commutation relations

$$\{\rho^p_{m+\eta-1/2}, \rho^q_{n+\eta-1/2}\} = \{\rho^p_{m+\eta-3/2}, \rho^q_{n+\eta-3/2}\}$$
$$= \{\mu^p_{m-\eta\pm 1/2}, (\mu^q_{n-\eta\pm 1/2})^\dagger\} = \delta_{mn}\delta^{pq}. \tag{10.70}$$

In the bosonic formulation of the heterotic string, the embedding of the point group in the gauge group is realized quite differently. Since, as discussed in §9.5, bosonic degrees of freedom are converted into complex fermionic degrees of freedom by exponentiation (as, for example, in (9.117)) a twist on the boundary conditions of a complex fermionic degree of freedom becomes a shift on the boundary conditions of the bosonic degree of freedom from which it is derived by fermionization. For the Z_3 orbifold, with the standard embedding of the point group, the twist on the fermionic boundary conditions $\Omega = \omega$, with ω as in (10.27), may be written as

$$\Omega = \exp\left(\frac{2\pi i}{3}(J_{12} + J_{34} + 2J_{56})\right) \tag{10.71}$$

where J_{12}, J_{34} and J_{56} are the generators of the SO(6) rotation group of which the SU(3) in which Ω is defined is a subgroup. From (10.71), the twist Ω on the boundary conditions of three complex fermionic degrees of freedom in the fermionic formulation becomes a shift πV^I on the boundary conditions of the bosonic degrees of freedom in the bosonic formulation, with

$$V^I = (\tfrac{1}{3}\tfrac{1}{3}\tfrac{2}{3}0^5)(0^8)' \tag{10.72}$$

where we have separated the bosonic degrees of freedom into those associated with E_8 and those associated with E_8'. (Notice the factor of 2 in the exponent in (9.117).) The role of p_L^I in (9.59) is now taken by $p_L^I - V^I$, so far as the boundary conditions are concerned, and the momentum lattice for the bosonic degrees of freedom becomes a shifted $E_8 \times E_8'$ lattice, with momenta shifted from the $E_8 \times E_8'$ lattice momenta by V^I.

Mass formulae for the orbifold twisted sector may be derived from the mode expansions following the steps described in Chapters 7 and 8 (Exercise 10.4). The contributions to $\frac{1}{4}M_R^2$ may be decomposed in the form

$$\tfrac{1}{4}M_R^2 = \tfrac{1}{4}M_R^2(\text{B}) + \tfrac{1}{4}M_R^2(\text{F}) - a \tag{10.73}$$

where $M_R^2(\text{B})$ and $M_R^2(\text{F})$ are the contributions of bosonic and fermionic degrees of freedom, and a is the normal-ordering constant. A bosonic degree of freedom Z_R^α with its boundary conditions twisted by $e^{-2\pi i\eta}$, with $0 < \eta < 1$, makes the following contribution to $M_R^2(\text{B})$:

$$\tfrac{1}{4}M_R^2(\text{B}) = \sum_{n=1}^{\infty} (\beta_{n-\eta}^\alpha)^\dagger \beta_{n-\eta}^\alpha + \sum_{n=0}^{\infty} (\gamma_{n+\eta}^\alpha)^\dagger \gamma_{n+\eta}^\alpha . \tag{10.74}$$

In the Ramond, sector, a fermionic degree of freedom with its boundary conditions twisted by $e^{-2\pi i\eta}$ makes the following contribution to $M_R^2(\text{F})$:

$$\tfrac{1}{4}M_R^2(\text{F}) = \sum_{n=1}^{\infty} (n-\eta)(c_{n-\eta}^\alpha)^\dagger c_{n-\eta}^\alpha + \sum_{n=0}^{\infty} (n+\eta)(e_{n+\eta}^\alpha)^\dagger e_{n+\eta}^\alpha$$

$$\text{R sector} \tag{10.75}$$

and in the Neveu–Schwarz sector a fermionic degree of freedom with a twist of $e^{-2\pi i\eta}$ superimposed on the underlying Neveu–Schwarz boundary condition contributes

$$\tfrac{1}{4}M_{\mathrm{R}}^2(\mathrm{F}) = \sum_{n=1}^{\infty}\Big((n-\eta-\tfrac{1}{2})(c^{\alpha}_{n-\eta-1/2})^{\dagger}c^{\alpha}_{n-\eta-1/2}$$
$$+(n+\eta-\tfrac{1}{2})(e^{\alpha}_{n+\eta-1/2})^{\dagger}e^{\alpha}_{n+\eta-1/2}\Big)$$
$$\text{NS sector, } 0<\eta<\tfrac{1}{2} \tag{10.76}$$

or

$$\tfrac{1}{4}M_{\mathrm{R}}^2(\mathrm{F}) = \sum_{n=1}^{\infty}\Big((n-\eta+\tfrac{1}{2})(c^{\alpha}_{n-\eta+1/2})^{\dagger}c^{\alpha}_{n-\eta+1/2}$$
$$+(n+\eta-\tfrac{3}{2})(e^{\alpha}_{n+\eta-3/2})^{\dagger}e^{\alpha}_{n+\eta-3/2}\Big)$$
$$\text{NS sector, } \tfrac{1}{2}<\eta<1. \tag{10.77}$$

As a consequence of the commutators and anti-commutators of the oscillators given above, $(\beta^{\alpha}_{n-\eta})^{\dagger}$, $(\gamma^{\alpha}_{n+\eta})^{\dagger}$, $(c^{\alpha}_{n-\eta})^{\dagger}$, $(e^{\alpha}_{n+\eta})^{\dagger}$, $(c^{\alpha}_{n-\eta-1/2})^{\dagger}$, $(e^{\alpha}_{n+\eta-1/2})^{\dagger}$, $(c^{\alpha}_{n-\eta+1/2})^{\dagger}$ and $(e^{\alpha}_{n+\eta-3/2})^{\dagger}$, acting on a string state, increase the value of $\frac{1}{4}M_{\mathrm{R}}^2$ for the state by $n-\eta$, $n+\eta$, $n-\eta$, $n+\eta$, $n-\eta-\frac{1}{2}$, $n+\eta-\frac{1}{2}$, $n-\eta+\frac{1}{2}$ and $n+\eta-\frac{3}{2}$ respectively.

The normal-ordering constant a in (10.73) may, as in Chapter 7 and Chapter 8, be fixed by using zeta-function regularization. Then a complex bosonic degree of freedom with boundary conditions twisted by $e^{-2\pi i\eta}$, with $0<\eta<1$, contributes

$$a = -\sum_{n=0}^{\infty}(n+\eta) = -\zeta(-1,\eta) = \tfrac{1}{12}-\tfrac{1}{2}\eta(1-\eta) \tag{10.78}$$

where

$$\zeta(z,a) = \sum_{n=0}^{\infty}(n+a)^{-z} \tag{10.79}$$

and a Ramond-sector complex fermionic degree of freedom contributes

$$a = -\tfrac{1}{12}+\tfrac{1}{2}\eta(1-\eta). \tag{10.80}$$

A complex fermionic degree of freedom in the Neveu–Schwarz sector with a twist of $e^{-2\pi i\eta}$, with $0<\eta<1$, superimposed upon the underlying Neveu–Schwarz boundary contribution, may be handled by replacing η by $\eta+\frac{1}{2}$ in (10.78), for $0<\eta<\frac{1}{2}$, and by replacing η by $\eta-\frac{1}{2}$ in (10.78) for $\frac{1}{2}<\eta<1$.

For the ω twisted sector of the Z_3 orbifold, we may construct the scalar super-partners of the fermions in which we are interested (such as quark and lepton generations) by considering the Neveu–Schwarz sector for the right movers. Then, in the light cone gauge, there are two real bosonic and two real fermionic degrees of freedom (or one complex bosonic and one complex fermionic degree of freedom) with untwisted boundary contributions. In addition, there are three complex bosonic degrees of freedom with boundary conditions twisted by $\eta = \frac{2}{3}$, and three complex fermionic degrees of freedom with boundary contributions twisted by $\eta = \frac{1}{6}$ (after subtracting $\frac{1}{2}$ for the underlying Neveu–Schwarz boundary condition). Thus

$$a = \tfrac{1}{12} + \tfrac{1}{24} - \tfrac{3}{36} - \tfrac{3}{72} = 0. \tag{10.81}$$

The massless right mover in the ω twisted Neveu–Schwarz sector is therefore the Neveu–Schwarz ground state $|0\rangle_{\mathrm{R}}$.

Similarly, in the fermionic formulation of the heterotic string, we may decompose $\frac{1}{4}M_{\mathrm{L}}^2$ in the form

$$\tfrac{1}{4}M_{\mathrm{L}}^2 = \tfrac{1}{4}M_{\mathrm{L}}^2(\mathrm{B}) + \tfrac{1}{4}M_{\mathrm{L}}^2(\mathrm{F}) - \tilde{a} \tag{10.82}$$

where $M_{\mathrm{L}}^2(\mathrm{B})$ and $M_{\mathrm{L}}^2(\mathrm{F})$ are the contributions of bosonic and fermionic degrees of freedom and $\tilde{a}$ is the normal-ordering constant. A bosonic degree of freedom Z_{L}^{α} with its boundary contributions twisted by $\mathrm{e}^{-2\pi\mathrm{i}\,\eta}$, with $0 < \eta < 1$, makes the following contribution to $M_{\mathrm{L}}^2(\mathrm{B})$:

$$\tfrac{1}{4}M_{\mathrm{L}}^2(\mathrm{B}) = \sum_{n=0}^{\infty} (\tilde{\beta}_{n+\eta}^{\alpha})^{\dagger}\tilde{\beta}_{n+\eta}^{\alpha} + \sum_{n=1}^{\infty} (\tilde{\gamma}_{n-\eta}^{\alpha})^{\dagger}\tilde{\gamma}_{n-\eta}^{\alpha} \tag{10.83}$$

and the oscillators $(\tilde{\beta}_{n+\eta}^{\alpha})^{\dagger}$ and $(\tilde{\gamma}_{n-\eta}^{\alpha})^{\dagger}$ acting on a string state increase the value of $\frac{1}{4}M_{\mathrm{L}}^2$ by $n + \eta$ and $n - \eta$ respectively. The contributions of the fermionic degrees of freedom in (10.66), (10.68) and (10.69) are analogous to those for the right movers. For a complex bosonic degree of freedom with boundary conditions twisted by $\mathrm{e}^{-2\pi\mathrm{i}\,\eta}$, with $0 < \eta < 1$, the contribution to $\tilde{a}$ is

$$\tilde{a} = \tfrac{1}{12} - \tfrac{1}{2}\eta(1 - \eta) \tag{10.84}$$

and for a Ramond-sector complex fermionic degree of freedom

$$\tilde{a} = -\tfrac{1}{12} + \tfrac{1}{2}\eta(1 - \eta). \tag{10.85}$$

For the Neveu–Schwarz sector the remarks following (10.80) again apply.

A discussion of the massless left movers for the ω twisted sector in the fermionic formation of the heterotic string requires the values for the normal-ordering constants $\tilde{a}(\alpha, \beta)$, with $\alpha, \beta = \mathrm{NS}$ or R, where the index α refers to the oscillators λ^A, $A = 1, \ldots, 16$, associated with E_8, and the index β refers to the oscillators $\bar{\lambda}^A$, $A = 1, \ldots, 16$, associated with E_8', as in §9.6. When the λ^A are assembled into eight complex fermionic degrees of freedom $\lambda^p, p = 1, \ldots, 8$, as in the discussion above, three of these complex degrees of freedom have boundary conditions twisted by $\eta = \frac{2}{3}$ in the Ramond sector, and, after subtracting $\frac{1}{2}$ for the underlying Neveu–Schwarz boundary condition, by $\eta = \frac{1}{6}$ in the Neveu–Schwarz sector. However, none of the $\bar{\lambda}^A$ have twisted boundary conditions because the embedding of the point group is entirely in the E_8 factor of $E_8 \times E_8'$. In addition, in the light cone gauge, there are two real bosonic degrees of freedom (or one complex bosonic degree of freedom) with untwisted boundary conditions, and three complex bosonic degrees of freedom with boundary conditions twisted by $\eta = \frac{2}{3}$. Consequently the normal-ordering constants are (Exercise 10.5)

$$\tilde{a}(\mathrm{R}, \mathrm{R}) = -1 \tag{10.86}$$

$$\tilde{a}(\mathrm{NS}, \mathrm{R}) = -\tfrac{1}{2} \tag{10.87}$$

$$\tilde{a}(\mathrm{NS}, \mathrm{NS}) = \tfrac{1}{2} \tag{10.88}$$

and

$$\tilde{a}(\mathrm{R}, \mathrm{NS}) = 0. \tag{10.89}$$

It follows that there are no massless left movers in the (R, R) or (NS, R) sectors, because of the positive definiteness of the oscillator terms $M_{\mathrm{L}}^2(\mathrm{B})$ and $M_{\mathrm{L}}^2(\mathrm{F})$ in M_{L}^2. In the (NS, NS) sector, if the twisted boundary conditions occur for $A = 11, \ldots, 16$, then we can make massless left movers $\lambda^A_{-1/2}|0\rangle_{\mathrm{L}}$, $A = 1, \ldots, 10$, using the oscillators of (9.120), and these 10 states constitute a **10** of an SO(10) subgroup of E_8. Also, using the degrees of freedom with twisted boundary conditions, we can construct a singlet of SO(10), using the oscillators of (10.69), $(\rho^p_{1/6})^\dagger(\rho^q_{1/6})^\dagger(\rho^r_{1/6})^\dagger|0\rangle_{\mathrm{L}}$, with p, q and r all different. In the (R, NS) sector, there are precisely 10 real fermionic degrees of freedom with (untwisted) periodic boundary conditions, and the massless ground state is then the 16-dimensional spinor **16** of SO(10). Taken together, the massless left movers make up the **27** of E_6, where, decomposed with respect to the SO(10) subgroup,

$$\mathbf{27} = \mathbf{16} + \mathbf{10} + \mathbf{1}. \tag{10.90}$$

In this way, we arrive at a (**27**, **1**) of the surviving gauge group $E_6 \times SU(3)$ of §10.4. There are 27 copies of (**27**, **1**) because there is an ω twisted sector associated with each of the fixed points (10.34).

In the bosonic formulation of the heterotic string, in the light cone gauge, the left movers consist of 24 real bosonic degrees of freedom, or 12 complex bosonic degrees of freedom with 3 of these complex degrees of freedom having boundary conditions twisted by $\eta = \frac{2}{3}$. In addition, the internal bosonic degrees of freedom have boundary conditions shifted by πV^I as in (10.72). (These shifts do not affect the normal-ordering constant.) The normal-ordering constant for the left movers is then

$$\tilde{a} = \tfrac{9}{12} - \tfrac{3}{36} = \tfrac{2}{3}. \tag{10.91}$$

The corresponding mass formula is

$$\tfrac{1}{4}M_L^2 = \frac{1}{2}\sum_{I=1}^{16}(p_L^I + V^I)^2 + \tilde{N} - \tfrac{2}{3} \tag{10.92}$$

where the shift V^I is as in (10.72), and $\tilde{N}$ is the oscillator term with bosonic degrees of freedom with twisted boundary conditions contributing as in (10.83) and bosonic degrees of freedom with untwisted boundary conditions contributing as in (9.74). The eigenvalues of $\tilde{N}$ are third integral and, in view of the positive definiteness of the first term in (10.92), the only possibilities for constructing massless left movers are for $\tilde{N} = 0$ or $\tilde{N} = \frac{1}{3}$. The momentum p_L^I is required to be on the $E_8 \times E_8'$ lattice of §9.4, where the lattice consists of all momenta of the form

$$p_L^I = (n_1, \ldots, n_8) \qquad \text{or} \qquad (n_1 + \tfrac{1}{2}, \ldots, n_8 + \tfrac{1}{2}) \tag{10.93}$$

where the integers n_i are constrained by

$$\sum_i n_i = \text{even integer}. \tag{10.94}$$

The only solutions of $M_L^2 = 0$ are for $\tilde{N} = 0$ and these are

$$p_L^I + V^I = (\tfrac{1}{3}\tfrac{1}{3} - \tfrac{1}{3}\ \underline{\pm 1\ 0^4}) \tag{10.95}$$

$$p_L^I + V^I = (-\tfrac{1}{6} - \tfrac{1}{6}\tfrac{1}{6}\,(\pm\tfrac{1}{2})^5) \tag{10.96}$$

and

$$p_L^I + V^I = (-\tfrac{2}{3} - \tfrac{2}{3}\tfrac{2}{3}\,0^5) \tag{10.97}$$

where in (10.95) the underlining signifies that all five permutations are to be included, and in (10.96) an even number of entries of $\frac{1}{2}$ is required. The states (10.95), (10.96) and (10.97) constitute **10**, **16** and **1** of SO(10) and, as in the fermionic formulation, we find 27 copies of (**27**, **1**) of $E_6 \times SU(3)$, one for each fixed point of ω.

10.6 Wilson lines

The orbifold theory discussed in §§10.3–10.5 has some shortcomings. The observable gauge group is $E_6 \times SU(3)$ which needs to be broken ultimately to $SU(3) \times SU(2) \times U(1)$, and, moreover there are too many generations of quarks and leptons. A mechanism that can be used to achieve some or all of the required gauge symmetry breaking[(1),(2),(3)] and at the same time to modify the matter-field content[(3),(4)] is the introduction of Wilson lines. The mechanism requires the existence in the theory of non-zero quantities U of the form

$$U \sim \exp \oint A_k \,\mathrm{d}x^k \tag{10.98}$$

referred to as Wilson lines, which are a generalization of integrals that occur in the Bohm–Aharonov effect in electrodynamics. In (10.98), $k = 3, \ldots, 8$ runs over compact manifold coordinates, the integral is round some closed loop not contractible to zero on the underlying torus of the orbifold, and the A_k are components of some 10-dimensional gauge field with zero field strength. The quantities (10.98) cannot be gauged to zero by an ordinary gauge transformation, but they can be gauged away by means of a non-single-valued gauge transformation. In this alternative formulation of the theory the Wilson lines (10.98) are no longer present. Instead, the fermionic degrees of freedom realizing the gauge group in the fermionic formulation of the heterotic string acquire extra phases upon a circuit of the torus as a result of the non-single-valued gauge transformation that has been performed. Equivalently, in the bosonic formulation, the bosonic degrees of freedom acquire extra shifts upon a circuit of the torus. As a consequence, the boundary conditions for the twisted sectors of the orbifold are modified, as we now discuss.

For the ω twisted sectors of the Z_3 orbifold associated with the various fixed points of ω, the boundary conditions are twisted by a space group element $(\omega, \boldsymbol{l})$, where, as discussed in §10.3, we may take the lattice vector $\boldsymbol{l}$ to be of the form

$$\boldsymbol{l} = \sum_{p=1}^{3} r_p \boldsymbol{e}_p \tag{10.99}$$

where $r_p = 0, \pm 1$. The basis vectors $\boldsymbol{e}_p$ of the torus lattice constitute single circuits of the torus in various 'directions' and should, according to the above discussion, be associated with additional twists on the boundary conditions of the fermionic degrees of freedom, in the fermionic formulation, or with additional shifts on the boundary conditions of the bosonic degrees of freedom, in the bosonic formulation of the heterotic string. Thus, not only is the point group rotation ω now embedded in the gauge group, but, in the theory in which non-trivial Wilson lines have been gauged away by a non-single-valued transformation, the discrete translations $\boldsymbol{e}_p$ are also embedded in the gauge group. (The lattice basis vectors $\boldsymbol{f}_p$ of (10.26) do *not* have independent Wilson lines because of the action of ω, as in (10.28), relating equivalent paths on the orbifold that are inequivalent on the torus.)

In the bosonic formulation, if the embedding of ω in the gauge group is represented by the shift πV^I on the boundary conditions of the bosonic degrees of freedom, and the embedding of $\boldsymbol{e}_p$ by the shift πa_p^I, then the momenta are shifted from the $E_8 \times E_8'$ lattice momenta p_L^I by $V^I + r_p a_p^I$. Then, the left-mover mass formula for the $(\omega, \boldsymbol{l})$ twisted sector of (10.92) is modified to

$$\tfrac{1}{4}M_L^2 = \tfrac{1}{2}\sum_{I=1}^{16}(p_L^I + V^I + r_\rho a_\rho^I)^2 + \tilde{N} - \tfrac{2}{3} \tag{10.100}$$

where $r_\rho = 0, \pm 1$, for the various fixed points of ω. The spectrum of massless states is now different for different fixed points because of the influence of the embeddings a_ρ^I of the lattice basis vectors $\boldsymbol{e}_\rho$. (The embeddings a_ρ^I themselves are often referred to as the Wilson lines.)

In addition, we must demand in the untwisted sector space group invariance rather than just point group invariance, i.e. invariance of the states under the action of space group elements $(\omega, \boldsymbol{l})$ and their embeddings rather than only under the action of point group elements ω and their embeddings. The extra invariance due to the embedding of $\boldsymbol{l}$ must reflect the fact that πa_ρ^I is a shift on the coordinates x^I and that $\exp(2\pi i\, p_L^I a_\rho^I)$ represents the effect of this shift on a state with momentum p_L^I because it is $2p_L^I$ that generates translations, as discussed in §9.4. Thus, space group invariance imposes the constraint that $\exp(2\pi i\, p_L^I a_\rho^I)$ should be 1, or equivalently

$$p_L^I a_\rho^I = \text{integer}. \tag{10.101}$$

(A similar condition is, in general, required for the twisted sectors where space group invariance is implemented in a slightly more subtle way as a generalized GSO projection. However, in the case of the Z_3 orbifold, all massless states in the twisted sectors survive the generalized GSO projection[5],[6] for arbitrary embedding of the point group and for arbitrary Wilson lines.)

Choices of the embedding of the space group in the gauge group are restricted by the need for the embedding to be a homomorphism and by modular invariance. In general, for space group elements (θ_1, l_1) and (θ_2, l_2)

$$(\theta_1, l_1)(\theta_2, l_2) = (\theta_1\theta_2, l_1 + \theta_1 l_2) \tag{10.102}$$

and consequently for the Z_3 point group generated by ω

$$(\omega, \boldsymbol{l})^3 = (I, \boldsymbol{0}). \tag{10.103}$$

For the embedding of the space group to be homomorphism it is therefore necessary to have

$$3(V^I + r_\rho a_\rho^I) \qquad \text{on an } E_8 \times E_8' \text{ lattice} \tag{10.104}$$

for $r_\rho = 0, \pm 1$. Thus, the embeddings of the point group element ω and the discrete translations $\boldsymbol{e}_\rho$ are constrained by

$$3V^I \qquad \text{on an } E_8 \times E_8' \text{ lattice} \tag{10.105}$$

and

$$3a_\rho^I \qquad \text{on an } E_8 \times E_8' \text{ lattice.} \tag{10.106}$$

Also, the fundamental modular invariance property of a consistent string theory (see Chapter 11) requires, for the Z_3 orbifold, that

$$3(V^I + r_\rho a_\rho^I)^2 = 0 \qquad (\text{mod } 2) \tag{10.107}$$

for $r_\rho = 0, \pm 1$.

A simple example (Exercise 10.6) of a Z_3-orbifold model with a single Wilson line that is consistent with the constraints (10.105)–(10.107) on the embeddings is obtained by taking V^I as in (10.72) and

$$a_1^I = a_2^I = a_3^I = (0^5\,\tfrac{1}{3}\tfrac{1}{3}\tfrac{2}{3})(0^8)'. \tag{10.108}$$

Retaining only space-group-invariant states in the untwisted sector the gauge group is reduced to $[\mathrm{SU}(3)]^4$, which we may interpret as $\mathrm{SU}_c(3) \times \mathrm{SU}_\mathrm{L}(3) \times \mathrm{SU}_\mathrm{R}(3) \times \mathrm{SU}(3)$ where the first three factors of SU(3) came from the E_6 factor of the original $\mathrm{E}_6 \times \mathrm{SU}(3)$ gauge group. The surviving massless matter fields in the untwisted sector constitute nine copies of $(\mathbf{1}, \mathbf{3}, \bar{\mathbf{3}}, \mathbf{1})$ of the $[\mathrm{SU}(3)]^4$ gauge group. The twisted sectors with $\Sigma_\rho r_\rho = 0$ (mod 3) provide nine complete $(\mathbf{27}, \mathbf{1})$ representations of the original $\mathrm{E}_6 \times \mathrm{SU}(3)$ group, which under $[\mathrm{SU}(3)]^4$ decompose as

$$(\mathbf{27}, \mathbf{1}) = (\mathbf{1}, \bar{\mathbf{3}}, \mathbf{3}, \mathbf{1}) + (\mathbf{3}, \mathbf{3}, \mathbf{1}, \mathbf{1}) + (\bar{\mathbf{3}}, \mathbf{1}, \bar{\mathbf{3}}, \mathbf{1}). \tag{10.109}$$

The twisted sectors with $\Sigma_\rho r_\rho = 1$ (mod 3) provide nine copies of

$$(\mathbf{3}, \bar{\mathbf{3}}, \mathbf{1}, \mathbf{1}) + (\mathbf{1}, \mathbf{3}, \mathbf{1}, \bar{\mathbf{3}}) + (\bar{\mathbf{3}}, \mathbf{1}, \mathbf{1}, \mathbf{3}) + 3(\mathbf{1}, \mathbf{1}, \mathbf{3}, \mathbf{1}) \tag{10.110}$$

and the twisted sectors with $\Sigma_\rho r_\rho = 2$ (mod 3) provide nine copies of

$$(\bar{\mathbf{3}}, \mathbf{1}, \mathbf{3}, \mathbf{1}) + (\mathbf{1}, \mathbf{1}, \bar{\mathbf{3}}, \bar{\mathbf{3}}) + (\mathbf{3}, \mathbf{1}, \mathbf{1}, \mathbf{3}) + 3(\mathbf{1}, \bar{\mathbf{3}}, \mathbf{1}, \mathbf{1}). \tag{10.111}$$

It can be seen that there is an exact cancellation of all non-abelian gauge anomalies amongst these surviving massless states, as required for a consistent gauge theory. It is a general result[(7)] that modular invariance of the string theory ensures this gauge anomaly cancellation. It will also be noticed that the twisted sectors in the presence of Wilson lines contain massless states in exotic representations of $\mathrm{SU}_c(3) \times \mathrm{SU}_\mathrm{L}(3) \times \mathrm{SU}_\mathrm{R}(3)$, some of which, with a standard definition of the electric charge Q_em in terms of the generators of the group

$$Q_\mathrm{em} = T_3^\mathrm{L} + T_3^\mathrm{R} + \tfrac{1}{2}Y_\mathrm{L} + \tfrac{1}{2}Y_\mathrm{R} \tag{10.112}$$

will be able to form fractionally charged colour singlets. This is a generic feature[(8)] of string theories. Fully realistic theories will either have to have hidden-sector non-abelian gauge group quantum numbers for these exotic states that confine them, or some spontaneous symmetry-breaking mechanism to provide them with large masses, in view of the tight cosmological bounds on the abundance of fractionally charged states.

10.7 Calabi–Yau manifolds

Throughout this chapter we have been focusing attention on orbifold compactifications of the string theory because for such compactifications the string field equations are the same as in the uncompactified theory. If one is prepared to pay the price of much more difficult field equations it is possible to compactify on a manifold[(9),(10)] rather than an orbifold. To ensure that (at least) $N = 1$ supersymmetry survives in four dimensions it is necessary to choose this six-dimensional manifold to be a manifold of SU(3) holonomy that is also Ricci flat, i.e. for which the Ricci tensor vanishes everywhere. These are the Calabi–Yau manifolds.

A modular invariant theory may be obtained by making an embedding in the gauge group analogous to the standard embedding of the point group in the gauge group for the orbifold case. For the Calabi–Yau compactification the analogous procedure is to identify the expectation value of the gauge fields in the SU(3) factor of an $E_6 \times SU(3)$ subgroup of the E_8 of $E_8 \times E_8'$ with the value of the spin connection for the compact manifold (of SU(3) holonomy).

Non-simply connected Calabi–Yau manifolds may be constructed by quotienting the original simply connected Calabi–Yau manifold by a freely acting (i.e. without fixed points) discrete symmetry group. Then gauge symmetry breaking may be achieved by introducing Wilson lines[(11)], using non-contractible closed loops on the non-simply connected space, in a similar fashion to the procedure in the orbifold case.

Calculation direct from the string theory is now very difficult because of the modification of the string field equations. However, topological methods[(9)] may be used to derive the spectrum of massless states, and topological methods together with discrete symmetries of the manifold may be used to derive some selection rules on Yukawa couplings. More detailed calculations may be carried out by exploiting the connection between Calabi–Yau manifolds and theories constructed using representations of the superconformal algebra (Gepner models)[(12)] for which explicit calculations of correlation functions are possible. (At least a subset of Calabi–Yau manifolds at special points in moduli space, i.e. for special choices of the 'radii' and 'angles' characterizing the manifold, are believed[(13)] to be equivalent to Gepner models.) It is beyond our scope here to pursue either the topological methods employed for Calabi–Yau manifolds, or the construction of Gepner models.

Exercises

10.1 Show that, except for special choices of the lattice and radii of the compact manifold, the massless states for an orbifold have no momentum or winding number on the compact manifold.

10.2 Check the action (10.28) and (10.29) of the point group element ω on the lattice basis vectors.

10.3 Show that the fixed points for the Z_3 orbifold may be written in the form (10.33).

10.4 Derive the mass formulae (10.73)–(10.77) and (10.82)–(10.83) for the orbifold twisted sector.

10.5 Calculate the normal-ordering constants of (10.86)–(10.89).

10.6 Construct the spectrum of massless states for the Z_3-orbifold model with point group embedding (10.72) and Wilson lines (10.108).

References

General references

Dixon L J 1987 Lectures given at the *1987 ICTP Summer Workshop in High Energy Physics and Cosmology*

Dixon L, Harvey J A, Vafa C and Witten E 1985 *Nucl. Phys.* B **261** 678; 1986 *Nucl. Phys.* B **274** 285

Font A, Ibanez L E, Quevedo F and Sierra A 1990 *Nucl. Phys.* B **331** 421

Ross G G 1988 Lectures given at the *Banff Summer Institute on Particles and Fields (Banff, UK, 1988)* (CERN-TH5109/88)

References in the text

1 Dixon L, Harvey J A, Vafa C and Witten E 1985 *Nucl. Phys.* B **261** 678; 1986 *Nucl. Phys.* B **274** 285

2 Bailin D, Love A and Thomas S 1987 *Nucl. Phys.* B **288** 431

3 Ibanez L E, Nilles H P and Quevedo F 1987 *Phys. Lett.* **187B** 25
Ibanez L E, Kim J E, Nilles H P and Quevedo F 1987 *Phys. Lett.* **191B** 3

4 Bailin D, Love A and Thomas S 1987 *Phys. Lett.* **188B** 193; 1987 *Phys. Lett.* **194B** 385

5 Bailin D, Love A and Thomas S 1988 *Mod. Phys. Lett.* A **3** 167

6 Ibanez L E, Jas J, Milles H P and Quevedo F 1988 *Nucl. Phys.* B **301** 157

7 Schellekens A N and Warner N 1986 *Phys. Lett.* **177B** 317
Gross D J and Mende P F 1987 *Nucl. Phys.* B **291** 653

8 Athanasiu G G, Atick J J, Dine M and Fischler W 1988 *Phys. Lett.* **214B** 55
Schellekens A N 1990 *Phys. Lett.* **237B** 363

9 Candelas P, Horowitz G, Strominger A and Witten E 1985 *Nucl. Phys.* B **258** 46

10 Greene B R, Kirklin K H, Miron P J and Ross G G 1986 *Phys. Lett.* **180B** 69; 1987 *Nucl. Phys.* B **292** 606

11 Hosotani Y 1983 *Phys. Lett.* **126B** 309; 1983 *Phys. Lett.* **129B** 193
Witten E 1984 *Phys. Lett.* **126B** 351

12 Gepner D 1988 *Nucl. Phys.* B **296** 757

13 Gepner D 1987 *Phys. Lett.* **199B** 380

11

DIRECT CONSTRUCTION OF FOUR-DIMENSIONAL HETEROTIC STRING THEORIES

11.1 Introduction

In Chapter 10 we saw how it was possible to construct heterotic string theories with only four space-time dimensions by compactifying six of the dimensions of the ten-dimensional heterotic string of Chapter 9 on an orbifold or Calabi–Yau manifold. It is also possible to construct four-dimensional heterotic string theories in a more direct fashion without the intermediate state of a ten-dimensional theory. One approach[1] returns to the original heterotic string, with 10 dimensions for the superstring right movers and 26 dimensions for the bosonic string left movers, bosonizes the fermionic degrees of freedom, other than those associated with four-dimensional space-time, and compactifies all the right- and left-moving bosonic degrees of freedom on a torus, with the exception of those associated with four-dimensional space-time. An alternative, essentially equivalent, approach[2],[3] which we shall pursue here exploits the possibility discussed in Chapter 9 of fermionizing toroidally compactified bosonic degrees of freedom. In this approach, all bosonic degrees of freedom other than the four-dimensional space-time degrees of freedom are replaced by fermionic degrees of freedom. The boundary conditions for all the internal fermionic degrees of freedom (i.e. other than four-dimensional space-time) are then chosen in such a way as to be consistent with the fundamental constraint of modular invariance. In the next two sections we shall describe the way in which modular invariance enters a consistent string theory.

11.2 Modular invariance and partition functions

In string theory, interactions involving strings may be described by world sheet diagrams such as figure 11.1 at tree level and such as figure 11.2 at one-loop level and so forth. (Interacting strings will be discussed in detail in Chapter 12.) The essential subtlety of the one-loop diagrams is contained in the vacuum-to-vacuum amplitude of figure 11.3 which is a toroidal world sheet. It is important not to double count contributions to the path integral for the amplitudes coming from equivalent tori. To discuss this point it is useful to introduce the modular parameter $\tilde{\tau}$ to characterize tori. If we

DOI: 10.1201/9780367805807-11

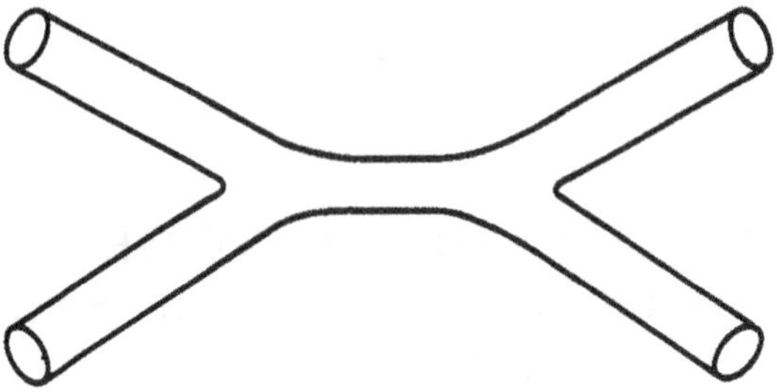

Figure 11.1 The string tree level amplitude.

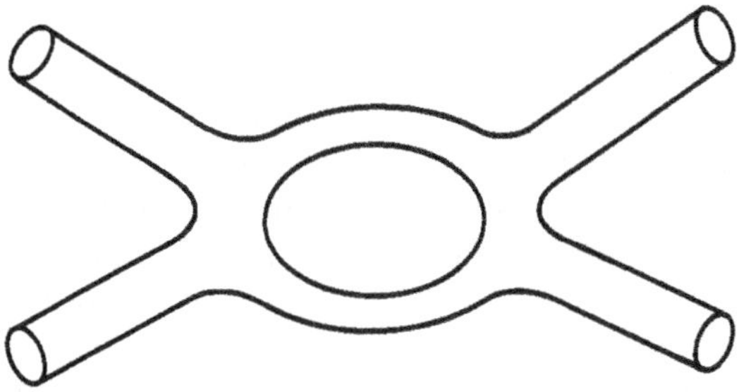

Figure 11.2 The one-loop string amplitude.

combine the world sheet coordinates τ and σ into a single complex coordinate

$$z = \sigma + \mathrm{i}\,\tau \tag{11.1}$$

then we can define a world sheet torus by making the identifications

$$z \equiv z + \pi\lambda_1 \tag{11.2}$$

and

$$z \equiv z + \pi\lambda_2 \tag{11.3}$$

so that

$$z \equiv z + \pi(n_1\lambda_1 + n_2\lambda_2) \tag{11.4}$$

where λ_1 and λ_2 are two complex numbers, and n_1 and n_2 are arbitrary integers. If we wish, the conformal invariance (7.8) may be used to rescale the world sheet metric for the torus so as to scale λ_1 to 1 while leaving the ratio fixed. It is therefore only the ratio

$$\tilde{\tau} = \lambda_2/\lambda_1 \tag{11.5}$$

that is of significance for characterizing tori. Points on the torus may be written as

$$z = \sigma_1\lambda_1 + \sigma_2\lambda_2 \qquad \text{with } 0 \leqslant \sigma_1, \sigma_2 < \pi \tag{11.6}$$

or, after rescaling λ_1 to 1, as

$$z = \sigma_1 + \tilde{\tau}\sigma_2. \tag{11.7}$$

However, not all values of $\tilde{\tau}$ describe inequivalent tori. Consider the so-called modular transformations

$$\begin{pmatrix} \lambda_2' \\ \lambda_1' \end{pmatrix} = \begin{pmatrix} a & b \\ c & d \end{pmatrix} \begin{pmatrix} \lambda_2 \\ \lambda_1 \end{pmatrix} \tag{11.8}$$

where a, b, c and d are integers with

$$ad - bc = 1. \tag{11.9}$$

These transformations induce the transformations on the modular parameter

$$\tilde{\tau}' = \frac{a\tilde{\tau} + b}{c\tilde{\tau} + d} \tag{11.10}$$

and form a group, the 'modular group', SL(2, Z). The inverse transformation to (11.8) is

$$\begin{pmatrix} \lambda_2 \\ \lambda_1 \end{pmatrix} = \begin{pmatrix} d & -b \\ -c & a \end{pmatrix} \begin{pmatrix} \lambda_2' \\ \lambda_1' \end{pmatrix} \tag{11.11}$$

and, consequently,

$$n_1\lambda_1 + n_2\lambda_2 = n_1'\lambda_1' + n_2'\lambda_2' \tag{11.12}$$

where, for arbitrary integers n_1 and n_2, n_1' and n_2' are also arbitrary integers. Then, when λ_1' and λ_2' are related to λ_1 and λ_2 by a modular transformation, they define the same torus, through (11.4). Consequently, tori with modular parameters related by (11.10) are equivalent.

Infinities in one-loop string amplitudes arising from including equivalent tori infinitely many times in the path integral over world sheet metrics may be avoided by restricting the integration to a finite range of modular parameters. With the aid of the transformation (11.10) relating equivalent modular parameters, this range is usually chosen to be

$$-\tfrac{1}{2} \leqslant \operatorname{Re}\tilde{\tau} < \tfrac{1}{2}, \qquad \operatorname{Im}\tilde{\tau} \geqslant 0, \qquad |\tilde{\tau}| \geqslant 1. \tag{11.13}$$

For this way of avoiding infinities to make sense it is necessary that the path integral over string degrees of freedom, which will depend on $\tilde{\tau}$, should be invariant under the modular transformation. (The dependence on $\tilde{\tau}$ will arise from the periodicity of the string degrees of freedom on the torus.) For the vacuum-to-vacuum amplitude what is required is that the so-called partition function

$$Z \sim \int \mathscr{D}X\,\mathscr{D}\Psi\,\exp(-S_{\mathrm{E}}) \tag{11.14}$$

should be modular invariant, where $\mathscr{D}X$ and $\mathscr{D}\Psi$ refer to path integrals over all bosonic and fermionic degrees of freedom of the string and S_{E} is the string action continued in Euclidean space. (The vacuum-to-vacuum amplitude itself involves a final integration over world sheet metrics.)

The evaluation of the partition function Z can be carried out by converting the Euclidean path integral to a determinant[(4)]. The form of the result may be made plausible by comparison with finite-temperature (T) field theory. In field theory at finite temperature [(5)], the path integral for bosonic theories is over fields that are periodic in the 'time' variable with periodicity β ($=1/kT$), and the vacuum-to-vacuum amplitude, which is the partition function, is given by $\mathrm{Tr}(\mathrm{e}^{-\beta\hat{H}})$ where $\hat{H}$ is the Hamiltonian operator. In the present situation, roughly speaking, the modular parameter $\tilde{\tau}$, which specifies the periodicity on the torus, plays the role of β and it is not too surprising that the contribution to the partition function (11.14) of free bosonic degrees of freedom is given by

$$Z = \mathrm{Tr}(q^{H_{\mathrm{L}}})\,\mathrm{Tr}(\bar{q}^{H_{\mathrm{R}}}) = \mathrm{Tr}(q^{H_{\mathrm{L}}}\bar{q}^{H_{\mathrm{R}}}) \tag{11.15}$$

where

$$q = \mathrm{e}^{\mathrm{i}\pi\tilde{\tau}} \tag{11.16}$$

and

$$\bar{q} = \mathrm{e}^{-\mathrm{i}\pi\tilde{\tau}^*}. \tag{11.17}$$

In (11.15), the Hamiltonian of (7.67) has been written as

$$H = H_{\mathrm{L}} + H_{\mathrm{R}} \tag{11.18}$$

where

$$H_{\mathrm{R}} = 2(L_0 - a) \tag{11.19}$$

and

$$H_{\mathrm{L}} = 2(\tilde{L}_0 - \tilde{a}) \tag{11.20}$$

where a and $\tilde{a}$ are the normal-ordering constants for right and left movers, respectively.

In the case of fermionic degrees of freedom, the path integral of finite-temperature field theory is over fields that are anti-periodic in the 'time' variable. We would therefore expect expression (11.15) for the partition function to apply in string theory for free fermionic degrees of freedom in the Neveu–Schwarz sector. To decide what happens for the Ramond sector, it is useful to think more generally about what happens when the boundary conditions for the fermionic degrees of freedom are twisted boundary conditions and to treat the transition from the Ramond to the Neveu–Schwarz sector as the special case of boundary conditions twisted by multiplying by -1.

For a complex fermionic left-moving degree of freedom $\Psi(\sigma_2, \sigma_1)$, the boundary conditions in the σ_1 and σ_2 directions of (11.6) on the torus may be specified by

$$\Psi(\sigma_2,\ \sigma_1 + \pi) = \mathrm{e}^{-2\pi\mathrm{i}\, v}\Psi(\sigma_2, \sigma_1) \tag{11.21}$$

and

$$\Psi(\sigma_2 + \pi, \sigma_1) = \mathrm{e}^{-2\pi\mathrm{i}\, u}\Psi(\sigma_2, \sigma_1) \tag{11.22}$$

with $0 \leqslant u, v \leqslant 1$. Then, the left-mover partition function factor for these boundary conditions takes the form

$$Z_u^v = \mathrm{Tr}(q^{H_{\mathrm{L}}(v)}\, \mathrm{e}^{2\pi\mathrm{i}(1/2 - u)\tilde{N}_{\mathrm{F}}(v)}) = \mathrm{Tr}(q^{H_{\mathrm{L}}(v)}\, (-1)^{\tilde{N}_{\mathrm{F}}(v)}\, \mathrm{e}^{-2\pi\mathrm{i}\, u\tilde{N}_{\mathrm{F}}(v)}) \tag{11.23}$$

where $H_{\mathrm{L}}(v)$ is the Hamiltonian for a complex fermionic left mover with boundary conditions twisted by $\mathrm{e}^{-2\pi\mathrm{i}\, v}$ (for example, $\eta = v$ in (10.64)) and $\tilde{N}_{\mathrm{F}}(v)$ is the fermionic number operator for left movers with these twisted boundary conditions. The $\tilde{N}_{\mathrm{F}}(v)$-dependent factor in (11.23) may be made plausible by the following heuristic argument. First notice that we can transform to a fermionic degree of freedom with anti-periodic boundary conditions in the σ_2 direction by making the change of variables

$$\Psi \to \hat{\Psi} = \mathrm{e}^{2\mathrm{i}(u - 1/2)\sigma_2}\Psi. \tag{11.24}$$

By construction, the effect of this over a period ($0 \leqslant \sigma_2 < \pi$) is $\mathrm{e}^{2\pi\mathrm{i}(u - 1/2)}$ for each occurrence of the fermionic degree of freedom. Starting from an expectation based on (11.15), and the following discussion for the fermionic Neveu–Schwarz sector, that for $\hat{\Psi}$ the contribution to Z would be given by

$$Z = \mathrm{Tr}(q^{H_{\mathrm{L}}(v)})$$

it might then be expected that upon returning to the original string degree of freedom $\hat{\Psi}$ we would have to introduce a factor

$$\mathrm{e}^{-2\pi\mathrm{i}(u - 1/2)\tilde{N}_{\mathrm{F}}(v)}.$$

In particular, if we take the boundary conditions in the σ_1 and σ_2 directions to be either Neveu–Schwarz (NS) or Ramond (R) then for a single left-moving real fermionic degree of freedom we should have

$$Z_{\mathrm{NS}}^{\mathrm{NS}} = \mathrm{Tr}(q^{H_{\mathrm{L}}(\mathrm{NS})}) \tag{11.25}$$

$$Z_{\mathrm{R}}^{\mathrm{NS}} = \mathrm{Tr}((-1)^F q^{H_{\mathrm{L}}(\mathrm{NS})}) \tag{11.26}$$

$$Z_{\mathrm{NS}}^{\mathrm{R}} = \mathrm{Tr}(q^{H_{\mathrm{L}}(\mathrm{R})}) \tag{11.27}$$

and

$$Z_{\mathrm{R}}^{\mathrm{R}} = \mathrm{Tr}((-1)^{F} q^{H_{\mathrm{L}}(\mathrm{R})}) \tag{11.28}$$

where the fermion number operator $\tilde{N}_{\mathrm{F}}(v)$ has been abbreviated to F. The left-mover Hamiltonians for a single real fermionic degree of freedom following from §8.5 with the normal-ordering constants as evaluated in §8.7 are

$$H_{\mathrm{L}}(\mathrm{NS}) = 2 \sum_{r>0,\, r\in\mathbb{Z}+1/2} r\tilde{b}_{-r}\tilde{b}_{r} - \tfrac{1}{24} \tag{11.29}$$

and

$$H_{\mathrm{L}}(\mathrm{R}) = 2 \sum_{n>0,\, n\in\mathbb{Z}} n\tilde{d}_{-n}\tilde{d}_{n} + \tfrac{1}{12}. \tag{11.30}$$

11.3 Partition functions and GSO projections

In §8.8, the GSO projections for the superstring were introduced *ad hoc* in order to delete tachyonic ground states and to obtain equal numbers of bosonic and fermionic degrees of freedom, as required for a space-time supersymmetric theory. In this section, it will be shown that projections of this type can be derived by demanding a modular invariant partition function, which as we saw in §11.2 is necessary to avoid infinities in one-loop string amplitudes.

For simplicity[(6)], we shall consider a group of eight left-moving real fermionic degrees of freedom with all eight degrees of freedom having the same boundary conditions on the world sheet torus. The left-mover Hamiltonian H_{L} following from (11.29) and (11.30) (for a single fermionic degree of freedom) takes the form

$$H_{\mathrm{L}}(\mathrm{NS}) = 2 \sum_{r>0,\, r\in\mathbb{Z}+1/2} r\tilde{b}^{i}_{-r}\tilde{b}^{i}_{r} - \tfrac{1}{3} \tag{11.31}$$

for Neveu–Schwarz boundary conditions, and

$$H_{\mathrm{L}}(\mathrm{R}) = 2 \sum_{n>0,\, n\in\mathbb{Z}} n\tilde{d}^{i}_{-n}\tilde{d}^{i}_{n} + \tfrac{2}{3} \tag{11.32}$$

for Ramond boundary conditions, where the sum over i runs over the eight degrees of freedom.

To construct a modular invariant partition function it is necessary in general to take a linear combination of terms with definite boundary conditions. To explore this, consider the effect of a modular transformation on a fermionic degree of freedom with boundary conditions as in (11.21) and (11.22), which we now write as

$$\Psi(\sigma_2, \sigma_1 + \pi) = h\Psi(\sigma_2, \sigma_1) \tag{11.33}$$

and

$$\Psi(\sigma_2 + \pi, \sigma_1) = g\Psi(\sigma_2, \sigma_1) \tag{11.34}$$

where

$$(h, g) = (\mathrm{e}^{-2\pi \mathrm{i}\, v}, \mathrm{e}^{-2\pi \mathrm{i}\, u}). \tag{11.35}$$

If points on the torus after the modular transformation (11.8) are written as

$$z = \sigma_1'\lambda_1' + \sigma_2'\lambda_2' \qquad \text{with } 0 \leqslant \sigma_1', \sigma_2' < \pi \tag{11.36}$$

the coordinates (σ_2', σ_1') are related to (σ_2, σ_1) by

$$\begin{pmatrix} \sigma_2' \\ \sigma_1' \end{pmatrix} = \begin{pmatrix} d & -c \\ -b & a \end{pmatrix} \begin{pmatrix} \sigma_2 \\ \sigma_1 \end{pmatrix}. \tag{11.37}$$

In terms of the new coordinates (σ_2', σ_1') the boundary conditions for $\Psi(\sigma_2', \sigma_1')$ may be written as

$$\Psi(\sigma_2', \sigma_1' + \pi) = h'\Psi(\sigma_2', \sigma_1') \tag{11.38}$$

and

$$\Psi(\sigma_2' + \pi, \sigma_1') = g'\Psi(\sigma_2', \sigma_1') \tag{11.39}$$

with

$$(h', g') = (h^d g^c, h^b g^a). \tag{11.40}$$

Put another way, if we write

$$(h', g') = (\mathrm{e}^{-2\pi \mathrm{i}\, v'}, \mathrm{e}^{-2\pi \mathrm{i}\, u'}) \tag{11.41}$$

the connection between the boundary conditions for Ψ on the torus defined by $\tilde{\tau}$ and the boundary conditions for Ψ on the torus defined by $\tilde{\tau}'$ of (11.10) is

$$\begin{pmatrix} u' \\ v' \end{pmatrix} = \begin{pmatrix} \mathrm{a} & \mathrm{b} \\ \mathrm{c} & \mathrm{d} \end{pmatrix} \begin{pmatrix} \mathrm{u} \\ v \end{pmatrix}. \tag{11.42}$$

Thus, the partition function contribution Z_u^v on the torus defined by $\tilde{\tau}$ is mapped to the partition function contribution $Z_{\mathrm{u}'}^{v'}$ on the torus defined by $\tilde{\tau}'$. Since, in general, the boundary conditions (v', u') differ from the boundary conditions (v, u), a modular invariant partition function must be constructed as a linear combination of terms of the type Z_u^v.

For the present case, if we consider the modular transformation

$$\tilde{\tau}' = -1/\tilde{\tau} \tag{11.43}$$

then the mapping between partition function contributions on the torus defined by $\tilde{\tau}$ and the torus defined by $\tilde{\tau}'$ is

$$Z^{1/2}_{1/2} \rightarrow Z^{1/2}_{1/2} \qquad Z^{1/2}_{0} \rightarrow Z^{0}_{1/2} \qquad Z^{0}_{1/2} \rightarrow Z^{1/2}_{0} \qquad Z^{0}_{0} \rightarrow Z^{0}_{0} \tag{11.44}$$

up to possible phase factors. For the modular transformation

$$\tilde{\tau}' = \frac{\tilde{\tau}}{\tilde{\tau}+1} \tag{11.45}$$

the corresponding mapping is

$$Z^{1/2}_{1/2} \rightarrow Z^{0}_{1/2} \qquad Z^{1/2}_{0} \rightarrow Z^{1/2}_{0} \qquad Z^{0}_{1/2} \rightarrow Z^{1/2}_{1/2} \qquad Z^{0}_{0} \rightarrow Z^{0}_{0} \tag{11.46}$$

up to possible phase factors. To determine the phase factors involved in these transformations we next construct the partition function terms explicitly.

Using the left-mover Hamiltonians (11.29) and (11.30), together with (11.25)–(11.28), and remembering that Z^{v}_{u} as defined in (11.23) is for a single *complex* fermionic degree of freedom, we find that (Exercise 11.1)

$$Z^{1/2}_{1/2} = q^{-1/12} \prod_{n=1}^{\infty} (1 + q^{2n-1})^2 \tag{11.47}$$

$$Z^{1/2}_{0} = q^{-1/12} \prod_{n=1}^{\infty} (1 - q^{2n-1})^2 \tag{11.48}$$

$$Z^{0}_{1/2} = 2q^{1/6} \prod_{n=1}^{\infty} (1 + q^{2n})^2 \tag{11.49}$$

and

$$Z^{0}_{0} = 0. \tag{11.50}$$

The factor of 2 in (11.49) and the vanishing of Z^{0}_{0} derive from the contributions of the two possible Ramond ground states for each pair of real fermions discussed in §8.8. For the Neveu–Schwarz and Ramond sectors under discussion here with eight real fermionic degrees of freedom or equivalently four complex fermionic degrees of freedom, the relevant partition function contributions are

$$Z^{\mathrm{NS}}_{\mathrm{NS}} = [Z^{1/2}_{1/2}]^4 \tag{11.51}$$

$$Z^{\mathrm{NS}}_{\mathrm{R}} = [Z^{1/2}_{0}]^4 \tag{11.52}$$

$$Z^{\mathrm{R}}_{\mathrm{NS}} = [Z^{0}_{1/2}]^4 \tag{11.53}$$

and

$$Z_{\mathrm{R}}^{\mathrm{R}} = [Z_0^0]^4. \tag{11.54}$$

The infinite products in (11.47)–(11.49) are special cases of the generalized θ-functions[7],

$$\theta\left(\begin{pmatrix}u\\v\end{pmatrix}, \bar{\tau}\right) = q^{v^2 - v + 1/6}\, \mathrm{e}^{\pi\mathrm{i}\, u(1-v)}$$

$$\times \prod_{n=1}^{\infty} (1 - q^{2(n-v)}\, \mathrm{e}^{2\pi\mathrm{i}\, u})(1 - q^{2(n+v-1)}\, \mathrm{e}^{-2\pi\mathrm{i}\, u}) \tag{11.55}$$

and in general (Exercise 11.2)

$$Z_u^v = \mathrm{e}^{-\pi\mathrm{i}\, u(1-v)}\, \theta\left(\begin{pmatrix}u\\v\end{pmatrix}, \bar{\tau}\right). \tag{11.56}$$

These generalized θ-functions possess the useful shift properties:

$$\theta\left(\begin{pmatrix}u+1\\v\end{pmatrix}, \bar{\tau}\right) = -\,\mathrm{e}^{-\pi\mathrm{i}\, v}\, \theta\left(\begin{pmatrix}u\\v\end{pmatrix}, \bar{\tau}\right) \tag{11.57}$$

and

$$\theta\left(\begin{pmatrix}u\\v+1\end{pmatrix}, \bar{\tau}\right) = -\,\mathrm{e}^{\pi\mathrm{i}\, u}\, \theta\left(\begin{pmatrix}u\\v\end{pmatrix}, \bar{\tau}\right). \tag{11.58}$$

They transform under modular transformations (11.10) as

$$\theta\left(\alpha\begin{pmatrix}u\\v\end{pmatrix}, \bar{\tau}\right) = \varepsilon_\alpha \theta\left(\begin{pmatrix}u\\v\end{pmatrix}, \alpha^{-1}\bar{\tau}\right) \tag{11.59}$$

where ε_α is a twelfth root of unity independent of u and v,

$$\alpha\colon \bar{\tau} \to \frac{a\bar{\tau} + b}{c\bar{\tau} + d} \tag{11.60}$$

and the action of α on

$$\begin{pmatrix}u\\v\end{pmatrix}$$

is

$$\alpha\begin{pmatrix}u\\v\end{pmatrix} = \begin{pmatrix}a & b\\c & d\end{pmatrix}\begin{pmatrix}u\\v\end{pmatrix}. \tag{11.61}$$

For the present case,

$$Z_{\mathrm{NS}}^{\mathrm{NS}} = -\left[\theta\left(\begin{pmatrix}\frac{1}{2}\\ \frac{1}{2}\end{pmatrix}, \bar{\tau}\right)\right]^4 \tag{11.62}$$

$$Z_{\mathrm{R}}^{\mathrm{NS}} = \left[\theta\left(\begin{pmatrix}0\\ \frac{1}{2}\end{pmatrix}, \bar{\tau}\right)\right]^4 \tag{11.63}$$

$$Z_{\mathrm{NS}}^{\mathrm{R}} = \left[\theta\left(\begin{pmatrix}\frac{1}{2}\\ 0\end{pmatrix}, \bar{\tau}\right)\right]^4 \tag{11.64}$$

and

$$Z_{\mathrm{R}}^{\mathrm{R}} = 0 \tag{11.65}$$

where we have used the identity

$$\prod_{n=1}^{\infty} (1 + q^{2n})(1 + q^{2(n-1)}) = 2 \prod_{n=1}^{\infty} (1 + q^{2n})^2$$

$$= \frac{1}{2} \prod_{n=1}^{\infty} (1 + q^{2(n-1)})^2. \tag{11.66}$$

We may now construct a modular invariant partition function by taking a suitable superposition of these terms

$$Z = \eta_{\mathrm{NS}}^{\mathrm{NS}} Z_{\mathrm{NS}}^{\mathrm{NS}} + \eta_{\mathrm{R}}^{\mathrm{NS}} Z_{\mathrm{R}}^{\mathrm{NS}} + \eta_{\mathrm{NS}}^{\mathrm{R}} Z_{\mathrm{NS}}^{\mathrm{R}} + \eta_{\mathrm{R}}^{\mathrm{R}} Z_{\mathrm{R}}^{\mathrm{R}}. \tag{11.67}$$

Relationships between the coefficients η required for modular invariance may be obtained by applying the transformation property (11.59). In the present discussion, employing only a group of eight left-moving real fermionic degrees of freedom, we should disregard the factors ε_α, which cancel between left and right movers in a complete model. Then, using the modular transformation

$$\bar{\tau} \rightarrow \bar{\tau}' = \bar{\tau}/(\bar{\tau} + 1) \tag{11.68}$$

we find that

$$Z_{\mathrm{NS}}^{\mathrm{NS}}(\bar{\tau}/(\bar{\tau} + 1)) = -Z_{\mathrm{NS}}^{\mathrm{R}}(\bar{\tau}) \tag{11.69}$$

and

$$Z_{\mathrm{NS}}^{\mathrm{R}}(\bar{\tau}/(\bar{\tau} + 1)) = -Z_{\mathrm{NS}}^{\mathrm{NS}}(\bar{\tau}) \tag{11.70}$$

and modular invariance of Z requires

$$\eta_{\mathrm{NS}}^{\mathrm{R}} = -\eta_{\mathrm{NS}}^{\mathrm{NS}}. \tag{11.71}$$

Also, using the modular transformation

$$\bar{\tau} \rightarrow \bar{\tau}' = -1/\bar{\tau} \tag{11.72}$$

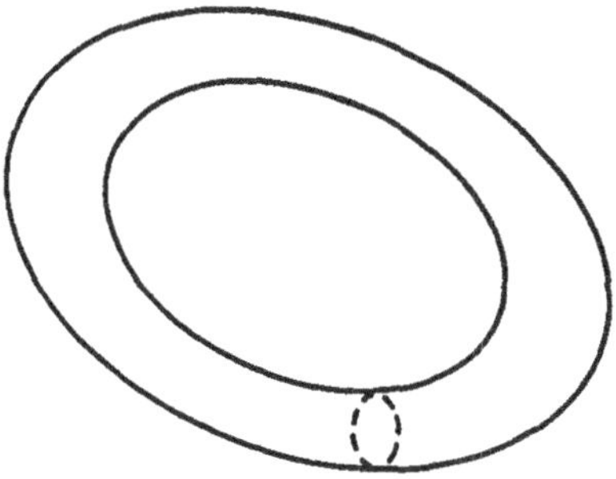

Figure 11.3 The vacuum-to-vacuum string amplitude.

we find that

$$Z_{\mathrm{R}}^{\mathrm{NS}}(-1/\bar{\tau}) = Z_{\mathrm{NS}}^{\mathrm{R}}(\bar{\tau}) \qquad (11.73)$$

and

$$Z_{\mathrm{NS}}^{\mathrm{R}}(-1/\bar{\tau}) = Z_{\mathrm{R}}^{\mathrm{NS}}(\bar{\tau}) \qquad (11.74)$$

and modular invariance of Z requires

$$\eta_{\mathrm{R}}^{\mathrm{NS}} = \eta_{\mathrm{NS}}^{\mathrm{R}}. \qquad (11.75)$$

Thus, the modular invariant partition function takes the form

$$(2\eta_{\mathrm{NS}}^{\mathrm{NS}})^{-1} Z = \mathrm{Tr}\left(\frac{(1+(-1)^{F+1})}{2} q^{H_{\mathrm{L}}(\mathrm{NS})}\right) - \mathrm{Tr}\left(\frac{(1+\eta(-1)^{F+1})}{2} q^{H_{\mathrm{L}}(\mathrm{R})}\right) \qquad (11.76)$$

where

$$\eta = \eta_{\mathrm{R}}^{\mathrm{R}}/\eta_{\mathrm{NS}}^{\mathrm{NS}}. \qquad (11.77)$$

Remembering that the partition function is essentially the vacuum-to-vacuum amplitude of the theory of figure 11.3, equation (11.76) implies projections (GSO projections) on the string states. For the Neveu–Schwarz sector, the GSO projection P is

$$P = \frac{1+(-1)^{F+1}}{2} \qquad \text{NS sector.} \qquad (11.78)$$

In agreement with (8.155). For the Ramond sector, the projection is

$$P = \frac{1+\eta(-1)^{F+1}}{2} \qquad \text{R sector} \qquad (11.79)$$

in agreement with (8.157). The coefficient η is not determined by modular invariance of the partition function.

However, consideration of unitarity shows that η should be ± 1. All this discussion has been for a group of eight real fermionic left movers. If instead we consider a group of sixteen real fermionic left movers, as appropriate for the internal degrees of freedom in the fermionic formulation of the heterotic string, the projection (11.78) is modified to $(1 + (-1)^F)/2$ in agreement with (9.134).

Strictly, we should be considering right and left movers simultaneously to impose modular invariance on the complete partition function of (11.15). For a complex fermionic right mover with boundary conditions twisted by $e^{-2\pi i\, v}$ and $e^{-2\pi i\, u}$ in the σ_1 and σ_2 directions on the world sheet torus, respectively, the partition function contribution is instead $\bar{Z}^v_u$, consistently with the replacement of q by $\bar{q}$ in (11.15). It is not difficult (Exercise 11.3) to repeat the arguments employed here retaining all right- and left-moving fermionic degrees of freedom of the superstring to arrive at the GSO projections (11.78) and (11.79), which now apply separately to left and right movers. As in §8.8, type-IIB and type-IIA superstring theories are possible depending on whether η in (11.79) has the same or opposite values for left and right movers. Thus, the GSO projections that were introduced *ad hoc* in §8.8 can now be derived from the fundamental requirement of modular invariance of the theory.

In the next section, we shall discuss the way in which the requirement of modular invariance can be used in the construction of consistent four-dimensional heterotic string theories.

11.4 Four-dimensional heterotic string theories

In this section, we shall develop the fermionic construction of four-dimensional heterotic string theories described in the introduction where all bosonic degrees of freedom other than those associated with four-dimensional space-time are fermionized[(2),(3)]. Before proceeding to the constraints imposed on such theories by modular invariance, we shall first establish notations for the boundary conditions for the fermionic degrees of freedom, and shall also consider the requirements for world sheet supersymmetry.

When all bosonic degrees of freedom other than those associated with four-dimensional space-time are fermionized the complete set of degrees of freedom of the heterotic string, in the light cone gauge, is as follows. Associated with four-dimensional space-time there are the bosonic degrees of freedom X^i, and the real fermionic right movers Ψ^i_R, $i = 1, 2$. It will be convenient to group these two real fermionic degrees of freedom into a single complex degree of freedom, which we denote by η. There are 18 real fermionic right-mover internal degrees of freedom λ^k_{1R}, λ^k_{2R} and λ^k_{3R}, $k = 3, \ldots, 8$, because for each of the 'original' real fermionic degrees of freedom there are two more arising from the fermionization of the corresponding bosonic degree of freedom.

We shall assume in what follows that these real fermionic degrees of freedom occur in pairs of triplets with the same boundary conditions, so that we can assemble the six triplets of real fermionic degrees of freedom into three triplets of complex fermionic degrees, which we denote by Λ^a_{1R}, Λ^a_{2R} and Λ^a_{3R}, $a = 1, 2, 3$. Finally, there are 44 real fermionic left-moving degrees of freedom, which we assume can be assembled into 22 complex fermionic degrees of freedom which we denote by ξ^A_L, $A = 1, \ldots, 22$. The action for all these string degrees of freedom is

$$S = \frac{1}{2\pi}\int d^2\sigma\,[\partial_\tau X^i\,\partial_\tau X^i - \partial_\sigma X^i\,\partial_\sigma X^i + 2i(\bar{\eta}\,\partial_+\eta + \text{HC}) + 2i(\bar{\Lambda}^a_{1R}\,\partial_+\Lambda^a_{1R} + \bar{\Lambda}^a_{2R}\,\partial_+\Lambda^a_{2R} + \bar{\Lambda}^a_{3R}\,\partial_+\Lambda^a_{3R} + \text{HC}) + 2i(\bar{\xi}^A_L\,\partial_-\xi^A_L + \text{HC})] \tag{11.80}$$

where

$$\partial_\pm = \tfrac{1}{2}(\partial_\tau \pm \partial_\sigma). \tag{11.81}$$

The fermionic degrees of freedom, both left and right movers, will be denoted collectively by $\Psi^l(\tau, \sigma)$, and twisted-sector boundary conditions will be written in the form

$$\Psi^l(\tau, \sigma + \pi) = e^{-2\pi i\, W^l}\Psi^l(\tau, \sigma) \tag{11.82}$$

with

$$0 \leqslant W^l < 1. \tag{11.83}$$

It is convenient to split the vector $\boldsymbol{W}$ of boundary conditions W^l into right- and left-mover boundary conditions $\boldsymbol{W}_R$ and $\boldsymbol{W}_L$ by writing

$$\boldsymbol{W} = (\boldsymbol{W}_R | \boldsymbol{W}_L). \tag{11.84}$$

It is also convenient to write $\boldsymbol{W}_R$ in a way that displays the triplets of fermionic degrees of freedom so that

$$\boldsymbol{W} = (s(a_1 b_1 c_1)\,(a_2 b_2 c_2)\,(a_3 b_3 c_3) | \boldsymbol{W}_L) \tag{11.85}$$

where the twists are for the degrees of freedom

$$(\eta(\Lambda^1_{1R}\Lambda^1_{2R}\Lambda^1_{3R})\,(\Lambda^2_{1R}\Lambda^2_{2R}\Lambda^2_{3R})\,(\Lambda^3_{1R}\Lambda^3_{2R}\Lambda^3_{3R}) | \xi^A_L).$$

It is important, for the elimination of ghosts from the theory to be possible, that the right movers (which derive from the superstring) should possess world sheet supersymmetry (unspoiled by the process of fermionization). With all the bosonic degrees of freedom other than those associated with four-dimensional space-time fermionized, the world sheet supersymmetry has to be realized in a non-linear way. To discover the form of the supersymmetry transformations in these circumstances it is necessary to find transformations that leave the action invariant and such that the commutator of two such transformations is a translation, as appropriate for the supersymmetry algebra. This program has been carried[8] through with the resulting world sheet supercurrent T_F (which is the analogue of (8.11))

$$T_F = 2\mathrm{i}\,\Psi^i_{\mathrm{R}}\,\partial_+ X^i + \mathrm{i}\sum_k \lambda^k_{1\mathrm{R}}\lambda^k_{2\mathrm{R}}\lambda^k_{3\mathrm{R}}. \tag{11.86}$$

For this supercurrent to be well defined, all terms in T_F must have the same boundary conditions, and so the boundary conditions for the product of $\lambda^k_{1\mathrm{R}}$, $\lambda^k_{2\mathrm{R}}$ and $\lambda^k_{3\mathrm{R}}$ must be the same as those for Ψ^i_{R}. If, as assumed above, the triplets $\lambda^k_{1\mathrm{R}}, \lambda^k_{2\mathrm{R}}$ and $\lambda^k_{3\mathrm{R}}$ of real fermionic degrees of freedom are assembled into triplets $\Lambda^a_{1\mathrm{R}}, \Lambda^a_{2\mathrm{R}}$ and $\Lambda^a_{3\mathrm{R}}$ of complex fermionic degrees of freedom, the resulting condition on the boundary conditions, referred to as the triplet constraint, is

$$a_a + b_a + c_a = s \,(\mathrm{mod}\ 1) \tag{11.87}$$

where a_a, b_a and c_a and s are half-integers or integers.

To construct a modular invariant partition function it is necessary to take a superposition of terms with the various allowed boundary conditions in the σ_1 and σ_2 directions on the world sheet torus for the vacuum-to-vacuum amplitude. In the present case, the generalization of (11.23) to include all left- and right-moving fermionic degrees of freedom is

$$Z^{\boldsymbol{W}}_{\boldsymbol{W}'} = \mathrm{Tr}(q^{H_{\mathrm{L}}(\boldsymbol{W})}\bar{q}^{H_{\mathrm{R}}(\boldsymbol{W})}\, \mathrm{e}^{2\pi\mathrm{i}(\boldsymbol{W}_0 - \boldsymbol{W}')\cdot\hat{\boldsymbol{N}}(\boldsymbol{W})}) \tag{11.88}$$

where $\boldsymbol{W}_0$ is defined to be

$$\boldsymbol{W}_0 = ((\tfrac{1}{2})^{10}|(\tfrac{1}{2})^{22}). \tag{11.89}$$

Scalar products are defined to have a minus sign for right movers so that, for example,

$$\boldsymbol{W}'\cdot\hat{\boldsymbol{N}} \equiv \boldsymbol{W}'_{\mathrm{L}}\cdot\hat{\boldsymbol{N}}_{\mathrm{L}} - \boldsymbol{W}'_{\mathrm{R}}\cdot\hat{\boldsymbol{N}}_{\mathrm{R}} \tag{11.90}$$

and

$$\hat{\boldsymbol{N}}(\boldsymbol{W}) = (\hat{\boldsymbol{N}}_{\mathrm{R}}(\boldsymbol{W})|\hat{\boldsymbol{N}}_{\mathrm{L}}(\boldsymbol{W})) \tag{11.91}$$

is the vector of fermionic number operators for right and left movers with boundary conditions twisted by $\boldsymbol{W}$.

$Z^{\boldsymbol{W}}_{\boldsymbol{W}'}$ factorizes as a product of partition function factors and their conjugates of the type (11.23) for the 22 complex left movers and 10 complex right movers.

$$Z_{W'}^{W} = \prod_{l=1}^{10} \bar{Z}_{W'^l}^{W^l} \prod_{l=11}^{32} Z_{W'^l}^{W^l}. \tag{11.92}$$

Now, a modular invariant partition function must be constructed as a superposition of the type

$$Z = \sum_{W, W'} C_{W'}^{W} Z_{W'}^{W} \tag{11.93}$$

for some coefficients $C_{W'}^{W}$, where the sum is over all allowed boundary conditions in the σ_1 and σ_2 directions.

Using the general expression (11.56) for the factors in $Z_{W'}^{W}$ in terms of generalized θ-functions and the modular transformation property (11.59), conditions for (11.93) to be modular invariant may be derived[(2),(3),(9)]. These conditions fall into two general types. First, when a modular transformation maps $Z_{W'}^{W}$ to itself, modular invariance requires that no phase factor should arise from the transformation. (This occurs when $(h', g') = (h, g)$ in (11.40).) This restricts the choices of boundary conditions W in a consistent theory. Second, when a modular transformation maps partition function terms $C_{W'}^{W} Z_{W'}^{W}$ into other such terms, relationships amongst the $C_{W'}^{W}$ are required for modular invariance. In the way discussed in §11.3, summing over σ_2 boundary conditions W for fixed σ_1 boundary conditions W' leads to generalized GSO projections.

The somewhat lengthy calculations involved are to be found given in some detail in the original literature[(2),(3)], to which we refer the reader. The results as regards the allowed boundary conditions in a modular invariant theory may be summarized as follows. All allowed boundary conditions can be written as linear combinations of a set of 'basis vectors' $\{W_i\}$, so the most general boundary conditions W are

$$W = \sum_i \overline{\alpha_i W_i}. \tag{11.94}$$

If the order m_i of W_i is defined to the smallest integer such that the components of $m_i W_i$ (no summation on i) are integers, then the α_i are integers taking values in the range 0 to $m_i - 1$. The bar over the top of $\alpha_i W_i$ means that one is to take the fractional part of its components so that the components of W satisfy (11.83). Basis vectors W_i consistent with modular invariance satisfy

$$m_i \sum_l W_{iL}^l = 0 \pmod 2 \tag{11.95}$$

$$m_i \sum_l W_{iR}^l = 0 \pmod 2 \tag{11.96}$$

where W has been separated into right- and left-mover boundary conditions as in (11.84),

$$m_i W_i \cdot W_i = 0 \pmod{2} \qquad m_i \text{ even} \tag{11.97}$$

$$m_i W_i \cdot W_i = 0 \pmod{1} \qquad m_i \text{ odd} \tag{11.98}$$

(no sum on i implied), and, for $i \neq j$,

$$m_{ij} W_i \cdot W_j = 0 \pmod{1} \qquad m_{ij} \text{ even} \tag{11.99}$$

and

$$2m_{ij} W_i \cdot W_j = 0 \pmod{1} \qquad m_{ij} \text{ odd} \tag{11.100}$$

where m_{ij} is the least common multiple of m_i and m_j, and the scalar products are again defined in the fashion of (11.90).

Once a set of basis vectors W_i has been chosen, all sectors of the theory corresponding to distinct choices of boundary conditions are obtained from (11.94). However, the theory is not completely specified until the generalized GSO projections have been determined. Modular invariance tightly constrains these projections for any choice of the W_i but there is still some freedom, which may be parametrized by certain parameters k_{ij} which are subject to the conditions

$$m_i k_{ij} = 0 \pmod{1} \tag{11.101}$$

$$k_{ij} + k_{ji} = W_i \cdot W_j \pmod{1} \tag{11.102}$$

and

$$k_{ii} + k_{i0} + s_i - \tfrac{1}{2} W_i \cdot W_i = 0 \pmod{1} \tag{11.103}$$

where s_i is the first entry of W_i as in (11.85), and no sum on i or j is implied in (11.101) or (11.103). The generalized GSO projections for the W twisted sector, with W given by (11.94), may be cast in the form

$$W_i \cdot N(W) = s_i - W_i \cdot W + \sum_j k_{ij} \alpha_j + k_{0i} \pmod{1} \tag{11.104}$$

where

$$N(W) = (N_{\mathrm{R}}(W) | N_{\mathrm{L}}(W)) \tag{11.105}$$

is the vector of eigenvalues of the fermionic number operator $\hat{N}(W)$ for the sector with boundary conditions twisted by W, and the scalar products are again defined as in (11.90).

There is an important subtlety when some of the fermionic degrees of freedom have Ramond boundary conditions. Then, $N(W)$ contains a contribution from the zero modes e_α of the subset Ψ^α with Ramond boundary conditions of the complete set of fermionic degrees of freedom (right- and left-moving) Ψ^l. These zero modes create spinor ground states

$$|n_\alpha\rangle = \prod_\alpha (e_\alpha^\dagger)^{n_\alpha}|0\rangle \qquad n_\alpha = 0 \text{ or } 1 \tag{11.106}$$

as in (8.159), and the contribution N_0 to $\Sigma_\alpha N^\alpha(W)$ is

$$N_0 = \sum_\alpha n_\alpha. \tag{11.107}$$

When $\Sigma_\alpha N^\alpha(W)$ is multiplied by $\frac{1}{2}$ in the generalized GSO projection (11.104) (as can occur when W_i has entries of $\frac{1}{2}$ in appropriate positions), the zero-mode contribution to (11.104) only depends on whether N_0 is even or odd, and this amounts to the chirality of the spinor ground state (for both right and left movers together) entering the projection, since the chirality of the ground state is given by

$$\chi = (-1)^{N_0} \tag{11.108}$$

as a consequence of (8.164).

11.5 Semi-realistic four-dimensional models

We shall now construct some examples[(2),(3)] of four-dimensional heterotic string theories using the approach of the previous section. The simplest example to be studied employs just two basis vectors w_0 and w_1 with w_0 (as before) given by

$$W_0 = ((\tfrac{1}{2})^{10}|(\tfrac{1}{2})^{22}) \tag{11.109}$$

and W_1 given by

$$W_1 = (0\,(0\tfrac{1}{2}\tfrac{1}{2})\,(0\tfrac{1}{2}\tfrac{1}{2})\,(0\tfrac{1}{2}\tfrac{1}{2})|(\tfrac{1}{2})^{22}). \tag{11.110}$$

(This choice of basis vectors is consistent with the modular invariance and world sheet supersymmetry constraints of §11.4.) Since both basis vectors have order two, the model has four sectors given by

$$W = \mathbf{0} \qquad W_0 \qquad W_1 \qquad W_0 + W_1. \tag{11.111}$$

The normal-ordering constants a_R and a_L for the $\mathbf{0}$ sector and $W_0 + W_1$ sector are negative and consequently these sectors contain no massless states. The normal-ordering constants for the other sectors are

$$(a_R, a_L) = (\tfrac{1}{2}, 1) \qquad \text{for the } W_0 \text{ sector}$$

$$(a_R, a_L) = (0, 1) \qquad \text{for the } W_1 \text{ sector}. \tag{11.112}$$

The parameters k_{ij} that enter the generalized GSO projections are restricted by (11.101)–(11.103) to take the values

$$k_{00} = 0 \quad \text{or} \quad \tfrac{1}{2} \pmod 1 \tag{11.113}$$

$$k_{11} = 0 \quad \text{or} \quad \tfrac{1}{2} \pmod 1 \tag{11.114}$$

and

$$k_{01} = k_{10} = k_{11} \pmod 1. \tag{11.115}$$

There are 32 complex fermionic degrees of freedom (both right and left movers) Ψ^l, $l = 1, \ldots, 32$, and there are transverse four-dimensional space-time real bosonic degrees of freedom X^j_{R} and X^j_{L}, $j = 1, 2$, whose oscillators can be used in the construction of string states. Let us consider the $\boldsymbol{W}_0$ sector first. Adopting the notations of (9.101) and (9.109) for complex fermionic degrees of freedom, with obvious modifications for right movers, the massless right movers are

$$(f^l_{1/2})^\dagger|0\rangle_{\mathrm{R}} \quad \text{and} \quad (g^l_{1/2})^\dagger|0\rangle_{\mathrm{R}} \qquad l = 1, \ldots, 10 \tag{11.116}$$

and the massless left movers are

$$\tilde{\alpha}^j_{-1}|0\rangle_{\mathrm{L}} \qquad j = 1, 2 \tag{11.117}$$

and

$$(\tilde{f}^m_{1/2})^\dagger(\tilde{f}^n_{1/2})^\dagger|0\rangle_{\mathrm{L}} \quad (\tilde{g}^m_{1/2})^\dagger(\tilde{g}^n_{1/2})^\dagger|0\rangle_{\mathrm{L}} \quad (\tilde{f}^m_{1/2})^\dagger(\tilde{g}^n_{1/2})^\dagger|0\rangle_{\mathrm{L}}$$

$$m, n = 11, \ldots, 32. \tag{11.118}$$

The $\boldsymbol{W}_0$ sector is subject to the generalized GSO projections (deriving from (11.104)),

$$N_{\mathrm{L}} - N_{\mathrm{R}} = 1 \pmod 2 \tag{11.119}$$

and

$$N_{\mathrm{L}} - \sum_{l=3,4,6,7,9,10} N^l_{\mathrm{R}} = 0 \pmod 2 \tag{11.120}$$

where N_{R} and N_{L} are the sums over all components of N_{R} and N_{L},

$$N_{\mathrm{R}} = \sum_{l=1,\ldots,10} N^l_{\mathrm{R}} \tag{11.121}$$

and

$$N_{\mathrm{L}} = \sum_{l=11,\ldots,32} N^l_{\mathrm{L}}. \tag{11.122}$$

Amongst other things, these projections delete potential tachyonic states with $m_R^2 = m_L^2 = -\frac{1}{2}$. The massless states in the W_0 sector surviving the generalized GSO projections are as follows. First there are the states

$$(\bar{f}^l_{1/2})^\dagger|0\rangle_R \bar{\alpha}^j_{-1}|0\rangle_L \tag{11.123}$$

and

$$(g^l_{1/2})^\dagger|0\rangle_R \bar{\alpha}^j_{-1}|0\rangle_L \tag{11.124}$$

which for $l = 1$ provide the graviton, dilaton and antisymmetric tensor. In the present four-dimensional case, the antisymmetric tensor is just a single (pseudo) scalar state, so the graviton is accompanied by two real scalars. For $l = 2$, 5 and 8, which is also consistent with the generalized GSO projections, we obtain instead six real vector fields. It may be conjectured at this stage that these are the six graviphotons that occur in the $N = 4$ supergravity multiplet of table 1.3, and we shall see shortly that the model does indeed contain all the massless states necessary to complete this supermultiplet.

Second, there are the states obtained by taking a right mover of the form (11.116) and a left mover of the form (11.118). Observing that $(\bar{f}^m_{1/2})^\dagger|0\rangle_L$ and $(\bar{g}^m_{1/2})^\dagger|0\rangle_L$ taken together provide the components of an SO(44) vector, we see that the left movers (11.118) are the antisymmetric part of the product of two SO(44) vectors and so transform as the adjoint representation of SO(44). Taking $l = 1$ for the right movers we obtain vector fields in the adjoint of SO(44) that can provide the gauge fields of SO(44). Taking instead $l = 2$, 5 or 8 (which also satisfies the generalized GSO projections) gives six multiplets of real scalars in the adjoint of SO(44) (as required for the $N = 4$ vector supermultiplet of table 1.3).

Next, let us consider the W_1 sector. In that case the massless right movers are just the ground state $|0\rangle_R$; and the massless left movers are (11.117) and (11.118), exactly as for the W_0 sector. Because the right movers Ψ^1, Ψ^2, Ψ^5 and Ψ^8 have periodic boundary conditions, the ground state $|0\rangle_R$ is an SO(8) spinor. In this case, the generalized GSO projections are

$$N_L - \sum_{l=3,4,6,7,9,10} N^l_R = 0 \pmod 2 \tag{11.125}$$

and

$$N_L - N_R = 2(k_{01} + k_{00}) \pmod 2. \tag{11.126}$$

It is important to notice that (11.126) contains contributions from the zero modes of Ψ^1, Ψ^2, Ψ^5 and Ψ^8, which means that the chirality of the SO(8) spinor ground state enters this projection in the way discussed at the end of §11.4. In consequence, the surviving massless states possess a single SO(8) chirality.

Taking $|0\rangle_R$ as the right mover and the left mover of (11.117) to obtain the states

$$|0\rangle_R \tilde{\alpha}^j_{-1} |0\rangle_L \qquad j = 1, 2 \tag{11.127}$$

provides four four-dimensional vector spinor states with helicity $\frac{3}{2}$ and four four-dimensional spinor states with helicity $\frac{1}{2}$ along with the same numbers of massless states with helicity $-\frac{3}{2}$ and $-\frac{1}{2}$, as required for $N = 4$ supergravity gravitinos and associated spin-$\frac{1}{2}$ components. Taken together with the massless states already obtained in the w_0 sector, these states complete the $N = 4$ supergravity multiplet of table 1.3.

Taking $|0\rangle_R$ as the right mover and the left movers of (11.118) provides four states with helicity $\frac{1}{2}$ and four with helicity $-\frac{1}{2}$ in the adjoint of SO(44) as required for the $N = 4$ gauginos. Together with the massless states in the adjoint of SO(44) from the $\boldsymbol{W}_0$ sector these states complete the $N = 4$ vector supermultiplet of table 1.3.

Thus, the model generated by the basis vectors $\boldsymbol{W}_0$ and $\boldsymbol{W}_1$ is an $N = 4$ supergravity theory with SO(44) gauge fields. More realistic models with $N = 1$ supergravity may be obtained by adding further basis vectors consistent with the modular invariance and world sheet supersymmetry constraints. We first add a single additional basis vector $\boldsymbol{W}_2$ where

$$\boldsymbol{W}_2 = (0\, (0\tfrac{1}{2}\tfrac{1}{2})\, (\tfrac{1}{2}0\tfrac{1}{2})\, (\tfrac{1}{2}0\tfrac{1}{2}) | (\tfrac{1}{2})^{14}\, 0^8). \tag{11.128}$$

Then, in addition to the massless states in the $\boldsymbol{W}_0$ and $\boldsymbol{W}_1$ sectors there are also massless states in the $\boldsymbol{W}_2$ sector with fermionic super-partners in the $\boldsymbol{W}_0 + \boldsymbol{W}_1 + \boldsymbol{W}_2$ sector. The parameters k_{ij} required for the generalized GSO projections satisfy (11.113)–(11.115) and additionally

$$k_{12} = 0 \quad \text{or} \quad \tfrac{1}{2} \pmod 1 \tag{11.129}$$

$$k_{22} = 0 \quad \text{or} \quad \tfrac{1}{2} \pmod 1 \tag{11.130}$$

$$k_{20} = k_{02} = k_{22} \pmod 1 \tag{11.131}$$

and

$$k_{21} = k_{12} + \tfrac{1}{2} \pmod 1. \tag{11.132}$$

The massless spectrum in the $\boldsymbol{W}_0$ sector is modified by the $\boldsymbol{W}_2$ generalized GSO projection

$$\sum_{l = 11, \ldots, 24} N^l_L - \sum_{l = 3,4,5,7,8,10} N^l_R = 0 \pmod 2. \tag{11.133}$$

The three complex gauge singlet vector fields are reduced to a single complex vector field or, equivalently, two real vector fields. For the gauge non-singlet vector fields, either *both* left-mover indices m and n of (11.118) have to belong to $\{11, \ldots, 24\}$ or *both* have to belong to $\{25, \ldots, 32\}$. In this way, the surviving gauge fields make up the adjoint of SO(28) × SO(16).

Moreover, the original three complex scalars in the adjoint of SO(44) are reduced to one complex scalar in the adjoint of SO(28), one complex scalar in the adjoint of SO(16) (each with right-mover index $l = 2$ in (11.116)) and two complex scalars which are vectors of both SO(28) and SO(16) (with right-mover index $l = 5$ or 8).

For the $\boldsymbol{W}_1$ sector, the $\boldsymbol{W}_2$ generalized GSO projection is

$$\sum_{l=11,\ldots,24} N_{\mathrm{L}}^{l} - \sum_{l=3,4,5,7,8,10} N_{\mathrm{R}}^{l} = 2(k_{12} + k_{22}). \qquad (11.134)$$

Noticing that N_{L}^{5} and N_{L}^{8} have zero-mode contributions, we see that for the gravitinos a definite chirality is chosen in the corresponding SO(4) subgroup of SO(8). In consequence, the number of gravitinos is reduced from four to two. Similarly, the number of massless spin-$\frac{1}{2}$ states associated with the gravitinos in the original $N = 4$ theory is reduced to two. In the case of the gauginos, the projection is also influenced by the representation of SO(28) × SO(16) to which the left movers belong. As a result, there are surviving gauginos in the adjoint of SO(28) and in the adjoint of SO(16) for one right-mover SO(4) chirality, and surviving erstwhile gauginos (not now associated with any gauge fields) transforming as a vector of both SO(28) and SO(16) for the other SO(4) chirality.

It can now be seen that the massless states correspond to an $N = 2$ supergravity theory with SO(28) × SO(16) gauge group. The $N = 2$ supergravity multiplet of table 1.2 is made up of the graviton and a single real gauge singlet vector from the $\boldsymbol{W}_0$ sector together with two gravitinos each with helicity $\pm\frac{3}{2}$ from the $\boldsymbol{W}_1$ sector. The remaining real gauge singlet vector from the $\boldsymbol{W}_0$ sector joins forces with two massless spin-$\frac{1}{2}$ states each with helicity $\pm\frac{1}{2}$ from the $\boldsymbol{W}_1$ sector and two real scalars from the $\boldsymbol{W}_0$ sector (originally the dilaton and antisymmetric tensor of the $N = 4$ theory) to form a gauge singlet $N = 2$ vector supermultiplet as in table 1.2. Moreover, the $\boldsymbol{W}_0$ and $\boldsymbol{W}_1$ sectors provide the gauge fields together with their gauginos and two real scalars, all in the adjoint of SO(28) × SO(16). In addition, the complex scalars and erstwhile gauginos in the vector representation of both SO(28) and SO(16) constitute an $N = 2$ hypermultiplet as in table 1.2.

In the $\boldsymbol{W}_2$ sector, the massless states in the first instance are

$$|0\rangle_{\mathrm{R}}|0\rangle_{\mathrm{L}} \qquad (11.135)$$

where because of the zero modes the right-mover ground state $|0\rangle_{\mathrm{R}}$ is an SO(8) spinor and the left-mover ground state is an SO(16) spinor. The $\boldsymbol{W}_0$, $\boldsymbol{W}_1$ and $\boldsymbol{W}_2$ generalized GSO projections on the $\boldsymbol{W}_2$ sector take the form

$$N_{\mathrm{L}} - N_{\mathrm{R}} = 1 + 2(k_{02} + k_{00}) \pmod 1 \qquad (11.136)$$

$$N_{\mathrm{L}} - \sum_{l=3,4,6,7,9,10} N_{\mathrm{R}}^{l} = 1 + 2(k_{11} + k_{12}) \pmod 1 \qquad (11.137)$$

and

$$\sum_{l=11,\ldots,24} N_L^l - \sum_{l=3,4,5,7,8,10} N_R^l = 0 \pmod 2). \tag{11.138}$$

After applying these generalized GSO projections, a definite right-mover SO(8) chirality is associated with a definite left-mover SO(16) chirality. However, since the SO(8) spinor contains both four-dimensional space-time chiralities the theory is non-chiral, as expected for $N = 2$ supergravity.

A chiral $N = 1$ supersymmetric theory may be obtained by the addition of one further basis vector

$$\boldsymbol{W}_3 = (0\,(\tfrac{1}{2}0\tfrac{1}{2})\,(0\tfrac{1}{2}\tfrac{1}{2})\,(\tfrac{1}{2}\tfrac{1}{2}0)|(\tfrac{1}{2})^7\,0^7\,(\tfrac{1}{2})^3\,0^5). \tag{11.139}$$

The parameters k_{ij} in the generalized GSO projections then satisfy (11.113)–(11.115), (11.129)–(11.132) and additionally

$$k_{30} = k_{03} = k_{33} + \tfrac{1}{2} \pmod 1) \tag{11.140}$$

$$k_{31} = k_{13} + \tfrac{1}{2} \pmod 1) \tag{11.141}$$

and

$$k_{32} = k_{23} \pmod 1). \tag{11.142}$$

The normal-ordering constants are such that no new sectors with massless states arise.

The $\boldsymbol{W}_3$ generalized GSO projection for the $\boldsymbol{W}_0$ sector takes the form

$$N_L(\mathrm{SO}(14)) + N_L(\mathrm{SO}(6)) - \sum_{l=2,4,6,7,8,9} N_R^l = 0 \pmod 2) \tag{11.143}$$

where

$$N_L(\mathrm{SO}(14)) = \sum_{l=11,\ldots,17} N_L^l \tag{11.144}$$

and

$$N_L(\mathrm{SO}(6)) = \sum_{l=25,\ldots,27} N_L^l. \tag{11.145}$$

This projection removes the remaining gauge singlet complex vector field and so removes the real vector in the $N = 2$ supergravity multiplet and the real vector in the gauge singlet $N = 2$ vector supermultiplet of which it is composed. The gauge fields are reduced to those of SO(14) × SO(14) × SO(10) × SO(6) and the model can be regarded as an SO(10) grand unified theory with horizontal symmetry group SO(6) and SO(14) × SO(14) hidden-sector gauge group.

In the W_1 sector, the W_3 projection takes the form

$$N_L(\mathrm{SO}(14)) + N_L(\mathrm{SO}(6)) - \sum_{l=2,4,6,7,8,9} N_R^l = 1 + 2(k_{13} + k_{33}) \quad (11.146)$$

where N_R^2 and N_R^8 have zero-mode contributions. The projection therefore selects a definite chirality in the SO(4) associated with N_R^2, and N_R^8 and halves the number of gravitinos leaving just a single gravitino as required for the $N = 1$ supergravity multiplet. The gauginos are unaffected. Thus, we end up with the supergravity and the gauge field vector supermultiplet of an $N = 1$ theory.

Finally, the W_3 generalized GSO projection for the W_2 sector takes the form

$$N_L(\mathrm{SO}(14)) + N_L(\mathrm{SO}(6)) - \sum_{l=2,4,6,7,8,9} N_R^l$$

$$= 1 + 2(k_{23} + k_{33}) \pmod 2 \quad (11.147)$$

where $N_L(\mathrm{SO}(6))$, N_R^2, N_R^6 and N_R^9 have a zero-mode contribution. Combining (11.147) with (11.136) which has zero-mode contributions from N_R^1, N_R^2, N_R^6, N_R^9 and

$$N_L(\mathrm{SO}(16)) \equiv \sum_{l=25,\ldots,32} N_L^l \quad (11.148)$$

we see that the four-dimensional space-time chirality for the right movers is correlated with the SO(10) chirality for the left movers. However, both SO(16) chiralities are allowed. Thus, after all projections, the W_2 sector contains massless SO(14) × SO′(14) singlet states in

$$2(\mathbf{16}, \mathbf{4})_L + 2(\mathbf{16}, \bar{\mathbf{4}})_L + 2(\overline{\mathbf{16}}, \bar{\mathbf{4}})_R + 2(\overline{\mathbf{16}}, \mathbf{4})_R \quad (11.149)$$

where $(\mathbf{a}, \mathbf{b})_{L,R}$ denotes the representation $\mathbf{a}$ of SO(10), $\mathbf{b}$ of SO(6) and four-dimensional space-time chirality L or R. The two copies of each representation occur because, as a result of the W_0, W_1 and W_3 projections, there is definite chirality in the SO(4) associated with N_R^6 and N_R^9 (and so two states) for a given space-time chirality. We now have a chiral theory, as is permitted for $N = 1$ supersymmetry, with 16 generations in **16** of SO(10), together with their anti-particles.

The discussion given in this chapter can be generalized slightly to allow for the possible presence of some real fermionic degrees of freedom[(2),(3)] which cannot be paired with other fermionic degrees of freedom with the same boundary conditions to form complex fermionic degrees of freedom. With this generalization, it has proved possible to construct potentially realistic three-generation models with flipped SU(5) × U(1) grand unification[(10)].

Exercises

11.1 Derive the partition function contributions for periodic and anti-periodic boundary conditions (11.47)–(11.50).

11.2 Derive the partition function Z_u^v of (11.56) for general v and u.

11.3 Use modular invariance to derive the GSO projections for the superstring.

References

General references

Antoniadis I, Bachas C P and Kounnas C 1987 *Nucl. Phys.* B **289** 87
Kawai H, Lewellen D C and Tye S-H 1987 *Nucl. Phys.* B **288** 1

References in the text

1 Lerche W, Lüst D and Schellekens A N 1987 *Nucl. Phys.* B **287** 677
2 Kawai H, Lewellen D C and Tye S-H 1987 *Nucl. Phys.* B **288** 1
3 Antoniadis I, Bachas C P and Kounnas C 1987 *Nucl. Phys.* B **289** 87
4 Polchinsky J 1986 *Commun. Math. Phys.* **104** 37
5 See, for example,
Bailin D and Love A 1986 *Introduction to Gauge Field Theory* (Bristol: Institute of Physics Publishing) Chapter 17
6 Seiberg N and Witten E 1986 *Nucl. Phys.* B **276** 272
7 Lang S 1973 *Elliptic Functions* (New York: Addison-Wesley)
8 Antoniadis I, Bachas C, Kounnas C and Windey P 1986 *Phys. Lett.* **171B** 51
9 Vafa C 1986 *Nucl. Phys.* B **273** 592
10 Antoniadis I, Ellis J, Hagelin J S and Nanopoulos D V 1989 *Phys. Lett.* **231B** 65

12

SUPERSTRING INTERACTIONS

12.1 Introduction

Apart from the intrinsic interest in formulating a consistent, relativistic quantum string theory, its importance stems from the fact (belief) that it may provide a finite quantum theory of all of the interactions in nature, including gravitation. The particles whose interactions we study are identified with single (massless) modes of the string. If we are to ascertain the consistency (or lack of it) of string theory with the real world, it is obviously essential that we can calculate scattering amplitudes whose external lines represent single-particle states, but whose internal structure includes all of the allowed string modes, not just a few single-particle states. Thus we are interested in calculating diagrams such as the closed-string diagrams shown in figure 12.1.

This task is made feasible by utilizing the invariance of the theory under a conformal rescaling of the world sheet metric $h_{\alpha\beta}(\tau, \sigma)$, described in §7.2. Under a finite such transformation,

$$h_{\alpha\beta} \to e^{\Lambda} h_{\alpha\beta} \tag{12.1}$$

with $\Lambda(\tau, \sigma)$ an arbitrary function of the world sheet coordinates τ, σ. By a suitable choice of Λ we can always arrange that the external lines 'puncture' the world sheet at *finite* points. The simplest illustration is provided by a single incoming particle and a single outgoing particle, as shown in figure 12.1(a), with a cylindrical world sheet, parametrized by z, φ with

$$-\infty < z < \infty \qquad 0 \leqslant \varphi < 2\pi \tag{12.2}$$

and having the metric

$$\mathrm{d}s^2 = \mathrm{d}z^2 + R^2\,\mathrm{d}\varphi^2. \tag{12.3}$$

Instead of z we may use the parameter θ defined by

$$z = 2R\ln(\tan\tfrac{1}{2}\theta) \qquad 0 < \theta < \pi. \tag{12.4}$$

Then

$$\mathrm{d}s^2 = R^2[(\sin\theta)^{-2}\,\mathrm{d}\theta^2 + \mathrm{d}\varphi^2] \tag{12.5}$$

and exploiting the conformal invariance allows us to rescale the metric with a factor

$$e^{\Lambda} = \sin^2\theta. \tag{12.6}$$

Then the new metric is

DOI: 10.1201/9780367805807-12

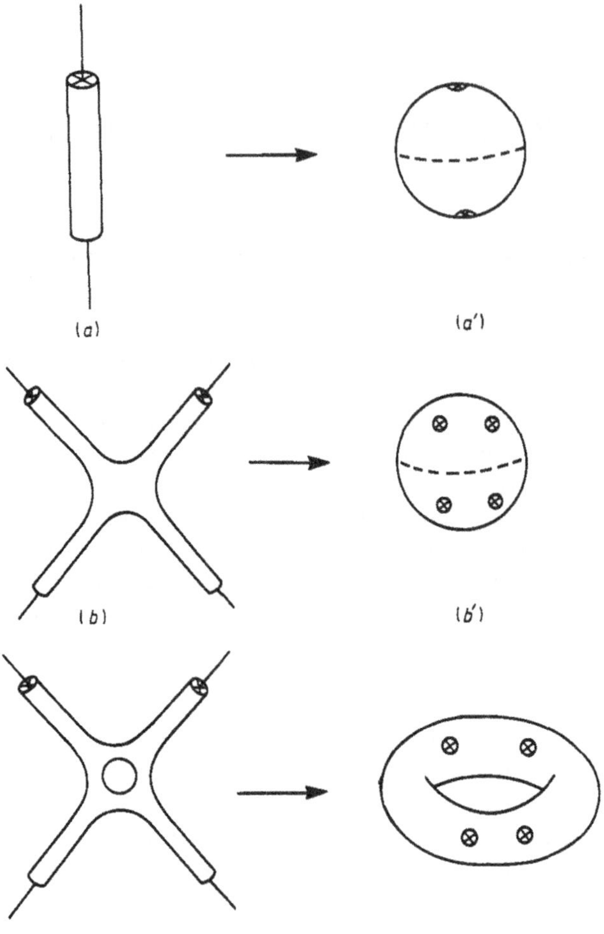

Figure 12.1 Conformal transformations of the closed-string world sheet mapping asymptotic states into finite points.

$$\tilde{d}s^2 = R^2(d\theta^2 + \sin^2\theta\, d\varphi^2) \tag{12.7}$$

which we recognize as the metric on a 2-sphere of radius R. The initial and final string states at $z = -\infty, +\infty$ correspond to $\theta = 0, \pi$, i.e. the north and south poles of the sphere, as shown in figure 12.1(a').

For more complicated processes, such as those shown in figure 12.1, (b) and (c), a suitable conformal factor e^{Λ} can always be found that maps each external string state onto a finite point on the world sheet, as shown in figure 12.1, (b') and (c'). This is possible because we only need to choose a

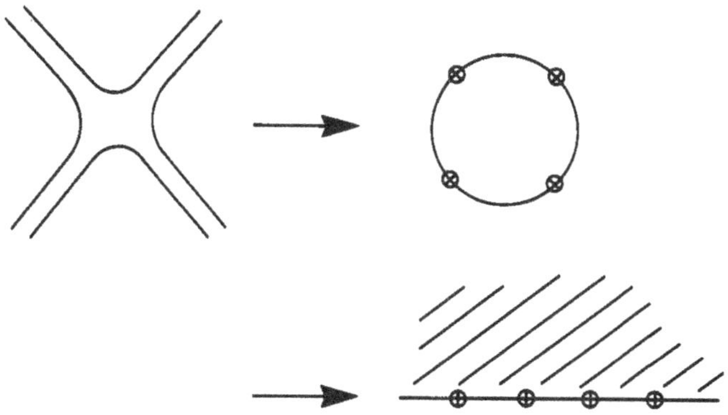

Figure 12.2 Conformal transformations of the open-string world sheet.

conformal factor e^{Λ} that has the desired *asymptotic* behaviour for each external string, and these can be chosen independently.

Similar remarks apply in the case of open-string scattering, as shown in figure 12.2(*a*). The world sheet can be conformally mapped onto a disk, or onto the upper half-plane, with the external states now appearing at finite points on the boundary, as shown in figure 12.2, (*b*) and (*c*).

In order to construct scattering amplitudes we must next deduce or define a vertex operator characterizing the particular particle being emitted or absorbed at this (finite) point on the world sheet. This is done for the lowest-lying states of the bosonic string in §12.2; the required operator is derived using the Lorentz properties of the particle, together with its 'conformal dimension', which describes its behaviour under world sheet reparametrizations. In §§12.2 and 12.3 we present the rules for calculating on-shell tree scattering amplitudes for the open (and closed) bosonic string, using the previously derived vertex operators and a propagator whose form is motivated by our experience of calculating scattering amplitudes in (bosonic) quantum field theory. We also verify that the (open-string) amplitudes possess the anticipated symmetry under a cyclic permutation of the external particles, and have poles corresponding to the known tachyon, vector states and the whole tower of massive string states. The generalization of these topics to the superstring is addressed in §12.5. In this case we need the vertex operators for the emission of a boson from a bosonic state of the string, the emission of a boson from a fermionic state of the string, and the emission of a fermion from a fermionic state of the string, turning it into a bosonic state. The form of these superstring operators is constrained by requiring invariance under supersymmetric reparametrizations of the world sheet. We also use the vertex operators to determine the three-graviton vertex and the graviton–gravitino–gravitino vertex in the context of the closed superstring.

The former reproduces the three-graviton vertex which may be derived from general relativity, while the latter reproduces the interaction vertex that was obtained previously in supergravity theory. Finally, in §12.6, we outline recent developments that are beyond the scope of this text. In particular we indicate how scattering amplitudes are calculated in (potentially) realistic theories such as the orbifold and fermionic models discussed in Chapters 10 and 11. We also give a short report on the current status of the most developed realistic model, the flipped SU(5) × U(1) (fermionic) model.

12.2 Bosonic string vertex operators and conformal dimensions

For closed strings, when we have mapped the external string states to puncture the world sheet at finite points, there must appear local operators that characterize the individual string states associated with each such point. We denote by $W_\Lambda(\tau, \sigma)$ the local operator that corresponds to the absorption of a string state $|\Lambda\rangle$ at the point (τ, σ) on the world sheet. W_Λ must carry the quantum numbers appropriate to the state $|\Lambda\rangle$ and must be constructed from the operator X^μ and its derivatives (in the case of the bosonic string). Thus in the case where $|\Lambda\rangle$ is the (tachyon) ground state, W_Λ must be a (D-dimensional) scalar, while if $|\Lambda\rangle$ is a graviton state, W_Λ must transform as a spin-2 Lorentz tensor. However, with only this information there is still considerable freedom in choosing an appropriate W_Λ.

Besides the Lorentz properties discussed above we must also ensure the correct behaviour under translations. A state $|k\rangle$ of momentum k^μ is multiplied by a factor $e^{-ik.a}$ when the translation

$$X^\mu \to X^\mu + a^\mu \tag{12.8}$$

is effected. Thus we expect $W_\Lambda(\tau, \sigma)$ to be multiplied by the operator $e^{-ik.X}$ if it is to describe the absorption of the state $|\Lambda, k\rangle$ having momentum k. Further, since any vertex operator $W_\Lambda(\tau, \sigma)$ may be inserted at any point on the world sheet, the quantity that is required for the calculation of scattering amplitudes is the operator

$$V_\Lambda(k) = \int d^2\sigma \sqrt{-h} W_\Lambda(\tau, \sigma)\, e^{-ik.X}. \tag{12.9}$$

Since (by construction) all string theories are invariant under reparametrizations of the world sheet, we should expect that $V_\Lambda(k)$ also has this property. In particular it should be invariant under rescaling of the world sheet parameters

$$\sigma \to \lambda\sigma \tag{12.10a}$$

$$\tau \to \lambda\tau. \tag{12.10b}$$

(Such a rescaling will not destroy the conformal gauge choice (7.13), (7.14).) Since the measure acquires a factor λ^2 as a result of the rescaling, we require that $W_\Lambda\, \mathrm{e}^{-\mathrm{i}k.X}$ acquires a (compensating) factor of λ^{-2}. Now, the invariance of the bosonic string action (7.4) under the above rescaling suggests that X^μ is invariant. (In that case the required λ^{-2} comes from the two derivatives ∂_α, ∂_β in (7.4).) Further, if X^μ is invariant, it would appear that $\mathrm{e}^{-\mathrm{i}k.X}$ is also invariant. But this is not correct, as we shall see shortly.

For open strings, the external string states are mapped onto the boundary of the world sheet, so there is a vertex operator $W_\Lambda(\tau)$ that is associated with the absorption of the state $|\Lambda\rangle$ at the point on the boundary parametrized by τ. In this case the quantity needed for the calculation of scattering amplitudes is

$$V_\Lambda(k) = \int \mathrm{d}\tau\, \sqrt{h_{\tau\tau}}\, W_\Lambda(\tau)\, \mathrm{e}^{-\mathrm{i}k.X(\tau)} \tag{12.11}$$

where k is the momentum of the absorbed state, and $X(\tau)$ is evaluated on the boundary ($\sigma = 0$ or 2π) of the world sheet. Then under the rescaling (12.10) of the parameter τ, by reasoning as we did above, we require that for the open string $W_\Lambda\, \mathrm{e}^{-\mathrm{i}k.X}$ acquires a (compensating) factor λ^{-1}; the difference arises because, since the external states are inserted on the boundary of the world sheet, there is only a single integration variable in (12.11).

The rescalings (12.10) are merely special cases of the reparametrizations

$$\sigma \to \sigma'(\sigma) \tag{12.12a}$$

$$\tau \to \tau'(\tau) \tag{12.12b}$$

which preserve the conformal gauge. The above observations about the behaviour of the local operator $W_\Lambda\, \mathrm{e}^{-\mathrm{i}k.X}$ under the rescalings can then be understood as statements about the 'conformal dimension' of the operator. This is defined as follows. Consider a local operator $A(\tau)$, such as arises in the open-string vertex operator for example. Under the reparametrizations (12.12*b*) $A(\tau)$ is transformed into $A'(\tau')$ and we say that $A(\tau)$ has 'conformal dimension J' when

$$A'(\tau') = \left(\frac{\mathrm{d}\tau}{\mathrm{d}\tau'}\right)^J A(\tau). \tag{12.13}$$

So if the open-string $W_\Lambda\, \mathrm{e}^{-\mathrm{i}k.X}$ has conformal dimension *one*, then (in particular) it will acquire the required λ^{-1} under the rescaling (12.10*b*).

The infinitesimal generators of the coordinate transformation (12.12*b*) are just the Virasoro operators introduced in §§7.5, 7.6. To see this consider an infinitesimal general coordinate transformation of the circle parametrized by θ, $0 \leqslant \theta < 2\pi$,

$$\theta \to \theta' = \theta + a(\theta). \tag{12.14}$$

Define

$$z = e^{i\theta} \tag{12.15}$$

then

$$z \to z' = z(1 + \epsilon(z)) = z + \sum_n \epsilon_n z^{n+1} \tag{12.16}$$

and

$$f(z') = f(z) + \sum_n \epsilon_n z^{n+1} \frac{df}{dz}. \tag{12.17}$$

Thus the operators (corresponding to $\epsilon(z) = -z^n$) are

$$L_n \equiv -z^{n+1} \frac{d}{dz} \tag{12.18a}$$

$$= i\,e^{in\theta} \frac{d}{d\theta} \tag{12.18b}$$

generate the reparametrization, and it is easy to check that they realize the Virasoro algebra (7.96) (without the central extension). The relevance of this to the case in hand is apparent when we note that although τ is not an angular variable it becomes such in the open-string mode expansion (7.43) since it arises only in the form $e^{in\tau}$ with n integral.

The definition (12.13) becomes

$$A'(z') = \left(\frac{z'}{z}\frac{dz}{dz'}\right)^J A(z) \tag{12.19}$$

in terms of the variable z, so for the infinitesimal transformation (12.16) we find

$$\delta A(z) = A'(z) - A(z) = -z\left[\epsilon \frac{dA}{dz} + JA\frac{d\epsilon}{dz}\right]. \tag{12.20}$$

Taking $\epsilon = -z^m$ we deduce

$$[L_m, A(z)] = z^m\left(z\frac{d}{dz} + mJ\right)A(z) \tag{12.21}$$

and expanding $A(z)$ in its moments

$$A(z) = \sum_n A_n z^{-n} \tag{12.22}$$

gives

$$[L_m, A_n] = [m(J-1) - n]A_{m+n}. \tag{12.23}$$

The Virasoro generators (7.92) for the open string are

$$L_m = -\tfrac{1}{2}\!:\sum_{p=-\infty}^{\infty} \alpha_{m-p}\alpha_p\!: \tag{12.24}$$

and with $\sigma = 0$ the moment X_n^μ of $X^\mu(\tau, 0) \equiv X^\mu(z)$ is (from (7.43))

$$X_n^\mu = \frac{\mathrm{i}}{n}\alpha_n^\mu \qquad (n \neq 0). \tag{12.25}$$

We leave it as an exercise (Exercise 12.2) to show that

$$[L_m, X_n^\mu] = -(m+n)X_{m+n}^\mu \tag{12.26}$$

thereby suggesting that $X^\mu(z)$ has conformal dimension $J = 0$; however, this does not work for $n = 0$, because the momentum part of X^μ in (7.43) involves $\tau = -\mathrm{i}\ln z$. Thus $X^\mu(z)$ is strictly *not* an operator of definite conformal dimension. On the other hand it is easy to check that $-\mathrm{i}\,\partial_\tau X^\mu(\tau, 0) = z\,\partial_z X^\mu(z)$ *does* have definite conformal dimension $J = 1$ (and this is essentially all that is needed for the invariance of the action), so in this weak sense X^μ has $J = 0$.

This also indicates why $\mathrm{e}^{\mathrm{i}k.X(z)}$ might have $J \neq 0$ in general; the exponentiation of the momentum term gives a power of z (see (12.27) below), and we can imagine that this might ensure that $\mathrm{e}^{\mathrm{i}k.X(z)}$ has definite conformal dimension. The verification that this is the case, and the determination of the J-value, is considerably more involved. First we have to be more precise about what we mean by $\mathrm{e}^{-\mathrm{i}k.X(x)}$. We shall study the normal-ordered form

$$A(z) \equiv :\mathrm{e}^{-\mathrm{i}k.X(z)}: = \exp\left(-\sum_{n=1}^{\infty}\frac{k.\alpha_{-n}}{n}z^n\right)\exp(-\mathrm{i}k.X - k.p\ln z)$$

$$\times\exp\left(\sum_{n=1}^{\infty}\frac{k.\alpha_n}{n}z^n\right). \tag{12.27}$$

Rather than determine the moments A_n it is easier to work with (12.21). The algebra is straightforward but tedious (Exercise 12.3); commuting L_m, given in (12.24), with $A(z)$ generates terms that are *not* in normal-ordered form whereas the right-hand side of (12.21) *is* normal ordered, and it is necessary to reorder carefully. The upshot is that the conformal dimension of $:\mathrm{e}^{-\mathrm{i}k.X(z)}:$ is

$$J = -k^2/2. \tag{12.28}$$

Let us return to the question of the vertex operators. It follows from (7.98) that the open-string ground state has a mass M satisfying

$$M^2 = -2 \tag{12.29}$$

(since $a = 1$). Thus the conformal dimension of $:e^{-i k.X(z)}:$ for the tachyonic ground state $|0, k\rangle$ having momentum k is $J = 1$. Since we require that $W_\Lambda\, e^{-i k.X}$ has conformal dimension *one*, it follows that the tachyon vertex operator W_0 has *zero* conformal dimension. Thus we may take

$$W_0 = 1. \tag{12.30}$$

The first excited state of the open string is the massless vector state obtained by operating on the state $|0, k\rangle$ with the α_{-1} operators. Since this is a massless vector ($k^2 = 0$) in D space-time dimensions, there are only $D - 2$ physical states corresponding to the $D - 2$ directions transverse to the vector k. A vector state $|v; \epsilon, k\rangle$ having polarization ϵ and momentum k is given by

$$|v; \epsilon, k\rangle = \epsilon_\mu \alpha^\mu_{-1}|0; k\rangle \tag{12.31a}$$

with

$$k^2 = 0 \qquad \epsilon.k = 0 \qquad \epsilon^2 = -1. \tag{12.31b}$$

Since the conformal dimension of $:e^{-i k.X}:$ is now zero, from (12.28), the corresponding vertex operator $W_{v,\epsilon}(\tau)$ must have conformal dimension $J = 1$. We therefore take

$$W_{v,\epsilon}(\tau) = \partial_\tau[\epsilon_\mu X^\mu(\tau, 0)] \tag{12.32a}$$

or

$$W_{v,\epsilon}(z) = i\, z \frac{d}{dz}[\epsilon_\mu X^\mu(z)]. \tag{12.32b}$$

(The overall normalization will be justified later.)

Things are only slightly more complicated when we address the closed-string vertex operators. In this case, since the external states are conformally mapped onto any point on the world sheet, we need first to know the conformal dimension of $:\exp(-i\, k.X(\tau, \sigma)):$ with

$$X^\mu(\tau, \sigma) = X^\mu_R(\tau - \sigma) + X^\mu_L(\tau + \sigma) \tag{12.33}$$

and $X^\mu_{R,L}$ having the mode expansions given in (7.26) and (7.27). The separation (12.32) implies that the exponential factorizes as

$$:\exp(-i\, k.X(\tau, \sigma)): = :\exp(-i\, k.X_R)::\exp(-i\, k.X_L): \tag{12.34}$$

and it is apparent that such an operator has conformal dimensions associated with each of the independent reparametrizations

$$\sigma_+ \to \sigma'_+ (\sigma_+) \tag{12.35a}$$

$$\sigma_- \to \sigma'_- (\sigma_-) \tag{12.35b}$$

where

$$\sigma_\pm \equiv \tau \pm \sigma \tag{12.35c}$$

parametrize the left- and right-moving modes respectively. Let us define J_L as the conformal dimension associated with (12.35*a*), and J_R that associated with (12.35*b*). For the operator (12.34) it is clear that

$$J_L = J_R. \tag{12.36}$$

The calculation of this common value proceeds very similarly to the open-string case with the result

$$J_L = J_R = -\tfrac{1}{8}k^2. \tag{12.37}$$

The factor of four difference between this result and the open-string result (12.28) arises from the overall factor of two difference in the mode expansions (7.26), (7.27) compared with the open-string expansion (7.43); effectively this replaces k by $k/2$ in the calculation of the conformal dimension.

We saw in §7.8 that the closed-string ground state is tachyonic with mass M satisfying

$$M^2 = -8 \tag{12.38}$$

so the conformal dimensions J_L and J_R of $:e^{-ik.X}:$ for the ground state are $J_L = J_R = 1$. Thus it acquires a factor of λ^{-1} under each of the independent rescalings

$$\sigma_+ \to \lambda\sigma_+ \tag{12.39a}$$

$$\sigma_- \to \lambda\sigma_- \tag{12.39b}$$

and consequently a factor λ^{-2} under the simultaneous rescaling. It follows that the (closed-string) tachyon vertex operator W_0 has zero conformal dimensions, and we may take

$$W_0 = 1 \tag{12.40}$$

as in the open string.

The first excited states of the closed string are the massless states given in (7.120), corresponding to the graviton, antisymmetric tensor, and dilaton states.

As in (12.31), the massless graviton state with polarization tensor $\epsilon_{\mu\nu}$ and momentum k is given by

$$|g; \epsilon, k\rangle = \epsilon_{\mu\nu}\alpha^{\mu}_{-1}|k\rangle_R\tilde{\alpha}^{\nu}_{-1}|k\rangle_L \tag{12.41a}$$

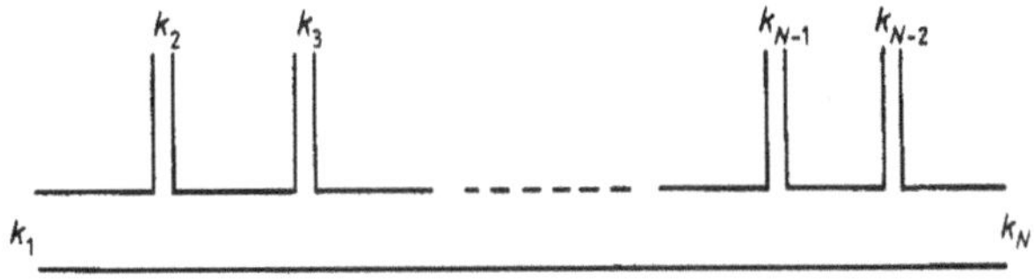

Figure 12.3 The N-particle open-string amplitude.

with

$$k^2 = 0 \qquad \epsilon_{\mu\nu} = \epsilon_{\nu\mu} \qquad \epsilon^{\mu}{}_{\mu} = 0 \qquad k^{\mu}\epsilon_{\mu\nu} = 0. \quad (12.41b)$$

The corresponding vertex operator $W_{g,\epsilon}$ must have $J_L = J_R = 1$ and is given by

$$W_{g,\epsilon}(\tau, \sigma) = \epsilon_{\mu\nu}\, \partial_+ X^{\mu}\, \partial_- X^{\nu} \qquad (12.42a)$$

that is

$$W_{g,\epsilon}(\tau, \sigma) = \tfrac{1}{4}\epsilon_{\mu\nu}\, \partial_{\alpha} X^{\mu}\, \partial^{\alpha} X^{\nu} \qquad (12.42b)$$

where, as in (8.38),

$$\partial_{\pm} \equiv \partial/\partial\sigma_{\pm} = \tfrac{1}{2}(\partial_{\tau} \pm \partial_{\sigma}). \qquad (12.42c)$$

The anti-symmetric tensor and dilaton states have vertex operators constructed in an analogous manner.

12.3 Bosonic open-string scattering amplitudes

To date there is no really satisfactory quantum field theory of strings. As a result we are not yet able to derive the rules for calculating amplitudes from a Lagrangian in the manner that we are accustomed to using for point particles. Instead we have to postulate certain rules for constructing diagrams, which have been found to yield scattering amplitudes with the features that we would expect in the light of our knowledge of point particle scattering amplitudes. At present the rules give satisfactory results for on-shell S-matrix elements only. In this section we shall address the calculation of tree amplitudes only. There are non-trivial complications in extending the techniques to loop amplitudes that are beyond the scope of this book.

Our experience of calculating Feynman diagrams leads us to expect that a (tree) scattering amplitude will have associated with it (i) vertex factors $V_{\Lambda_i}(k_i)$, in general momentum dependent, characterizing the absorption of string states $|\Lambda_i; k_i\rangle$ of momentum k_i, and (ii) propagator factors Δ associated with the propagation of the string between two vertices[1]. Then the amplitude for the N-particle process shown in figure 12.3 is given by

$$A_N = g^{N-2}\langle \Lambda_1 ; k_1 | V_{\Lambda_2}(k_2)\, \Delta V_{\Lambda_3}(k_3) \cdots \Delta V_{\Lambda_{N-1}}(k_{N-1}) | \Lambda_N ; k_N \rangle \qquad (12.43)$$

where g is the string coupling constant. The vertex factors are just the products $:W_\Lambda\, \mathrm{e}^{-\mathrm{i}\, k.X}:$, introduced in §12.2, but evaluated at $\tau = 0$ ($z = 1$). The propagators Δ are analogous to the familiar scalar-field Feynman propagator $(p^2 - m^2 + \mathrm{i}\,\epsilon)^{-1}$; the (open-string) mass formula is given in (7.98):

$$\tfrac{1}{2}M^2 = -\sum_{n=1}^{\infty} \alpha^{\mu}_{-n}\alpha_{\mu n} - 1 \qquad (12.44)$$

and we see from (7.91) and (7.92) that the Hamiltonian H, given in (7.97), is

$$H = L_0 - 1 = -\tfrac{1}{2}(p^2 - M^2). \qquad (12.45)$$

We therefore take the propagator to be

$$\Delta = (L_0 - 1 - \mathrm{i}\,\epsilon)^{-1} \qquad (12.46)$$

not worrying too much at present with the normalization. We write each propagator Δ in (12.43) using the integral representation

$$\Delta = \int_0^1 \mathrm{d}z\, z^{L_0 - 2}. \qquad (12.47)$$

Then

$$\begin{aligned} A_N = g^{N-2} \int_0^1 \frac{\mathrm{d}z_3}{z_3} \int_0^1 \frac{\mathrm{d}z_4}{z_4} \cdots \int_0^1 \frac{\mathrm{d}z_{N-1}}{z_{N-1}} \langle \Lambda_1; k_1 | V_{\Lambda_2}(k_2) z_3^{L_0 - 1} V_{\Lambda_3} z_3^{-L_0 + 1} \\ \times (z_3 z_4)^{L_0 - 1} V_{\Lambda_4}(k_4)(z_3 z_4)^{-L_0 + 1} \cdots (z_3 z_4 \cdots z_{N-1})^{L_0 - 1} \\ \times V_{\Lambda_{N-1}}(k_{N-1})(z_3 z_4 \cdots z_{N-1})^{-L_0 + 1} \\ \times (z_3 z_4 \cdots z_{N-1})^{L_0 - 1} | \Lambda_N; k_N \rangle. \end{aligned} \qquad (12.48)$$

The operator L_0 generates τ translations, so for any local operator $A(\tau)$ we have

$$A(\tau) = \mathrm{e}^{\mathrm{i}\,\tau L_0} A(0)\, \mathrm{e}^{-\mathrm{i}\,\tau L_0} = \mathrm{e}^{\mathrm{i}\,\tau(L_0 - 1)} A(0)\, \mathrm{e}^{-\mathrm{i}\,\tau(L_0 - 1)}. \qquad (12.49)$$

In terms of the variable introduced in the previous section

$$z = \mathrm{e}^{\mathrm{i}\,\tau} \qquad (12.50)$$

this gives

$$A(z) = z^{L_0 - 1} A(1) z^{-L_0 + 1}. \qquad (12.51)$$

Thus we may rewrite (12.48) as

$$A_N = g^{N-2} \int_0^1 \frac{dz_3}{z_3} \int_0^1 \frac{dz_4}{z_4} \cdots \int_0^1 \frac{dz_{N-1}}{z_{N-1}} \langle \Lambda_1; k_1 | V_{\Lambda_2}(k_2, z_2)$$
$$\times V_{\Lambda_3}(k_3, z_3) V_{\Lambda_4}(k_4, z_3 z_4) \cdots$$
$$\times V_{\Lambda_{N-1}}(k_{N-1}, z_3 z_4 \cdots z_{N-1}) | \Lambda_N; k_N \rangle \tag{12.52a}$$

where

$$z_2 \equiv 1 \tag{12.52b}$$

and we have used

$$(L_0 - 1) | \Lambda_N; k_N \rangle = 0. \tag{12.52c}$$

Changing variables to

$$y_i = z_2 z_3 z_4 \cdots z_i \qquad (i = 2, 3, \ldots, N-1) \tag{12.53}$$

gives

$$\prod_{i=3}^{N-1} \frac{dy_i}{y_i} = \prod_{i=3}^{N-1} \frac{dz_i}{z_i} \tag{12.54}$$

and the domain of integration is

$$\mathfrak{D}: \quad 0 < y_{N-1} < y_{N-2} \cdots < y_3 < y_2 = 1. \tag{12.55}$$

Then

$$A_N = g^{N-2} \int_{\mathfrak{D}} \prod_{i=3}^{N-1} \frac{dy_i}{y_i} \langle \Lambda_1; k_1 | V_{\Lambda_2}(k_2, y_2) V_{\Lambda_3}(k_3, y_3) \cdots$$
$$\times V_{\Lambda_{N-1}}(k_{N-1}, y_{N-1}) | \Lambda_N; k_N \rangle. \tag{12.56}$$

We may view the above manipulations as follows. The integral representation of the propagator, used in (12.47), requires the integration of the variable τ in (12.50) to be along the *imaginary* axis $\tau \equiv i\tau'$, $0 < \tau' < \infty$. It is therefore natural to associate the initial state $|\Lambda_N, k_N\rangle$ with $\tau'_N = \infty$, corresponding to $y_N = 0$, and the final state $|\Lambda_1\ k_1\rangle$ with $\tau'_1 = -\infty$, $y_1 = \infty$.

However, we observed in §12.1 that for open-string scattering we can always choose a mapping of the world sheet onto the upper half-plane, or onto a disk, with the external states appearing at finite points, as in figure 12.2, (*b*) and (*c*). Thus the scattering amplitude ought to be invariant under the cyclic transformation of the N external particles:

$$(\Lambda_1, k_1; \Lambda_2, k_2; \cdots; \Lambda_N, k_N) \rightarrow (\Lambda_N, k_N; \Lambda_1, k_1; \cdots; \Lambda_{N-1}, k_{N-1}). \tag{12.57}$$

Obviously to prove this statement, we will need to perform a conformal transformation in which

$$\tau'_i \to \tau'_{i+1} \qquad (i = 1, \ldots, N-1) \tag{12.58a}$$

$$\tau'_N \to \tau'_1 \tag{12.58b}$$

but to do this we need to associate vertex operators $V_{\Lambda_N}(k_N, y_N)$ and $V_{\Lambda_1}(k_1, y_1)$ with the initial and final states in (12.56).

Consider first the case where $|\Lambda\rangle$ is a tachyon state, so $W_\Lambda = 1$. Then

$$|\Lambda; k\rangle = \lim_{z \to 0} z^{k^2/2} V_\Lambda(k, z)|0, 0\rangle \tag{12.59}$$

where $|0, 0\rangle$ is the zero-momentum Fock space ground state. The reason for inclusion of the factor $z^{k^2/2} = z^{-1}$ is that we may write $V_\Lambda(k, z)$ in the form (12.27) and only the zero-mode piece

$$Z_0 = \mathrm{e}^{-\mathrm{i}\,k.x - k.p \ln z} \tag{12.60}$$

contributes in the tachyon state. It is easy to see (Exercise 12.3) that we may rewrite this as

$$Z_0 = \mathrm{e}^{-\mathrm{i}\,k.x} z^{-k.p - k^2/2} = z^{-k.p + k^2/2}\,\mathrm{e}^{-\mathrm{i}\,k.x} \tag{12.61}$$

so

$$Z_0|0, 0\rangle = z^{-k^2/2}\,\mathrm{e}^{-\mathrm{i}\,k.x}|0, 0\rangle = z^{-k^2/2}|\Lambda, k\rangle \tag{12.62}$$

and it is this $z^{-k^2/2}$ ($= z$ for the tachyon) that must be cancelled in (12.59). Similarly

$$\langle\Lambda; k| = \lim_{z \to \infty} \langle 0, 0|V_\Lambda(k, z) z^{-k^2/2}. \tag{12.63}$$

Thus we may rewrite (12.56) as

$$A_N = \lim_{y_1 \to \infty,\, y_N \to 0} g^{N-2} y_1{}^{-k_1^2/2} y_N{}^{k_N^2/2} \int_{\mathscr{D}} \prod_{i=3}^{N-1} \frac{\mathrm{d}y_i}{y_i} \langle 0, 0|V_{\Lambda_1}(k_1, y_1)$$
$$\times V_{\Lambda_2}(k_2, y_2) \cdots V_{\Lambda_{N-1}}(k_{N-1}, y_{N-1}) V_{\Lambda_N}(k_N, y_N)|0, 0\rangle \tag{12.64}$$

with the domain $\mathscr{D}$ given in (12.55)

Using this form for the amplitude the cyclic invariance may be readily established. It is instructive to do this, since the proof utilizes the residual symmetry (SL(2, R) in fact), which preserves the conformal gauge choice made in (7.13), (7.14). However, it is not needed for the calculation of the simple string amplitudes that constitute the main objective of this chapter. We therefore refer the interested reader to Appendix B for the details.

The simplest amplitudes to calculate are the three-point functions A_3, since it is apparent from (12.56) that no integrations are needed. The three-tachyon vertex is

$$A_3 = g\langle 0; k_1|V_0(k_2, 1)|0; k_3\rangle \tag{12.65}$$

with

$$V_0(k, z) \equiv :W_0\, e^{-ik.X(z)}: \tag{12.66}$$

and $W_0 = 1$, as given in (12.30). It is clear from (12.27) that only the zero-mode part of $V_0(k, 1)$ contributes:

$$\begin{aligned}A_3 &= g\langle 0; k_1|e^{-ik_2.X}|0; k_3\rangle = g\langle 0; k_1|0; k_2 + k_3\rangle \\ &= g(2\pi)^D\delta^{(D)}(k_1 + k_2 + k_3).\end{aligned} \tag{12.67}$$

Evidently such an energy–momentum-conserving δ-function, multiplied by $(2\pi)^D$, will appear in all amplitudes, and we shall only exhibit the remainder of the amplitude:

$$A_3 \sim g. \tag{12.68}$$

Next consider the tachyon–tachyon–vector vertex:

$$A_3 = g\langle 0; k_1|V_{v,\epsilon_2}(k_2, 1)|0; k_3\rangle \tag{12.69a}$$

where

$$V_{v,\epsilon}(k, z) = W_{v,\epsilon}(z)\, e^{-ik.X(z)} \tag{12.69b}$$

with $W_{v,\epsilon}(z)$ given in (12.32). Thus using (7.43) (and setting $l = 1$)

$$W_{v,\epsilon}(z) = \epsilon.p + \sum_{n\neq 0} \epsilon.\alpha_n z^{-n}. \tag{12.70}$$

Hence

$$A_3 = g\langle 0; k_1|\epsilon_2.p|0; k_2 + k_3\rangle \sim g\epsilon_2.(k_2 + k_3) \tag{12.71}$$

with ϵ_2 the polarization of the external vector state. Since ϵ_2 is transverse to k_2, as given in (12.31b),

$$A_3 \sim g\epsilon_2.k_3 = -g\epsilon_2.k_1 = \tfrac{1}{2}g\epsilon_2.(k_3 - k_1) \tag{12.72}$$

using energy–momentum conservation. The same result must of course be obtained if we cyclically permute the states so that the vector state is the initial state and we insert a tachyon vertex operator:

$$\begin{aligned}A_3 &= g\langle 0; k_3|V_0(k_1, 1)|V; \epsilon_2, k_2\rangle = g\langle 0; k_3|V_0(k, 1)\epsilon_2.\alpha_{-1}|0; k_2\rangle \\ &= g\langle 0; k_3|e^{-ik_1.x}\, e^{k_1.\alpha_1}\epsilon_2.\alpha_{-1}|0; k_2\rangle \\ &= g\langle 0; k_3 + k_1|[k_1.\alpha_1, \epsilon_2.\alpha_{-1}]|0; k_2\rangle \sim -g\epsilon_2 k_1\end{aligned} \tag{12.73}$$

in agreement with (12.72). Incidentally, this agreement provides the promised justification of the overall normalization of the vector emission vertex operator (12.32).

The triple-vector vertex is given by

$$
\begin{aligned}
A_3 &= g\langle \nu; \epsilon_1, k_1 | V_{\nu,\epsilon_2}(k_2, 1) | \nu; \epsilon_3, k_3 \rangle \\
&= g\langle 0; k_1 | \epsilon_1 . \alpha_1 [(\epsilon_2 . \alpha_{-1} + \epsilon_2 . p + \epsilon_2 . \alpha_1) \, \mathrm{e}^{-k_2 . \alpha_{-1}} \\
&\quad \times \mathrm{e}^{-\mathrm{i} k_2 . x} \, \mathrm{e}^{k_2 . \alpha_1}] \epsilon_3 . \alpha_{-1} | 0; k_3 \rangle \\
&\sim - g[\epsilon_1 . \epsilon_2 \epsilon_3 . k_1 + \epsilon_2 . \epsilon_3 \epsilon_1 . k_2 + \epsilon_3 . \epsilon_1 \epsilon_2 . k_3 \\
&\quad - (\epsilon_1 . k_2)(\epsilon_2 . k_3)(\epsilon_3 . k_1)] \\
&= \tfrac{1}{2} g[\epsilon_1 . \epsilon_2 \epsilon_3 . (k_2 - k_1) + \epsilon_2 . \epsilon_3 \epsilon_1 . (k_3 - k_2) + \epsilon_3 . \epsilon_1 \epsilon_2 . (k_1 - k_3) \\
&\quad - \epsilon_1 . (k_3 - k_2) \epsilon_2 . (k_1 - k_3) \epsilon_3 . (k_2 - k_1)]
\end{aligned} \tag{12.74}
$$

and we have used the fact that only the $n = 0, \pm 1$ modes are active in this case. The final form shows that this amplitude is antisymmetric:

$$A_3(1, 2, 3) = - A_3(2, 1, 3) \tag{12.75}$$

so the full (Bose symmetric) amplitude for the process, obtained by adding these two amplitudes, for example, is zero. Thus, as in QED, the triple-(massless-) vector amplitude is zero in an abelian gauge theory. However, it survives in a non-abelian theory, in which there is a compensating totally anti-symmetric group theory factor f^{abc} associated with the vertex. We note too that the first three terms of (12.74) give precisely the momentum dependence of the triple-vector vertex in non-abelian theories, as can be verified by comparison with equation (10.75) of Bailin and Love[(2)]. (Of course in that case the vertex is for general off-shell vector particles, and derives from the cubic terms of the Yang–Mills Lagrangian $-\frac{1}{4}F_a{}^{\mu\nu}F_{a\mu\nu}$.) The last term in (12.74) is actually of order α', and amounts to an additional term proportional to $f^{abc}F_{a\mu}{}^{\nu}F_{b\nu}{}^{\rho}F_{c\rho}{}^{\mu}$ in the effective (Yang–Mills) Lagrangian.

We turn now to the simplest four-particle amplitude, involving four tachyons. From (12.56) we have

$$A_4 = g^2 \int_0^1 \frac{\mathrm{d}y}{y} \langle 0; k_1 | V_0(k_2, 1) V_0(k_3, y) | 0; k_4 \rangle. \tag{12.76}$$

Using (12.27) and (12.61) the zero-mode part of $V_0(k_3, y)$ gives

$$
\begin{aligned}
&\mathrm{e}^{-\mathrm{i} k_3 . x} y^{-k_3 . p - k_3^2/2} | 0; k_4 \rangle = \mathrm{e}^{-\mathrm{i} k_3 . x} y^{-k_3 . k_4 + 1} | 0; k_4 \rangle \\
&\quad = y^{-s/2 - 1} | 0; k_3 + k_4 \rangle
\end{aligned} \tag{12.77a}
$$

since $k_3^2 = -2$ for a tachyon state and we have defined

$$s \equiv (k_1 + k_2)^2 = (k_3 + k_4)^2 = 2k_3 . k_4 - 4. \tag{12.77b}$$

Similarly the zero-mode part of $V_0(k_2, 1)$ converts $\langle 0; k_1|$ to $\langle 0; k_1 + k_2|$. The contribution of the non-zero modes is then determined by

$$\langle 0, 0 | \exp\left(\sum_{m=1}^{\infty} \frac{k_2 . \alpha_m}{m} \right) \exp\left(- \sum_{n=1}^{\infty} \frac{k_3 . \alpha_{-n}}{n} y^n \right) | 0, 0 \rangle \tag{12.78}$$

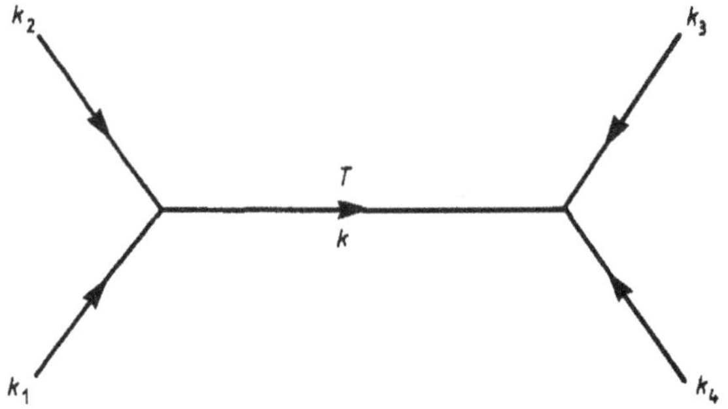

Figure 12.4 *s*-channel exchange contribution to A_4.

and this is most easily evaluated using the 'coherent state' methods, described in Appendix C. Using the result

$$\langle :e^A::e^B: \rangle = e^{\langle AB \rangle} \tag{12.79}$$

it is apparent that we only need

$$\langle AB \rangle = \langle 0, 0| \left(\sum_{m=1}^{\infty} \frac{k_2 . \alpha_m}{m} \right) \left(- \sum_{n=1}^{\infty} \frac{k_3 . \alpha_{-n}}{n} y^n \right) |0, 0\rangle$$

$$= \sum_{n=1}^{\infty} \frac{k_2 . k_3}{n} y^n = - k_2 . k_3 \ln(1 - y). \tag{12.80}$$

Putting these results together we get (the Veneziano[3] amplitude)

$$A_4 \sim g^2 \int_0^1 dy\, y^{-s/2 - 2}(1 - y)^{-t/2 - 2} = g^2 B\left(- \frac{s}{2} - 1, - \frac{t}{2} - 1 \right) \tag{12.81a}$$

where

$$t \equiv (k_1 + k_4)^2 = (k_2 + k_3)^2 = 2k_2 . k_3 - 4 \tag{12.81b}$$

and B is the beta function

$$B(a, b) \equiv \int_0^1 dx\, x^{a-1}(1 - x)^{b-1} = \frac{\Gamma(a)\Gamma(b)}{\Gamma(a + b)}. \tag{12.81c}$$

Note that because the gamma functions have poles where their arguments are zero or negative integers, the amplitude A_4 has poles when $s, t = -2, 0, 2, 4, \ldots$. This result is in accord with what we would expect from field theory considerations. We have already established in (12.68) the existence of a non-zero three-tachyon vertex, when all three tachyons are on-shell. Thus there ought to be a contribution to A_4 arising from the exchange of a tachyon (T) between the pairs k_1, k_2 and k_3, k_4, as shown in figure 12.4.

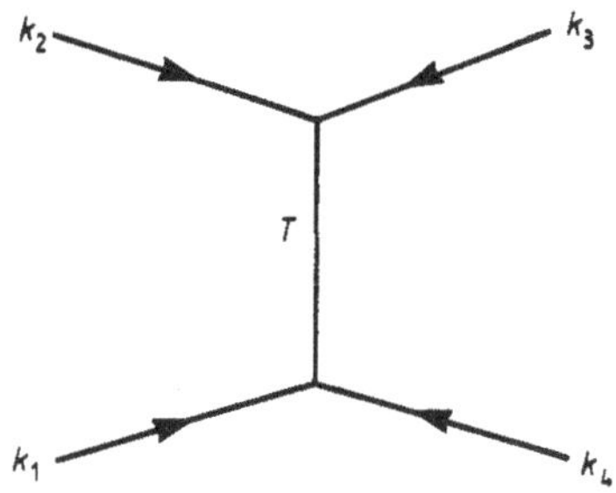

Figure 12.5 t-channel exchange contribution to A_4.

Our experience of field theory leads us to expect a pole in the amplitude from the propagator of T which contributes $(k^2 - m_T^2)^{-1}$, where k is the momentum of T, and m_T is its mass. Thus using energy–momentum conservation, the pole is when $k^2 = (k_1 + k_2)^2 = s = -2$, in accordance with the vanishing of the argument of the gamma function $\Gamma(-s/2 - 1)$ in A_4. The other s-channel poles arise from couplings of the tachyon to the massless vector, as suggested by (12.73), and to the whole tower of massive string states. In the same way the t-channel poles may be understood to arise from exchanges between the pairs k_2, k_3 and k_1, k_4.

The fact that this amplitude (12.81) is *symmetric* under the interchange of s and t is, however, quite amazing from a field theory viewpoint. It amounts to the statement that the diagrams in figure 12.4 are *equal* to the diagrams in figure 12.5. More generally, for N-point amplitudes, A_N is invariant under cyclic interchange of the momenta $k_1, k_2, \ldots, k_N$.

12.4 Bosonic closed-string amplitudes

The evaluation of closed-string amplitudes proceeds analogously to that of the open-string amplitudes. The principal difference arises because the world sheet for a closed string is (topologically) a sphere, and the external particles puncture the sphere (generally) at finite points. As explained in §12.1, the world sheet can be transformed to an ordinary round sphere, and then mapped onto the entire (complex) place. Then after the Wick rotation $\tau = \mathrm{i}\,\tau'$ the right-moving part of the world sheet $X_{\mathrm{R}}(\tau - \sigma)$ becomes a function $X_{\mathrm{R}}(z)$ of the complex variable

$$z = \mathrm{e}^{2\mathrm{i}(\tau - \sigma)} = \mathrm{e}^{-2(\tau' + \mathrm{i}\,\sigma)} \tag{12.82a}$$

while the left-moving piece $X_{\mathrm{L}}(\tau + \sigma)$ becomes a function $X_{\mathrm{L}}(\bar{z})$ of the complex conjugate variable

$$\bar{z} = \mathrm{e}^{2\mathrm{i}(\tau + \sigma)} = \mathrm{e}^{-2(\tau' - \mathrm{i}\,\sigma)}. \tag{12.82b}$$

There is a difference too in the propagator. The closed-string mass formulae are given in (7.84), (7.85), (7.86), and using the expressions for L_0, $\tilde{L}_0$ given in (7.65), (7.66) we see that the Hamiltonian H, given in (7.67), is

$$H = 2(L_0 + \tilde{L}_0 - 2) = -\tfrac{1}{4}(p^2 - M^2) \tag{12.83}$$

so, with the same normalization as (12.46), we take the propagator to be

$$\Delta = \tfrac{1}{2}(L_0 + \tilde{L}_0 - 2)^{-1} = \tfrac{1}{2}\int_0^1 \mathrm{d}\rho\, \rho^{L_0 + \tilde{L}_0 - 3} \tag{12.84}$$

where we have introduced the integral representation analogous to that given in (12.47). However, we also have the constraint (7.87) which requires

$$(L_0 - \tilde{L}_0)|\varphi\rangle = 0 \tag{12.85}$$

for any physical state $|\varphi\rangle$. Since this constraint has not been incorporated into (12.84), it is clear that non-physical states will be propagated by Δ. To restrict the propagation to physical states obeying (12.85) we modify Δ to

$$\Delta = \frac{1}{4\pi}\int_0^1 \mathrm{d}\rho \int_0^{2\pi} \mathrm{d}\varphi\, \rho^{L_0 + \tilde{L}_0 - 3}\, \mathrm{e}^{\mathrm{i}\varphi(L_0 - \tilde{L}_0)} = \frac{1}{4\pi}\int_{|z| \leqslant 1} \frac{\mathrm{d}^2 z}{|z|^2}\, z^{L_0 - 1}\bar{z}^{\tilde{L}_0 - 1} \tag{12.86a}$$

where

$$z = \rho\,\mathrm{e}^{\mathrm{i}\varphi} \tag{12.86b}$$

$$\mathrm{d}^2 z = \rho\, \mathrm{d}\rho\, \mathrm{d}\varphi. \tag{12.86c}$$

Just as in the open-string case (12.43) the amplitude for a general (tree) scattering process is given by

$$A_N = g^{N-2}\langle \Lambda_1; k_1 | V_{\Lambda_2}(k_2)\, \Delta V_{\Lambda_3}(k_3) \cdots \Delta V_{\Lambda_{N-1}}(k_{N-1}) | \Lambda_N; k_N\rangle + \text{permutations} \tag{12.87}$$

with g the string coupling constant, and the vertex factors $V_{\Lambda_i}(k_i)$ characterize the absorption of a state $|\Lambda_i\rangle$ of momentum k_i (at the point $(\tau, \sigma) = (0, 0)$). Since there is no well-defined order for the $N-2$ emitted particles, the amplitude A_N includes a sum over all permutations of the vertex operators. Then, just as the open-string amplitude is symmetric under cyclic reordering of the N external particles, so the closed-string amplitude includes all possible orderings of the N external states.

The separation of the world sheet $X^\mu(\tau, \sigma)$ into right- and left-moving pieces also entails a similar separation of the vertex factors $V_\Lambda(k)$, so

$$V_\Lambda(k) = V_{\Lambda\mathrm{R}}(k, \sigma_- = 0) V_{\Lambda\mathrm{L}}(k, \sigma_+ = 0) \tag{12.88}$$

with $\sigma_\pm$ defined in (12.35c). L_0, $\tilde{L}_0$ generate translations of σ_-, σ_+ for the right- and left-moving pieces, so

$$V_{\Lambda R}(k, \tau - \sigma) = e^{2i(\tau - \sigma)L_0} V_{\Lambda R}(k, 0)\, e^{-2i(\tau - \sigma)L_0} \tag{12.89a}$$

$$V_{\Lambda L}(k, \tau + \sigma) = e^{2i(\tau + \sigma)\tilde{L}_0} V_{\Lambda L}(k, 0)\, e^{-2i(\tau + \sigma)\tilde{L}_0} \tag{12.89b}$$

which, as in (12.51), can be rewritten in terms of the variables z, $\bar{z}$ introduced in (12.82):

$$V_{\Lambda R}(k, z) = z^{L_0 - 1} V_{\Lambda R}(k, 1) z^{-L_0 + 1} \tag{12.90a}$$

$$V_{\Lambda L}(k, \bar{z}) = \bar{z}^{\tilde{L}_0 - 1} V_{\Lambda L}(k, 1) \bar{z}^{-\tilde{L}_0 + 1}. \tag{12.90b}$$

As before these are precisely the factors that appear in the propagator Δ given in (12.86). For this reason the final expression for A_N is very similar to (12.56):

$$A_N = \left(\frac{g}{4\pi}\right)^{N-2} 4\pi \int_{\mathscr{D}} \prod_{i=3}^{N-1} \frac{d^2 z_i}{|z_i|^2} \langle \Lambda_1; k_1 | V_{\Lambda_2}(k_2, 1, 1) V_{\Lambda_3}(k_3, z_3, \bar{z}_3) \cdots$$

$$\times V_{\Lambda_{N-1}}(k_{N-1}, z_{N-1}, \bar{z}_{N-1}) | \Lambda_N; k_N \rangle + \text{permutations} \tag{12.91a}$$

with

$$\times V_{\Lambda}(k, z, \bar{z}) = V_{\Lambda R}(k, z) V_{\Lambda L}(k, \bar{z}) \tag{12.91b}$$

as in (12.88). The domain of integration is

$$\mathscr{D}: \quad 0 < |z_{N-1}| < |z_{N-2}| < \cdots < |z_3| < 1. \tag{12.91c}$$

In this case the simplest amplitude, the three-tachyon vertex is

$$A_3 = g\langle 0; k_1 | e^{-ik_2 \cdot X_R(1)} e^{-ik_2 \cdot X_L(1)} | 0; k_3 \rangle = g\langle 0; k_1 | e^{-ik_2 \cdot x} | 0; k_3 \rangle \tag{12.92}$$

since only the zero-mode part of the vertex operator is activated. The external tachyon states are defined by

$$|0; k\rangle \equiv e^{-ik \cdot x} |0\rangle_R |0\rangle_L \tag{12.93}$$

(with $k^2 = -8$). Then as in (12.68)

$$A_3 = g\langle 0; k_1 | 0; k_2 + k_3 \rangle \tag{12.94a}$$

that is

$$A \sim g \tag{12.94b}$$

omitting the energy–momentum-conserving δ-function multiplied by $(2\pi)^D$.

Similarly the tachyon–tachyon–graviton vertex may be derived in a way that closely parallels the tachyon–tachyon–vector vertex in open-string theory:

$$A_3 = g\langle 0; k_1 | V_{g, \epsilon_2}(k_2, 1, 1) | 0; k_3 \rangle \tag{12.95a}$$

where

$$V_{g,\epsilon}(k, z, \bar{z}) = W_{g,\epsilon}(z, \bar{z})\, \mathrm{e}^{-\mathrm{i}k.X(z,\bar{z})} \tag{12.95b}$$

with $W_{g,\epsilon}(\tau, \sigma)$ given in (12.42). Using the expansions (7.26), (7.27) we find

$$W_{g,\epsilon}(1, 1) = \epsilon^{\mu\nu}(\tfrac{1}{2}p_\mu + \Sigma\alpha_\mu)(\tfrac{1}{2}p_\nu + \Sigma\tilde{\alpha}_\nu). \tag{12.96}$$

Hence

$$\begin{aligned} A_3 &= \tfrac{1}{4}g\langle 0; k_1|\epsilon_2^{\mu\nu} p_\mu p_\nu|0; k_2 + k_3\rangle \sim \tfrac{1}{4}g\epsilon_2^{\mu\nu}k_{3\mu}k_{3\nu} \\ &= \tfrac{1}{4}g\epsilon_2^{\mu\nu}k_{1\mu}k_{1\nu} = -\tfrac{1}{4}g\epsilon_2^{\mu\nu}k_{3\mu}k_{1\nu} \end{aligned} \tag{12.97}$$

using energy–momentum conservation and the transverse properties (12.41*b*) of $\epsilon_2^{\mu\nu}$.

We leave the other closed-string three-point vertices as exercises and turn now to the four-tachyon scattering amplitude. Equation (12.87) gives

$$\begin{aligned} A_4 &= g^2\langle 0; k_1|V_0(k_2)\,\Delta V_0(k_3)|0; k_4\rangle \\ &\quad + g^2\langle 0; k_1|V_0(k_3)\,\Delta V_0(k_2)|0; k_4\rangle \end{aligned} \tag{12.98}$$

and, using (12.86), we can write the first term as in (12.91) as

$$\frac{g^2}{4\pi}\int_{|z|\leqslant 1}\frac{\mathrm{d}^2z}{|z|^2}\langle 0; k_1|V_0(k_2, 1, 1)V_0(k_3, z, \bar{z})|0; k_4\rangle. \tag{12.99}$$

Similarly, we can use the propagator in the second term to write it as

$$\frac{g^2}{4\pi}\int_{|z|\leqslant 1}\frac{\mathrm{d}^2z}{|z|^2}\langle 0; k_1|V_0(k_3, z^{-1}, \bar{z}^{-1})V_0(k_2, 1, 1)|0; k_4\rangle. \tag{12.100}$$

Now change variables in this term to

$$w = z^{-1} \qquad \bar{w} = \bar{z}^{-1} \tag{12.101}$$

so that

$$\frac{\mathrm{d}^2z}{|z|^2} = \frac{\mathrm{d}^2w}{|w|^2}. \tag{12.102}$$

Then

$$A_4 = \frac{g^2}{4\pi}\int\frac{\mathrm{d}^2z}{|z|^2}\langle 0; k_1|P[V_0(k_2, 1, 1)V_0(k_3, z, \bar{z})]|0; k_4\rangle \tag{12.103}$$

where the region of integration is the whole of the z-plane and P is an ordering operation, defined by

$$P[V_0(k_2, 1, 1)V_0(k_3, z, \bar{z})] = \begin{cases} V_0(k_2, 1, 1)V_0(k_3, \bar{z}, \bar{z}) & |z| \leqslant 1 \\ V_0(k_3, z, \bar{z})V_0(k_2, 1, 1) & |z| > 1. \end{cases} \tag{12.104}$$

The factorization property (12.88) of the vertex operator means that the integrand in (12.103) also factorizes into left- and right-moving pieces. The contributions of each of these is readily evaluated as in the open-string case (see (12.77) to (12.80)). First note that using the notation (12.82)

$$X^\mu(z,\bar{z}) = X^\mu_{\mathrm{R}}(z) + X^\mu_{\mathrm{L}}(\bar{z}) \tag{12.105a}$$

with

$$X^\mu_{\mathrm{R}}(z) = \tfrac{1}{2}x^\mu - \frac{\mathrm{i}}{4}p^\mu \ln z + \frac{\mathrm{i}}{2}\sum_{n\neq 0}\alpha^\mu_n z^{-n} \tag{12.105b}$$

$$X^\mu_{\mathrm{L}}(\bar{z}) = \tfrac{1}{2}x^\mu - \frac{\mathrm{i}}{4}p^\mu \ln \bar{z} + \frac{\mathrm{i}}{2}\sum_{n\neq 0}\tilde{\alpha}^\mu_n \bar{z}^{-n}. \tag{12.105c}$$

Thus the zero-mode part of $V_0(k_3, z, \bar{z})$ gives

$$\begin{aligned}\exp(-\mathrm{i}\,k_3.x - \tfrac{1}{4}k_3.p\ln z\bar{z})|0;k_4\rangle &= \mathrm{e}^{-\mathrm{i}\,k_3.x}(z\bar{z})^{-k_3.k_4/4 - k_3^2/4}\\ &= |z|^{-k_3.k_4/2+2}|0;k_3+k_4\rangle\end{aligned} \tag{12.106}$$

since for the *closed*-string tachyon state $k_3^2 = -8$. As in (12.80), the non-zero mode contribution is

$$(1-z)^{-k_2.k_3/4}(1-\bar{z})^{-k_2.k_3/4} \tag{12.107}$$

so

$$A_4 = \frac{g^2}{4\pi}\int \mathrm{d}^2z\,|z|^{-k_3.k_4/2}|1-z|^{-k_2.k_3/2}. \tag{12.108}$$

The general expression for integrals of this and more general forms is given in Appendix D (see (D11)).

Thus[4]

$$\begin{aligned}A_4 &= \frac{g^2}{4}\frac{\Gamma(1-k_3.k_4/4)\Gamma(1-k_2.k_3/4)\Gamma(1-k_1.k_3/4)}{\Gamma(k_3.k_4/4)\Gamma(k_2.k_3/4)\Gamma(k_1.k_3/4)}\\ &= \frac{g^2}{4}\frac{\Gamma(-1-s/8)\Gamma(-1-t/8)\Gamma(-1-u/8)}{\Gamma(2+s/8)\Gamma(2+t/8)\Gamma(2+u/8)}\end{aligned} \tag{12.109a}$$

where

$$s \equiv (k_1+k_2)^2 = 2k_1.k_2 - 16 \tag{12.109b}$$

$$t \equiv (k_2+k_3)^2 = 2k_2.k_3 - 16 \tag{12.109c}$$

$$u = (k_1+k_3)^2 = 2k_1.k_3 - 16 \tag{12.109d}$$

now, because of the different tachyon mass. As expected this has poles at

$$s, t, u = 8(n-1) \qquad n = 0, 1, 2, \ldots \tag{12.110}$$

corresponding to the tachyon, (massless) graviton, anti-symmetric tensor, dilaton, and the whole tower of massive string states.

The factorization (12.88) of the closed-string vertex operator into vertex operators (12.89*a*) and (12.89*b*) associated with the right- and left-moving modes, suggests that the result (12.109) might also be separable. Also, since both of the vertex operators (12.89*a*) and (12.89*b*) are similar to the open-string vertex operator, we might suspect that our result is expressible in terms of the open-string result (12.81) for the four-tachyon scattering amplitude. It is easy to verify that this is the case. Using the elementary property

$$\Gamma(x)\Gamma(1-x) = \pi/\sin \pi x \tag{12.111}$$

we see that[5]

$$g^2 A_4^{\text{closed}}(s, t, u) = \frac{\sin \pi t/8}{4\pi} A_4^{\text{open}}\left(\frac{s}{4}, \frac{t}{4}\right) A_4^{\text{open}}\left(\frac{t}{4}, \frac{u}{4}\right) \tag{12.112}$$

where $A_4^{\text{open}}(s, t)$ is given in (12.81).

12.5 The superstring vertex operator

In order to calculate scattering amplitudes for a superstring theory, it is obviously essential to construct vertex operators for three basic processes: (i) the emission of an on-shell bosonic state from a bosonic string; (ii) the emission of an on-shell bosonic state from a fermionic string; and (iii) the emission of an on-shell fermionic state from a bosonic string, changing it to a fermionic string. (The latter will then fix the emission of an on-shell fermionic state from a fermionic string, changing it to a bosonic string.) One might suspect that the first of these, involving only bosons and the bosonic string, would be covered by our earlier treatment in §12.2, and indeed it *is* true that we require the vertex operator $V_\Lambda(0)$, describing the emission of the bosonic state Λ from the point $\tau = 0$ on the edge of an open string, to have conformal dimension

$$J = 1. \tag{12.113}$$

However, in a superstring theory $V_\Lambda(0)$ is constrained by the Virasoro super-algebra; the rescalings (12.10) are now merely special cases of the world sheet super-reparametrizations that preserve the super-conformal gauge (8.28), (8.29). In the (open-) superstring theory we now require the existence of an operator $W_\Lambda(0)$ such that

$$V_\Lambda(0) = [G_r, W_\Lambda(0)]_\pm \tag{12.114}$$

for all $r \in \mathbb{Z} + \frac{1}{2}$. We take the commutator $[\ ,\ ]_-$ or anti-commutator

$[\ ,\ ]_+$ depending upon whether $W_\Lambda(0)$ is a world sheet bosonic or fermionic operator.

Then it can be shown (Exercise 12.10) that $V_\Lambda(0)$ has the required conformal dimension $J = 1$ if and only if $W_\Lambda(0)$ has conformal dimension $J = \frac{1}{2}$. We can verify this as follows. Take

$$W_\Lambda(0) = :\exp(-\mathrm{i}\, k\,.\,X(0)): \tag{12.115}$$

with $X(z)$ having the usual (open-string) expansion, as in (12.27). Then using

$$G_r = -\sum_{n\in\mathbb{Z}} b^\mu_{r-n}\alpha_{\mu n} \tag{12.116}$$

which is the open-string analogue of (8.76), we find[6]

$$[G_r, W_\Lambda(0)] = -\sqrt{2}k\,.\,\psi(0) :\mathrm{e}^{-\mathrm{i}\,k\,.\,X(0)}: \tag{12.117a}$$

where

$$\psi^\mu(\tau) = \frac{1}{\sqrt{2}}\sum_{r\in\mathbb{Z}+1/2} b^\mu_r\,\mathrm{e}^{-\mathrm{i}\,r\tau} \tag{12.117b}$$

is the (NS) world sheet fermion field. Now, the invariance of the superstring action (8.1) under the rescalings (12.10) requires that ψ^μ (and $\bar{\psi}^\mu$) acquire a factor of $\lambda^{-1/2}$ under the rescaling, which is consistent with ψ having conformal dimension $J = \frac{1}{2}$. Thus in this case $V_\Lambda(0)$ defined in (12.114) will have the required conformal dimension $J = 1$ if and only if the conformal dimension of W_Λ in (12.115) is $J = \frac{1}{2}$. We have seen already, in (12.28), that the conformal dimension of W_Λ is $-k^2/2$, so we only get the required value for $k^2 = -1$, which corresponds to the tachyonic state that is removed by the GSO projection.

The (undeleted) massless vector state $|v; \epsilon, k\rangle$ having momentum k and polarization vector ϵ is given by

$$|v; \epsilon, k\rangle = -\epsilon_\mu b^\mu_{-1/2}|0; k\rangle. \tag{12.118}$$

This suggests that to construct the vertex operator we take

$$W_v = -\sqrt{2}\epsilon\,.\,\psi(0) :\mathrm{e}^{-\mathrm{i}\,k\,.\,X(0)}: \tag{12.119}$$

which has $J = \frac{1}{2}$, since $k^2 = 0$. Then

$$V_v = \{G_r, W_v\} \tag{12.120}$$

in which we take the *anti*-commutator of G_r with W_v, since both are fermionic, has the required conformal dimension $J = 1$. This may also be readily verified by calculating V_v explicitly, which gives[6]

$$V_v = :[-\epsilon\,.\,\partial_\tau X(0) - 2\epsilon\,.\,\psi(0)k\,.\,\psi(0)]\,\mathrm{e}^{-\mathrm{i}\,k\,.\,X(0)}:. \tag{12.121}$$

The general rule is that if W is a bosonic operator (on the world sheet), such as (12.115), involving an even number of ψs, then V is constructed using the commutator with G_r, as in (12.114). However, if W is a fermionic operator on the world sheet, such as (12.119), involving an odd number of ψs then V is constructed using the anti-commutator with G_r, as in (12.120). In fact we saw in §8.8 that the GSO projection deletes states involving an even number of ψs in the NS sector, so we will only ever need the anti-commutator (12.120) for physical processes in this sector.

In the Ramond sector, describing space-time fermions, the bosonic emission vertex operators have essentially the same form as when emitted in the Neveu–Schwarz sector; a vertex operator is associated with emission at a specific point on the world sheet, and this should not be affected by the difference in boundary conditions which is the only distinction between the space-time bosonic (NS) and the space-time fermionic (R) sectors. The formulae (12.114) relating V_Λ to W_Λ must have G_r replaced by F_m with

$$F_m = -\sum_{n \in \mathbb{Z}} d^\mu_{m-n}\alpha_{\mu n} \tag{12.122}$$

which is the open-string analogue of (8.78), and the upshot is that one simply uses the expansion (12.117*b*) of $\psi^\mu(\tau)$ in half-integral modes for emission from a bosonic (NS) string or

$$\psi^\mu(\tau) = \frac{1}{\sqrt{2}}\sum_{n \in \mathbb{Z}} d^\mu_n \mathrm{e}^{-\mathrm{i}n\tau} \tag{12.123}$$

for emission from a fermionic (R) string.

The vertex operator (V_F) that describes fermion emission[7] from a bosonic string is altogether more complicated. It must change the incoming bosonic string into an outgoing fermionic string. In the R–NS formulation that we are using this entails changing the boundary conditions of the ψ^μ field. Thus V_F must be associated with a cut on the world sheet with a branch point at the specific point where the emission occurs, and it is difficult, at first sight, to see how to construct such an operator from X^μ, ψ^μ. The trick is to bosonize the R–NS fermions in pairs:

$$\psi^{2m-1} \pm \mathrm{i}\,\psi^{2m} = \mathrm{e}^{\pm\mathrm{i}\varphi_m} \tag{12.124}$$

and then construct the spin operators

$$D^m_{\pm 1/2} = \mathrm{e}^{\pm\mathrm{i}\varphi_m/2}. \tag{12.125}$$

Then

$$\Theta_\alpha \equiv D^1_{\pm 1/2}D^2_{\pm 1/2}D^3_{\pm 1/2}D^4_{\pm 1/2}D^5_{\pm 1/2} \tag{12.126}$$

has 32 (=2^5) components and transforms as an SO(10) spinor $\mathbf{16}+\overline{\mathbf{16}}$ representation, and creates the required cuts in all of the ψ^μ. However, Θ_α

has conformal dimension $\frac{5}{8}$ and has to be augmented by a (ghost spin) operator $\Sigma_{+1/2}$ having dimension $\frac{3}{8}$ in order to obtain the requisite $J = 1$. It is beyond the scope of this book to give further details of this, and we refer the interested reader to the extensive literature on this topic[8]. Suffice it to say that the vertex operator

$$V_F = \Sigma_{+1/2}\bar{u}^{\alpha}\Theta_{\alpha}\, e^{-ik.X} \tag{12.127}$$

relates to the emission of a massless fermion of momentum k described by the spinor u^{α}, and so

$$\Gamma^{\mu}k_{\mu}u = 0 \tag{12.128}$$

with Γ^{μ} the 10-dimensional gamma matrices. It is also true, but not obvious, that the appearance of the super-conformal ghost operator $\Sigma_{+1/2}$ is compatible with the decoupling of these ghosts, as well as the time-like ghosts discussed in §8.6.

The extension of all of these considerations to the *closed* superstring is straightforward, and very similar to that given in §12.2 for bosonic string. This is because the left- and right-moving components are essentially independent in the closed string, and each behaves very much like the open string. As in (12.88), the vertex operator factorizes into left- and right-moving pieces:

$$V_{\Lambda} = V_{\Lambda L}V_{\Lambda R} \tag{12.129}$$

and we require that the conformal dimensions satisfy

$$J_L = 1 = J_R. \tag{12.130}$$

Consider first the massless (ten-dimensional) graviton bosonic states (8.168) that survive the GSO projection with momentum k and polarization tensor $\epsilon_{\mu\nu}$. Then the associated vertex operator is

$$V_{g,\epsilon}(k) = \epsilon_{\mu\nu}[\partial_- X^{\mu}_{R} + \tfrac{1}{2}\psi^{\mu}_{R}k.\psi_R][\partial_+ X^{\nu}_{L} + \tfrac{1}{2}\psi^{\nu}_{L}k.\psi_L]\, e^{-ik.X} \tag{12.131}$$

in which the right- and left-moving contributions are obtained by computing the anti-commutators of G_r and $\bar{G}_r$ with

$$\psi^{\mu}_{L,R}\, e^{-ik.X_{L,R}},$$

analogously to (12.121); $\partial_{\pm}$ are defined in (12.42*c*).

The vertex operators for the other massless states in (8.166), (8.167) and (8.169) that survive the GSO projection are constructed similarly. Thus for the (ten-dimensional) gravitino fermionic states (8.166) having momentum k and vector spinor u^{α}_{μ} of definite chirality, the vertex operator is

$$V_{\tilde{g}}(\tilde{u}^{\alpha}_{\mu}, k) = \bar{u}^{\alpha}_{\mu}\Sigma^{R}_{+1/2}\Theta^{R}_{\alpha}[\partial_+ X^{\mu}_{L} + \tfrac{1}{2}\psi^{\mu}_{L}k.\psi_L]\, e^{-ik.X} \tag{12.132}$$

where $\Sigma^{R}_{+1/2}$ is the right movers'conformal ghost operator, and Θ^{R}_{α} is defined as in (12.126) for the right-moving R–NS fermions.

12.6 Superstring scattering amplitudes

As in the bosonic string we consider only tree scattering amplitudes. We start with the N-boson amplitude A_N. As before, we expect that the superstring amplitude has the form (12.43)

$$A_N = g^{N-2}\langle\varphi_1|V_{\Lambda_2}(k_2)\,\Delta V_{\Lambda_3}(k_3)\cdots V_{\Lambda_{N-1}}(k_{N-1})|\varphi_N\rangle \tag{12.133}$$

and that for suitably chosen vertex operators V_{Λ_i}, and propagator Δ, this will possess the usual properties of unitarity and, in the case of open strings, cyclic symmetry. The propagator is determined, as before, from the Hamiltonian. For the open string we have

$$H = -\tfrac{1}{2}(p^2 - M^2) = L_0 - a \tag{12.134}$$

where, in the bosonic (NS) sector, L_0 is given by (8.93) and $a_{\rm NS} = \frac{1}{2}$, as given in (8.154). Thus we take the superstring propagator to be

$$\Delta = (L_0 - \tfrac{1}{2})^{-1} = \int_0^1 \mathrm{d}z\, z^{L_0 - 3/2}. \tag{12.135}$$

Then proceeding as in §12.3 it is easy to show that A_N is given precisely as in (12.56) by

$$A_N = g^{N-2}\int_{\mathscr{D}} \prod_{i=3}^{N-1} \frac{\mathrm{d}y_i}{y_i} \langle\varphi_1|V_{\Lambda_2}(k_2, 1)V_{\Lambda_3}(k_3, y_3)\cdots$$

$$\times V_{\Lambda_{n-1}}(k_{N-1}, y_{N-1})|\varphi_N\rangle. \tag{12.136}$$

However, it is not possible to proceed as we did before to obtain the form (12.64) which facilitates the proof of cyclic symmetry. The problem is best illustrated by considering an N-tachyon amplitude, even though the GSO projection deletes this state. We see from (12.62) that the incoming tachyon state $|0; k\rangle$ is obtained from

$$|0; k\rangle = \lim_{z\to 0} z^{k^2/2}\,\mathrm{e}^{-\mathrm{i}kX}|0, 0\rangle = \lim_{z\to 0} z^{-1/2}W_0(k, z)|0, 0\rangle. \tag{12.137}$$

Then the analogue of (12.64) will involve $N-2$ V-operators and two W-operators, rather than the N V-operators we had in the bosonic string; this derives from the conformal dimensions of the operator W_0 being $-k^2/2 = \frac{1}{2}$ rather than the $J = 1$ of the vertex operators V_0. For this reason the proof of cyclic symmetry is most readily given in a different 'picture' to the one we have discussed so far. It is beyond the scope of this book to give details of the alternative (F_1) picture or its equivalence to the (F_2) picture we have used hitherto. The interested reader is referred to other texts[(8)] for this.

The simplest amplitude is the three-point function A_3. Since the tachyon state is deleted by the GSO projection, the lowest-mass bosonic state is the massless vector state. The three-vector vertex[7] is

$$A_3 = g\langle v; \epsilon_1, k_1 | V_{v,\epsilon_2}(k_2) | v; \epsilon_3, k_3\rangle$$
$$= g\langle 0; k_1 | \epsilon_1 . b_{1/2} V_{v,\epsilon_2}(k_2) \epsilon_3 . b_{-1/2} | 0; k_3\rangle. \quad (12.138)$$

Using the vertex (12.121), and exhibiting only the active modes in this amplitude, we find

$$V_{v,\epsilon_2}(k_2) = [-\epsilon_2 . p - \epsilon_2 . b_{-1/2} k_2 . b_{1/2} + k_2 . b_{-1/2} \epsilon_2 . b_{1/2}]\, \mathrm{e}^{-\mathrm{i} k_2 . x}. \quad (12.139)$$

Thus, using the anti-commutation relations, we find

$$A_3 = g(\epsilon_1 . \epsilon_3 \epsilon_2 . k_3 - \epsilon_1 . \epsilon_2 \epsilon_3 . k_2 + \epsilon_1 . k_2 \epsilon_2 . \epsilon_3) = \tfrac{1}{2} g[\epsilon_1 . \epsilon_2 \epsilon_3 . (k_1 - k_2)$$
$$+ \epsilon_2 . \epsilon_3 \epsilon_1 . (k_2 - k_3) + \epsilon_3 . \epsilon_1 \epsilon_2 . (k_3 - k_1)]. \quad (12.140)$$

Just like the (open-) bosonic string amplitude (12.74), this amplitude is Bose anti-symmetric, and survives only in a non-abelian theory. However, unlike in (12.74), there are no additional $\mathrm{O}(\alpha')$ terms; superstring theory delivers precisely the three-vector interaction vertex of the Yang–Mills Lagrangian.

For the *closed* superstring, the analogous process is the three-graviton vertex. We leave it as an exercise to show that using the graviton emission vertex (12.131) we get

$$A_3 = \tfrac{1}{16} g \epsilon_1^{\rho\sigma} \epsilon_2^{\mu\nu} \epsilon_3^{\alpha\beta} [\eta_{\rho\mu}(k_1 - k_2)_\alpha + \eta_{\mu\alpha}(k_2 - k_3)_\rho + \eta_{\alpha\rho}(k_3 - k_1)_\mu]$$
$$\times [\eta_{\sigma\nu}(k_1 - k_2)_\beta + \eta_{\nu\beta}(k_2 - k_3)_\sigma + \eta_{\beta\sigma}(k_3 - k_1)_\nu]. \quad (12.141)$$

Thus if we write the open-string three-vector amplitude (12.140) as

$$A_3^{\text{open}} = g \epsilon_1^\rho \epsilon_2^\mu \epsilon_3^\alpha V_{\rho\mu\alpha}(k_1, k_2, k_3) \quad (12.142)$$

we see that the closed-string three-graviton amplitude (12.141) is given by

$$A_3^{\text{closed}} = g \epsilon_1^{\rho\sigma} \epsilon_2^{\mu\nu} \epsilon_3^{\alpha\beta} V_{\rho\mu\alpha}(\tfrac{1}{2}k_1, \tfrac{1}{2}k_2, \tfrac{1}{2}k_3) V_{\sigma\nu\beta}(\tfrac{1}{2}k_1, \tfrac{1}{2}k_2, \tfrac{1}{2}k_3). \quad (12.143)$$

This gives precisely the same three-graviton vertex as that implied by the Einstein–Hilbert action of general relativity:

$$\mathcal{L}_{\mathrm{GR}} = \sqrt{-g}\mathcal{R} \quad (12.144a)$$

where

$$g = \det(g_{\mu\nu}) \quad (12.144b)$$

and $\mathcal{R}$ is the curvature scalar. In the weak-field limit, in which $g_{\mu\nu}(x)$ is expanded about the flat metric $\eta_{\mu\nu}$:

$$g_{\mu\nu}(x) = \eta_{\mu\nu} - \kappa h_{\mu\nu}(x)$$

$\mathcal{L}_{\mathrm{GR}}$ generates trilinear terms in the graviton field $h_{\mu\nu}$ which reproduce the vertex (12.142) for on-shell gravitons (Exercise 12.14). This is one of the string theory 'miracles'. Nothing like *space-time* reparametrization invariance was input into the string action, and yet it has emerged in the effective action.

As before, the first non-trivial amplitudes are the four-point functions. For the open-string four-vector amplitude we need

$$A_4 = g^2 \int \frac{dz}{z} \langle v; \epsilon_1, k_1 | V_{v,\epsilon_2}(k_2, 1) V_{v,\epsilon_3}(k_3, z) | v; \epsilon_4, k_4 \rangle \tag{12.145a}$$

where

$$|v; \epsilon, k\rangle = \epsilon . b_{-1/2} |0; k\rangle \tag{12.145b}$$

and

$$V_{v,\epsilon}(k, z) = - :[\mathrm{i}\, z\epsilon\, \partial_z X(z) + 2\epsilon . \psi(z) k . \psi(z)]\, \mathrm{e}^{-\mathrm{i}\, kX(z)} \tag{12.145c}$$

with $z = \mathrm{e}^{\mathrm{i}\tau}$, as in the bosonic string. The existence of the fermionic modes, as well as the additional complication of the vector vertex, make this calculation considerably more involved than that for the four-tachyon amplitude that was carried out in §12.4. It is nevertheless straightforward and we merely outline the calculation. First we use the zero-mode contributions to the $\exp(-\mathrm{i}\, kX)$ factors to write

$$\begin{aligned} A_4 = g^2 \int dz\, z^{-s/2-1} \langle 0; k_1 + k_2 | \epsilon_1 . b_{1/2} :\Big[-\epsilon_2 . k_1 + \sum_{m \neq 0} \epsilon_2 . \alpha_m \\ + 2k_2 . \psi(1) \epsilon_2 . \psi(1) \Big] W_0(k_2, 1) :: \Big[\epsilon_3 . k_4 + \sum_{n \neq 0} \epsilon_3 . \alpha_n z^{-n} \\ + 2k_3 . \psi(z) \epsilon_3 . \psi(z) \Big] W_0(k_3, z) : \epsilon_4 . b_{-1/2} | 0; k_3 + k_4 \rangle \end{aligned} \tag{12.146a}$$

where

$$W_0(k, z) = \exp\left(- \sum_{p=1}^{\infty} \frac{k . \alpha_{-p}}{p} z^p \right) \exp\left(\sum_{q=1}^{\infty} \frac{k . \alpha_q}{q} z^{-q} \right) \tag{12.146b}$$

and because the vector particles are massless

$$\begin{aligned} s &\equiv (k_1 + k_2)^2 = 2k_1 . k_2 \\ t &\equiv (k_2 + k_3)^2 = 2k_2 . k_3 \\ u &\equiv (k_3 + k_1)^2 = 2k_3 . k_1 . \end{aligned} \tag{12.146c}$$

This splits into four pieces as follows. First, there is the contribution proportional to the bosonic piece of both square brackets. This may be evaluated using the previous results in (12.78) and (12.80), as well as

$$\left[\sum_{m>0} \epsilon_2 . \alpha_m, W_0(k_3, z)\right] = -\epsilon_2 . k_3 \frac{z}{1-z} W_0(k_3, z) \qquad (12.147a)$$

$$\left[W_0(k_2, 1), \sum_{n<0} \epsilon_3 . \alpha_n z^{-n}\right] = \epsilon_3 . k_2 \frac{z}{1-z} W_0(k_2, 1) \qquad (12.147b)$$

$$\left[\sum_{m} \epsilon_2 . \alpha_m, \sum_{n<0} \epsilon_3 . \alpha_n z^{-n}\right] = -\epsilon_2 . \epsilon_3 \frac{z}{(1-z)^2} . \qquad (12.147c)$$

Next there is the contribution proportional to the fermionic piece $k_2 . \psi(1)\epsilon_2 . \psi(1)$ of the first bracket and the bosonic piece of the second bracket. The fermionic operators generate a factor $\epsilon_1 . k_2 \epsilon_2 . \epsilon_4 - \epsilon_1 . \epsilon_2 k_2 \cdot \epsilon_4$, and the remaining bosonic piece can be determined as above. Similarly the bosonic piece of the first bracket multiplied by the fermionic piece of the second generates a factor $\epsilon_1 . k_3 \epsilon_3 . \epsilon_4 - \epsilon_1 . \epsilon_3 k_3 . \epsilon_4$. Finally, there is the contribution proportional to the fermionic piece of both square brackets. This requires the evaluation of

$$\langle 0|\epsilon_1 . b_{1/2} k_2 . \psi(1)\epsilon_2 . \psi(1) k_3 . \psi(z)\epsilon_3 . \psi(z)\epsilon_4 . b_{-1/2}|0\rangle. \qquad (12.148)$$

This too is straightforward (but tedious) and uses the contraction

$$\langle 0|\psi^\mu(1)\psi^\nu(z)|0\rangle = -\frac{\sqrt{z}}{1-z}\eta^{\mu\nu} \qquad (12.149)$$

extensively. The final result is that

$$\begin{aligned} A_4 = -\tfrac{1}{2}g^2 \frac{\Gamma(-s/2)\Gamma(-t/2)}{\Gamma(1+u/2)} &\{-\tfrac{1}{4}[st\epsilon_{13}\epsilon_{24} + su\epsilon_{23}\epsilon_{14} + tu\epsilon_{12}\epsilon_{34}] \\ &+ \tfrac{1}{2}s[k_{14}k_{32}\epsilon_{24} + k_{23}k_{41}\epsilon_{13} + k_{13}k_{42}\epsilon_{23} + k_{24}k_{31}\epsilon_{14}] \\ &+ \tfrac{1}{2}t[k_{21}k_{43}\epsilon_{31} + k_{34}k_{12}\epsilon_{24} + k_{24}k_{13}\epsilon_{34} + k_{31}k_{42}\epsilon_{21}] \\ &+ \tfrac{1}{2}u[k_{12}k_{43}\epsilon_{32} + k_{34}k_{21}\epsilon_{14} + k_{14}k_{23}\epsilon_{34} + k_{32}k_{41}\epsilon_{12}]\} \end{aligned} \qquad (12.150a)$$

where

$$\epsilon_{ij} \equiv \epsilon_i . \epsilon_j \qquad k_{ij} \equiv \epsilon_i . k_j . \qquad (12.150b)$$

The calculation of the analogous process for the closed superstring, namely the four-graviton amplitude[5], proceeds similarly, and, as in (12.112), the final result is expressible in terms of the open-string four-vector amplitude (12.150). If we write the latter as

$$A_4^{\text{open}} = g^2 \frac{\Gamma(-s/2)\Gamma(-t/2)}{\Gamma(1+u/2)} \epsilon_1^{\alpha} \epsilon_2^{\beta} \epsilon_3^{\gamma} \epsilon_4^{\delta} V_{\alpha\beta\gamma\delta}(k_1, k_2, k_3, k_4) \tag{12.151}$$

then the required closed-string four-graviton amplitude is given by

$$A_4^{\text{closed}} = g^2 C(s,t,u) \epsilon_1^{\alpha\alpha'} \epsilon_2^{\beta\beta'} \epsilon_3^{\gamma\gamma'} \epsilon_4^{\delta\delta'} V_{\alpha\beta\gamma\delta}(\tfrac{1}{2}k_1, \tfrac{1}{2}k_2, \tfrac{1}{2}k_3, \tfrac{1}{2}k_4)$$
$$\times V_{\alpha'\beta'\gamma'\delta'}(\tfrac{1}{2}k_1, \tfrac{1}{2}k_2, \tfrac{1}{2}k_3, \tfrac{1}{2}k_4) \tag{12.152a}$$

where

$$C(s,t,u) = -\pi \frac{\Gamma(-s/8)\Gamma(-t/8)\Gamma(-u/8)}{\Gamma(1+s/8)\Gamma(1+t/8)\Gamma(1+u/8)}. \tag{12.152b}$$

As expected, both open- and closed-string amplitudes have no tachyon poles, and possess the poles from the massless and massive modes in the superstring spectrum.

The only other amplitudes that we address are those (tree) processes in which bosons are emitted from a fermionic string. Thus there are two external fermion lines, and, if bosons are emitted, the required amplitude is given by[9] an expression analogous to (12.133):

$$A_{2,N} = g^N \langle \psi_1 | W_1 S W_2 S \cdots S W_N | \psi_2 \rangle \tag{12.153}$$

where $|\psi_1\rangle$ and $|\psi_2\rangle$ are physical *fermion* states. For the open string these satisfy

$$F_m |\psi\rangle = 0 \qquad m > 0 \tag{12.154}$$

analogously to the closed-string constraint (8.104). Actually the constraint equations also require

$$F_0 |\psi\rangle = 0 \tag{12.155}$$

since there is no normal-ordering ambiguity. Also, as in (8.111) and (8.115), we see that

$$F_0^2 = L_0 \tag{12.156}$$

so F_0 plays the role of the Dirac operator, and

$$S = F_0^{-1} = F_0 L_0^{-1} \tag{12.157}$$

is then the analogue of the fermionic propagator that is needed in (12.153). The factors W_i are precisely those discussed earlier in this section. So for vector emission we use the vertex operator given in (12.119), but *now*, since we are concerned with emission from a fermionic string, we use the Ramond sector expansion of

$$\psi^\mu(\tau) = \frac{1}{\sqrt{2}} \sum_{n \in \mathbb{Z}} d_n^\mu \, e^{-in\tau}. \tag{12.158}$$

In the case of the open string the simplest example is the emission of a vector particle from the Ramond (spinor) ground state

$$|0; k\rangle_\alpha u^\alpha(k) \tag{12.159}$$

where α is the (summed) spinor index and $u^\alpha(k)$ is a (massless) spinor. Then (suppressing the spinor indices)

$$\begin{aligned} A_{2,1} &= g\bar{u}(k_1)\langle 0; k_1|W_{v,\epsilon_2}(k_2)|0; k_3\rangle u(k_3) \\ &= -g\bar{u}(k_1)\langle 0; k_1|\epsilon_2 d_0|0; k_2 + k_3\rangle u(k_3) \\ &\sim \frac{1}{\sqrt{2}} \mathrm{i}\, g\epsilon_2^\mu \bar{u}(k_1)\gamma_\mu u(k_3) \end{aligned} \tag{12.160}$$

where the last step follows because the operators

$$\gamma^\mu \equiv \mathrm{i}\sqrt{2} d_0^\mu \tag{12.161}$$

satisfy the Clifford algebra

$$\{\gamma^\mu, \gamma^\nu\} = 2\eta^{\mu\nu} \tag{12.162}$$

as in (8.124) and (8.125), and therefore are represented by Dirac matrices (also denoted by γ^μ) acting on the spinor indices. We note that this is precisely the gaugino vertex contained in the (non-abelian) vector superfield Lagrangian (3.138).

The analogous process for the closed string is graviton emission from a gravitino (ground) state. For example we consider the gravitino state of momentum k described by the vector spinor u^μ which is constructed from the right-moving Neveu–Schwarz ground state and the left-moving Ramond ground state:

$$|\tilde{g}; u^\mu, k\rangle = -b^\mu_{-1/2}|0; k\rangle_\mathrm{R}|0; k\rangle_{\mathrm{L}\alpha} u^\alpha_\mu(k) \tag{12.163a}$$

$$\gamma^\rho k_\rho u_\mu(k) = 0 = k^\mu u_\mu(k) \tag{12.163b}$$

The required graviton vertex operator is then constructed using the bosonic right-moving prescription tensored with the fermionic left-moving recipe:

$$\epsilon_{\mu\nu}[\partial_\tau X^\mu_\mathrm{R}(0) + \tfrac{1}{2}\psi^\mu_\mathrm{R} k\,.\,\psi_\mathrm{R}(0)]\psi^\nu_\mathrm{L}(0)\,\mathrm{e}^{-\mathrm{i}k.X}. \tag{12.164}$$

The amplitude for graviton emission from a gravitino then factorizes to give

$$\begin{aligned} A_{2,1} &= g\epsilon_{2\rho\sigma}\,{}_\mathrm{R}\langle 0; k_1|b^\mu_{1/2}{:}[\partial_\tau X^\rho_\mathrm{R}(0) + \tfrac{1}{2}\psi^\rho_\mathrm{R}(0)\, k_2\,.\,\psi_\mathrm{R}(0)]b^\lambda_{-1/2}|0; k_2 + k_3\rangle_\mathrm{R} \\ &\quad \times \bar{u}_\mu(k_1)\,{}_\mathrm{L}\langle 0; k_1|\psi^\sigma_\mathrm{L}(0)|0; k_2 + k_3\rangle_\mathrm{L}\, u_\lambda(k_3) \\ &= \tfrac{1}{2} g\epsilon_{2\rho\sigma}[-\eta^{\mu\lambda}(k_2 + k_3)^\rho + \eta^{\mu\rho}k_2^\lambda - \eta^{\rho\lambda}k_2^\mu]\bar{u}_\mu(k_1)\gamma^\sigma u_\lambda(k_3) \\ &= \tfrac{1}{4} g\epsilon_{2\rho\sigma}[\eta^{\mu\lambda}(k_1 - k_3)^\rho + \eta^{\mu\rho}(k_2 - k_1)^\lambda + \eta^{\rho\lambda}(k_3 - k_2)^\mu] \\ &\quad \times \bar{u}_\mu(k_1)\Gamma^\sigma u_\lambda(k_3). \end{aligned} \tag{12.165}$$

We leave it as an exercise to verify that this is precisely the interaction vertex that arises in the supergravity Lagrangian discussed in Chapter 4.

12.7 A review of further developments

It is beyond the scope of a book at this level to take the development of string theory much further. In particular we shall *not* construct the vertex operators or scattering amplitudes for the ten-dimensional heterotic string theory that was described in Chapter 9. We trust that a conscientious reader who has followed the developments thus far should have little difficulty in sewing together the bosonic string vertex operators, appropriate to the left-moving modes of the heterotic string, and the superstring vertex operators, appropriate to the right movers. It is, though, perhaps of some interest to comment on how calculations are performed in the potentially realistic cases, such as when the ten-dimensional heterotic string is compactified to four dimensions on an orbifold, as discussed in Chapter 10, or when a four-dimensional heterotic string theory is constructed directly, as discussed in Chapter 11.

The purpose of all of these calculations is to determine the effective supergravity (grand unified?) theory that emerges at the string scale, and then to use the renormalization group techniques discussed in Chapter 6 to confront the low-energy (TeV scale) experimental data. We have seen in Chapter 5 that in general a supergravity theory involving chiral and vector superfields is characterized by the superpotential $W(\Phi_i)$, the Kahler potential $G(\varphi_i^*, \varphi_i)$, and the gauge kinetic function $f_{ab}(\Phi_i)$; as before, Φ_i denotes the chiral superfields, and φ_i their scalar-field components. The most immediately accessible of these is the superpotential W. Its form determines the (renormalizable and non-renormalizable) point interactions of the fields, so the calculation of these interactions enables one to infer W.

We start with the orbifold compactifications discussed in Chapter 10. Physical states arise in both the untwisted sector (U) as well as twisted sectors (T) and the allowed couplings are constrained by the requirement of point group, or more generally space group, invariance. Suppose we consider a trilinear (Yukawa) coupling of three twisted-sector states. Each is associated with the string centre-of-mass coordinates at a fixed point (f) of the orbifold satisfying

$$(\theta, l)f \equiv \theta f + l = f \tag{12.166}$$

where θ is a point group element and l a lattice vector. Since f is only defined up to a lattice vector (λ)

$$f \sim f + \lambda \tag{12.167}$$

the lattice vector l associated with f is only defined up to $(1-\theta)\lambda$:

$$l \sim l + (1 - \theta)\lambda. \tag{12.168}$$

Let us denote the space group elements associated with the three fixed points by (θ_i, l_i) $(i = 1, 2, 3)$. Then point group invariance requires

$$\theta_1\theta_2\theta_3 = I \tag{12.169}$$

the identity element of the point group. Thus there is no constraint on UUU couplings from point group invariance and TUU couplings are always forbidden. For the Z_3 orbifold the only (left chiral) twisted-sector states T are in the $\theta = \omega$ sector, so TTU couplings are also forbidden. The allowed TTT couplings are further constrained by space group invariance. We require that

$$(\theta_1, l_1)(\theta_2, l_2)(\theta_3, l_3) \sim (I, 0) \tag{12.170}$$

using the equivalences (12.168) for each i. It is easy to see that this entails

$$l_1 + l_2 + l_3 \sim 0. \tag{12.171}$$

In the case of the Z_3 orbifold the associated lattice vectors (l_i) are given in (10.35) and space group invariance requires the indices $p_\rho^{(i)}$ to satisfy

$$p_\rho^{(1)} + p_\rho^{(2)} + p_\rho^{(3)} = 0 \pmod 3. \tag{12.172}$$

There remains the problem of calculating the non-zero couplings that *are* allowed by space group invariance. The simplest case is when all states are untwisted; these correspond to states that are already present in the Hilbert space before the orbifold construction, i.e. states from strictly periodic loops. In this case the construction of the vertex operator is a straightforward application of (12.129) but *now* only $V_{\Lambda R}$ is a superstring vertex operator; $V_{\Lambda L}$ is built using the bosonic string results of §12.2. The Yukawa couplings can then be calculated, for example, by evaluating the matrix element of the vertex operator for the emission of an (untwisted) boson between two (untwisted) fermionic states. In fact this prescription works also for the TTU Yukawa couplings, when they exist. The two external fermionic states are taken to be the twisted-sector fermions, and the vertex operator for the emission of the untwisted bosonic state is evaluated using the expansions of $X(z, \bar{z})$, $\psi_R(z)$ appropriate to the twisted sector to which the incoming fermion belongs. (Point group invariance requires that the outgoing fermion belongs to the *same* twisted sector.) In this case it is clear that the coupling is independent of the particular fixed point associated with the twisted-sector states. Because of this the TTU and UUU Yukawa couplings have a universal strength and there is no chance of obtaining a hierarchy of Yukawa coupling strengths. Such a hierarchy is desirable phenomenologically, since when the electroweak symmetry breaking occurs it will convert into a hierarchy of fermion masses, thereby explaining the huge disparities observed in the quark and lepton mass spectrum. (It was for this reason that in §6.3 as a first approximation we ignored all Yukawa couplings *except* the coupling of the top quark to the electroweak Higgs.)

The evaluation of the TTT couplings, in which all three states belong to twisted sectors, is more subtle and the outcome has more chance of phenomenological success. This is because the strength of the coupling *does* now depend upon the fixed points associated with each twisted state. We shall not give much detail, but we shall endeavour to explain how the fixed-point dependence enters.

We consider a Z_N point group with elements θ^r $(r = 0, 1, \ldots, N-1)$ satisfying

$$\theta^N = 1. \tag{12.173}$$

As before the twisted-sector states are associated with fixed points f satisfying

$$(1 - \theta^r)(f + \lambda) = l \tag{12.174}$$

where l, λ are lattice vectors. The twisted-sector ground states are created by 'twist fields' $\sigma_{\theta^r, f}(z, \bar{z})$ analogous to the spin field introduced in §12.5:

$$\sigma_{\theta^r, f}(z, \bar{z})|0\rangle = |0_{\theta^r, f}\rangle. \tag{12.175}$$

The required Yukawa coupling acquires its dependence on the fixed points via the three-point correlation function

$$Z \equiv \langle 0|\sigma_{\theta^k, f_a}(z_a, \bar{z}_a)\sigma_{\theta^l, f_b}(z_b, \bar{z}_b)\sigma_{\theta^m, f_c}(z_c, \bar{z}_c)|0\rangle \tag{12.176a}$$

where

$$k + l + \mathrm{m} = 0 \pmod{N} \tag{12.176b}$$

is necessary to satisfy point group invariance (12.169) and the lattice vectors $l_{a,b,c}$, defined in (12.174), must satisfy (12.171). The correlation function may be calculated using the path integral method mentioned in §7.9, in which we perform a functional integral over the string coordinate fields $X^i(z, \bar{z})$. To do this we split $X^i(z, \bar{z})$ into a classical piece (X_{cl}) with quantum excitations (X_{q}):

$$X^i(z, \bar{z}) = X^i_{\mathrm{cl}}(z, \bar{z}) + X^i_{\mathrm{q}}(z, \bar{z}) \tag{12.177}$$

where only the classical piece feels the lattice shift when taken around the twist field $\sigma_{\theta^k, f}$. The action S is given by

$$S = \frac{1}{4\pi}\int \mathrm{d}^2 z\,(\partial_z X\,\partial_{\bar{z}}\bar{X} + \partial_{\bar{z}} X\,\partial_z \bar{X}) \tag{12.178}$$

where X^α, $\bar{X}^\alpha$ are three complex coordinates (in which the twist acts diagonally, as in (10.27), for example). Since S is quadratic the required correlation function Z factorizes into a quantum and classical part

$$Z = Z_q \sum_{X_{cl}} e^{-S_{cl}} \tag{12.179}$$

and the fixed-point dependence enters via S_{cl}. To see how, consider a loop $\mathscr{C}$ enclosing the two twist fields at z_a, z_b *with net zero twist*. Thus if p and q are the smallest integers such that

$$pk = ql, \tag{12.180}$$

$\mathscr{C}$ encircles z_a p times in an anti-clockwise sense and z_b l times in a clockwise sense. Then the shift in X_{cl} around this closed path is

$$\Delta X_{cl} \equiv \oint_{\mathscr{C}} dz\, \partial_z X_{cl} + \oint_{\mathscr{C}} d\bar{z}\, \partial_{\bar{z}} X_{cl} \tag{12.181}$$

and this is required to be the lattice vector v arising in the product of the space group elements

$$(\theta^k, l_a)^p(\theta^l, l_b)^{-q} = [\theta^{pk}, (1 - \theta^{pk})(f_a + \lambda_a)]$$
$$\times [\theta^{-ql}, (1 - \theta^{-ql})(f_b + \lambda_b)] = (1, v) \tag{12.182a}$$

where

$$v = (1 - \theta^{pk})(f_a - f_b + \lambda) \tag{12.182b}$$

is related to the difference between the fixed points. This determines the overall normalization of $\partial_z X_{cl}$ (or $\partial_{\bar{z}} X_{cl}$), and hence the value of S_{cl}, the required correlation function, and the Yukawa coupling. Because of the *exponential* suppression there is the possibility that for suitable values of the scale factors (moduli) of the orbifold and the other (angular) deformation parameters we can achieve the required hierarchy of Yukawa couplings at the string scale needed to generate the observed hierarchy of fermion masses. Recent work[(10)] suggests that a reasonable fit to the physical fermion masses *is* feasible for some orbifold models. However, the Yukawa couplings also determine the mixing of the electroweak eigenstates to form the mass eigenstates. This is encoded in the Cabibbo–Kobayashi–Maskawa matrix, and there is no possibility of fitting its parameters, at least at the renormalizable level.

In fact, the calculation of the interaction terms in the Lagrangian (or the superpotential $W(\Phi_i)$) is not sufficient to determine the coupling strength of the *physical* fields because, until the Kähler potential $G(\varphi, \varphi_i^*)$ is known, we do not know that the string states have canonical kinetic terms; diagonalizing and normalizing these could enhance or weaken any hierarchy emerging from the superpotential. The derivation of the Kähler potential from string amplitudes can be done, at least for untwisted moduli fields ($\tilde{U}_a$). The upshot is that the quantity

$$G^i_j \equiv \frac{\partial^2 G}{\partial \varphi_i \, \partial \varphi^{j^*}} \tag{12.183}$$

defined in (5.40), which multiplies the kinetic term $D_\mu \varphi_i \, D^\mu \varphi^{j^*}$ in (5.38), has the generic form

$$G^i_j \sim \delta^i_j \prod_{a=1}^{3} (\tilde{U}_a + \tilde{U}^+_a)^{-p_{ai}} \tag{12.184}$$

with the 'modular weights' p_{ai} fractional numbers typically in the range $(-1, 5)$. Thus there is a power law hierarchy as well as the exponential hierarchy already discussed.

We saw in Chapter 3 that the convergence of the gauge coupling constants in the supersymmetric standard model provides the best (circumstantial) evidence so far for supersymmetry and grand unification. String theory also requires the gauge coupling constants to have a common value[(11)], so one might also construe the above convergence as evidence for string theory. However, the unification scale from the string is at 10^{18} GeV, significantly higher than the energy scale (10^{16} GeV) at which the coupling constants are 'observed' to converge. So the convergence is also the best evidence *against* string theory. It may be that there is additional so-far-unobserved matter, beyond the reach of current accelerators, but lighter than the unification scale. This matter would affect the running of the coupling constants and could delay the unification. There are certainly models in which this occurs[(12)]. Alternatively, it may be that the massive string modes generate threshold corrections that have the same effect, although this has not so far been achieved in model orbifold calculations[(13)]. At any rate, the aesthetic and theoretical arguments for string theory, that it is the only known theory that can provide a consistent *quantum* theory of all of the interactions observed in nature, remain compelling. The 'evidence' against it is recognized as a problem, but it is not (yet?) regarded as fatal.

The fact that the gauge coupling constants unify at the string scale is not necessarily evidence of non-abelian unification. However, the fact that the known matter is organized into (three complete generations of) a few representations ($\bar{\mathbf{5}}$, **10**, **1**) of SU(5) or a single representation (**16**) of SO(10) certainly points towards a grand unification group that contains something like these groups. If we assume, as we shall, that this (supergravity) GUT originates in the string theory, then the possible models are constrained by a (fairly) general theorem that excludes the existence of matter in the adjoint or higher representations[(14)]. This means that the minimal SU(5) theory, which we discussed in Chapter 6, *cannot* emerge from string theory; the adjoint scalar multiplet Σ, introduced in (6.16) in order to break the SU(5), does not occur. The allowed matter representations are **5**, **10**, **1** and their conjugates. Thus we require a GUT with electroweak and GUT Higgs particles in one or more of these representations. The flipped SU(5) $\times$ U(1) model is then a prime candidate in this context, and realistic models have been constructed using the direct construction discussed in Chapter 11, in which all degrees of freedom other than these relating to the four-dimensional space-time are fermionized[15]. The model is called 'flipped' because, compared with the ordinary (minimal) SU(5), lepton and quark flavour assignments to the representations are interchanged (or flipped). Thus the assignments are now

$$\bar{\mathbf{5}}: L, u^c \tag{12.185a}$$

$$\mathbf{10}: Q, d^c, \nu^c \tag{12.185b}$$

$$\mathbf{1}: e^c \tag{12.185c}$$

where L, Q denote the (left chiral) lepton and quark doublets, and u^c, d^c, ν^c, e^c the corresponding singlets. Although the $SU(3)_c$ and $SU(2)_L$ gauge groups are (still) embedded in SU(5), the $U(1)_{em}$ is not, as is immediately apparent from the assignment of e^c to a singlet representation. This is why the GUT must be enlarged to SU(5) × U(1). The appearance of the electroweak singlet ν_c in the **10** also indicates that the SU(5) can be broken using this representation for the GUT scalars, rather than the adjoint (**24**) scalars of the minimal SU(5). It is also natural that besides the (three generations of) chiral matter there is additional 'vector-like' matter, in this case in $\mathbf{5} + \bar{\mathbf{5}}$ and $\mathbf{10} + \overline{\mathbf{10}}$ representations, besides the Higgs scalars required to ensure the spontaneous breaking of the GUT and electroweak symmetry breaking. This additional matter removed the gap between the 'observed' unification scale of 10^{16} GeV and the string unification scale of 10^{18} GeV, as discussed earlier.

As already mentioned, any specific string model generates a unique supergravity theory that can in principle be compared with experiment. In practice, though, it is necessary to make additional assumptions. For example, besides the 'observable' sector of the theory, which houses the GUT that we have been discussing, there is a 'hidden' sector, with respect to whose gauge group all observable matter is a singlet representation. The hidden-sector gauge couplings are assumed to become large at an intermediate scale and trigger supersymmetry breaking via hierarchically small, soft, supersymmetry-breaking parameters. In principle what happens is fully determined, but in practice these non-perturbative effects are not really calculable, and we have to supplement the model with assumptions about the precise nature of the soft supersymmetry-breaking parameters. Even so, the number of parameters is considerably less than the 20 or so of the minimal SU(5) supergravity GUT discussed earlier. Such models are at the level of making falsifiable predictions[16] for the particle and sparticle spectra, which should be tested soon at the Tevatron and LEP 200. This interplay between string-inspired models and experimental data is the only method currently available for determining which particular string theory really is the 'theory of everything'.

Exercises

12.1 Check that the operators L_n in (12.18) obey the Virasoro algebra (7.96) without the central extension.

12.2 Verify (12.26) for $n \neq 0$. Hence show that $\partial_\tau X$ has conformal dimension $J = 1$.

12.3 Show that $:\exp(-\mathrm{i}\,k.X(z)):$ has conformal dimension $J = -\frac{1}{2}k^2$ directly from (12.21).

12.4 Verify (12.62).

12.5 Calculate the open-bosonic-string vector–vector–tachyon vertex.

12.6 Show that the $\mathrm{O}(k^3)$ term in the three-vector vertex (12.74) would arise from an effective Lagrangian $\mathrm{tr}(F_\mu{}^\nu F_\nu{}^\rho F_\rho{}^\mu)$.

12.7 Calculate the closed-bosonic-string graviton–graviton–tachyon vertex.

12.8 Calculate the closed-bosonic-string three-graviton vertex.

12.9 Verify (12.112).

12.10 Show that if the superstring vertex operators V_Λ, W_Λ are related as in (12.114) then V_Λ has conformal dimension $J = 1$ if and only if W_Λ has conformal dimension $J = \frac{1}{2}$.

12.11 Show that the invariance of the superstring action (8.1) under the rescalings (12.10) requires that ψ^μ acquires a factor $\lambda^{-1/2}$ under the rescaling.

12.12 Show that if the open-string state $|v;\, \epsilon, k\rangle$ is a physical state, obeying $G_{1/2}|\psi\rangle = 0$ then $\epsilon.k = 0$.

12.13 Show that the closed-superstring three-graviton vertex is given by (12.141), and verify (12.143).

12.14 Show that the (Einstein–Hilbert) action (12.144) generates the three-graviton vertex given in (12.141).

12.15 Verify (12.147) and (12.149).

12.16 Calculate the gravitino–gravitino–graviton vertex deriving from the supergravity action (4.65).

References

General references

The books and references that we have found most useful in preparing this chapter are as follows.

Brink L and Hennaux M 1988 *Principles of String Theory* (New York: Plenum)
Dixon L, Friedan D, Martinec E and Shenker S 1987 *Nucl. Phys.* B **282** 13
Green M B, Schwarz J H and Witten E 1987 *Superstring Theory* vol 1 (Cambridge: Cambridge University Press)
Hamidi S and Vafa C 1987 *Nucl. Phys.* B **279** 465
Kaku M 1988 *Introduction to Superstrings* (New York: Springer)
Kaplunovsky V S 1988 *Nucl. Phys.* B **307** 145
Scherk J 1975 *Rev. Mod. Phys.* **47** 123

References in the text

1 Fubini S and Veneziano G 1970 *Nuovo Cimento* A **67** 29
2 Bailin D and Love A 1993 *Introduction to Gauge Field Theory* (Bristol: Institute of Physics Publishing)
3 This form for the four-scalar scattering amplitude was first conjectured by Veneziano G 1968 *Nuovo Cimento* A **57** 190
4 Virasoro M A 1969 *Phys. Rev.* **177** 2309
5 Kawai H, Lewellen D D and Tye S-H H 1986 *Nucl. Phys.* B **269** 1
6 Neveu A and Schwarz J H 1971 *Nucl. Phys.* B **31** 86
Neveu A, Schwarz J H and Thorn C B 1971 *Phys. Rev. Lett.* **27** 1758
7 Thorn C B 1971 *Phys. Rev.* D **4** 1112
Schwarz J H 1971 *Phys. Lett.* **37B** 315
Corrigan E and Olive D 1972 *Nuovo Cimento* A **11** 749
8 Friedan D, Martinec E and Shenker S 1985 *Phys. Lett.* **160B** 55; 1986 *Nucl. Phys.* B **271** 93
9 Neveu A and Schwarz J H 1971 *Phys. Rev.* D **4** 1104
Thorn C B 1971 *Phys. Rev.* D **4** 1112
10 Casas J A, Gomez F and Muñoz C 1992 *Phys. Lett.* **292B** 42
11 Ginsparg P 1987 *Phys. Lett.* **197B** 139
12 Antoniadis I, Ellis J, Kelley S and Nanopoulos D V 1991 *Phys. Lett.* **272B** 31
Kelley S, Lopez J L and Nanopoulos D V 1992 *Phys. Lett.* **278B** 140
Bailin D and Love A 1992 *Phys. Lett.* **280B** 26; 1992 *Mod. Phys. Lett.* **7A** 1485
13 Antoniadis I, Ellis J, Lacaze R and Nanopoulos D V 1991 *Phys. Lett.* **268B** 188
Kalara S, Lopez J L and Nanopoulos D V 1991 *Phys. Lett.* **269B** 84
Bailin D and Love A 1992 *Phys. Lett.* **278B** 125
Ibanez L E, Lust D and Ross G G 1991 *Phys. Lett.* **272B** 251
14 Ellis J, Lopez J L and Nanopoulos D V 1990 *Phys. Lett.* **245B** 375
Font A, Ibãnez L and Quevedo F 1990 *Nucl. Phys.* B **345** 389
15 Antoniadis I, Ellis J, Hagelin J S and Nanopoulos D V 1989 *Phys. Lett.* **231B** 65
16 See, for example,
Lopez J L, Nanopoulos D V and Zichichi A 1993 *CERN Preprint* CERN-TH-6934/93, *Proc. Int. Sch. Subnuclear Physics, 30th Course, 'From Superstrings to the Real Superworld' (Erice, 14–22 July 1992)*; *CERN Preprint* CERN-TH-6926/93 *Proc. INFN Eloisatron Project, 26th Workshop, 'From Superstrings to Supergravity' (Erice, 5–12 December 1992)*

APPENDIX A

WEYL SPINOR FIERZ IDENTITIES

We summarize here the complete set of Fierz identities for Weyl spinors. They may all be derived from the matrix identities (1.85) and (1.88), which in an obvious notation may be written as

$$1 \times 1 = \tfrac{1}{2}\sigma^{\mu} \otimes \bar{\sigma}_{\mu} \tag{A1}$$

or

$$1 \times 1 = \tfrac{1}{2}[1 \otimes 1 - \sigma^{\mu\nu} \otimes \bar{\sigma}_{\mu\nu}]. \tag{A2}$$

Then besides the immediate identities (1.87) and (1.89):

$$(\theta\varphi)(\chi\eta) = -\tfrac{1}{2}[(\theta\eta)(\chi\varphi) - (\theta\sigma^{\mu\nu}\eta)(\chi\sigma_{\mu\nu}\varphi)] \tag{A3}$$

$$(\theta\varphi)(\bar{\chi}\bar{\eta}) = -\tfrac{1}{2}(\theta\sigma^{\mu}\eta)(\bar{\chi}\bar{\sigma}_{\mu}\varphi) \tag{A4}$$

we have

$$(\theta\varphi)(\chi\sigma^{\mu}\bar{\eta}) = -\tfrac{1}{2}[(\theta\sigma^{\mu}\bar{\eta})(\chi\varphi) + 2(\theta\sigma_{\nu}\bar{\eta})(\chi\sigma^{\mu\nu}\varphi)] \tag{A5}$$

$$(\theta\varphi)(\bar{\chi}\bar{\sigma}^{\mu}\eta) = -\tfrac{1}{2}[(\theta\eta)(\bar{\chi}\bar{\sigma}^{\mu}\varphi) - 2(\theta\sigma^{\mu\nu}\eta)(\bar{\chi}\bar{\sigma}_{\nu}\varphi)] \tag{A6}$$

$$\begin{aligned}(\theta\sigma^{\mu}\bar{\varphi})(\chi\sigma^{\nu}\bar{\eta}) = -\tfrac{1}{2}[&(\theta\sigma^{\mu}\bar{\eta})(\chi\sigma^{\nu}\bar{\varphi}) + (\theta\sigma^{\nu}\bar{\eta})(\chi\sigma^{\mu}\bar{\varphi}) \\ &- \eta^{\mu\nu}(\theta\sigma^{\lambda}\bar{\eta})(\chi\sigma_{\lambda}\bar{\varphi}) - \mathrm{i}\,\epsilon^{\mu\nu\kappa\lambda}(\theta\sigma_{\kappa}\bar{\eta})(\chi\sigma_{\lambda}\bar{\varphi})]\end{aligned} \tag{A7}$$

$$\begin{aligned}(\theta\sigma^{\mu}\bar{\varphi})(\bar{\chi}\bar{\sigma}^{\nu}\eta) = -\tfrac{1}{2}[&\eta^{\mu\nu}(\theta\eta)(\bar{\chi}\bar{\varphi}) + 2(\theta\sigma^{\mu\nu}\eta)(\bar{\chi}\bar{\varphi}) \\ &- 2(\theta\eta)(\bar{\chi}\bar{\sigma}^{\mu\nu}\bar{\varphi}) - 4(\theta\sigma^{\nu\lambda}\eta)(\bar{\chi}\bar{\sigma}_{\lambda}{}^{\mu}\bar{\varphi})]\end{aligned} \tag{A8}$$

$$\begin{aligned}(\theta\varphi)(\chi\sigma^{\mu\nu}\eta) = -\tfrac{1}{2}[&(\theta\eta)(\chi\sigma^{\mu\nu}\varphi) + (\theta\sigma^{\mu\nu}\eta)(\chi\varphi) \\ &- (\theta\sigma^{\mu\lambda}\eta)(\chi\sigma_{\lambda}{}^{\nu}\varphi) + (\theta\sigma^{\nu\lambda}\eta)(\chi\sigma_{\lambda}{}^{\mu}\varphi)]\end{aligned} \tag{A9}$$

$$\begin{aligned}(\theta\varphi)(\bar{\chi}\bar{\sigma}^{\mu\nu}\bar{\eta}) = -\tfrac{1}{4}[&(\theta\sigma^{\nu}\bar{\eta})(\bar{\chi}\bar{\sigma}^{\mu}\varphi) - (\theta\sigma^{\mu}\bar{\eta})(\bar{\chi}\bar{\sigma}^{\nu}\varphi) \\ &+ \mathrm{i}\,\epsilon^{\mu\nu\kappa\lambda}(\theta\sigma_{\kappa}\bar{\eta})(\bar{\chi}\bar{\sigma}_{\lambda}\varphi)]\end{aligned} \tag{A10}$$

$$\begin{aligned}(\theta\sigma^{\mu\nu}\varphi)(\chi\sigma^{\lambda}\bar{\eta}) = \tfrac{1}{4}[&\eta^{\mu\lambda}(\theta\sigma^{\nu}\bar{\eta})(\chi\varphi) - \eta^{\nu\lambda}(\theta\sigma^{\mu}\bar{\eta})(\chi\varphi) \\ &+ \mathrm{i}\,\epsilon^{\mu\nu\lambda\rho}(\theta\sigma_{\rho}\bar{\eta})(\chi\varphi)] + \tfrac{1}{2}[(\theta\sigma^{\nu}\bar{\eta})(\chi\sigma^{\lambda\mu}\varphi) \\ &- (\theta\sigma^{\mu}\bar{\eta})(\chi\sigma^{\lambda\nu}\varphi) + \mathrm{i}\,\epsilon^{\mu\nu\kappa\rho}(\theta\sigma_{\kappa}\bar{\eta})(\chi\sigma_{\rho}{}^{\lambda}\varphi)]\end{aligned} \tag{A11}$$

$$\begin{aligned}(\theta\sigma^{\mu\nu}\varphi)(\bar{\chi}\bar{\sigma}^{\lambda}\eta) = -\tfrac{1}{4}[&\eta^{\mu\lambda}(\theta\eta)(\bar{\chi}\bar{\sigma}^{\nu}\varphi) - \eta^{\nu\lambda}(\theta\eta)(\bar{\chi}\bar{\sigma}^{\mu}\varphi) \\ &+ \mathrm{i}\,\epsilon^{\mu\nu\lambda\rho}(\theta\eta)(\bar{\chi}\bar{\sigma}_{\rho}\varphi)] - \tfrac{1}{2}[(\theta\sigma^{\mu\lambda}\eta)(\bar{\chi}\bar{\sigma}^{\nu}\varphi) \\ &- (\theta\sigma^{\nu\lambda}\eta)(\bar{\chi}\bar{\sigma}^{\mu}\varphi) + \mathrm{i}\,\epsilon^{\mu\nu\kappa\rho}(\theta\sigma_{\kappa}{}^{\lambda}\eta)(\bar{\chi}\bar{\sigma}_{\rho}\varphi)]\end{aligned} \tag{A12}$$

$$
\begin{aligned}
(\theta\sigma^{\mu\nu}\varphi)(\bar{\chi}\bar{\sigma}^{\kappa\lambda}\bar{\eta}) = -\tfrac{1}{8}[&(\eta^{\mu\lambda}\eta^{\nu\kappa} - \eta^{\mu\kappa}\eta^{\nu\lambda})(\theta\sigma^{\rho}\bar{\eta})(\bar{\chi}\bar{\sigma}_{\rho}\varphi) \\
&+ \eta^{\mu\kappa}(\theta\sigma^{\nu}\eta)(\bar{\chi}\bar{\sigma}^{\lambda}\varphi) + \eta^{\mu\kappa}(\theta\sigma^{\lambda}\eta)(\bar{\chi}\bar{\sigma}^{\nu}\varphi) \\
&- \eta^{\nu\kappa}(\theta\sigma^{\mu}\eta)(\bar{\chi}\bar{\sigma}^{\lambda}\varphi) - \eta^{\nu\kappa}(\theta\sigma^{\lambda}\eta)(\bar{\chi}\bar{\sigma}^{\mu}\varphi) \\
&+ \eta^{\nu\lambda}(\theta\sigma^{\mu}\eta)(\bar{\chi}\bar{\sigma}^{\kappa}\varphi) + \eta^{\nu\lambda}(\theta\sigma^{\kappa}\eta)(\bar{\chi}\bar{\sigma}^{\mu}\varphi) \\
&- \eta^{\mu\lambda}(\theta\sigma^{\nu}\eta)(\bar{\chi}\bar{\sigma}^{\kappa}\varphi) - \eta^{\mu\lambda}(\theta\sigma^{\kappa}\eta)(\bar{\chi}\bar{\sigma}^{\nu}\varphi) \\
&+ \mathrm{i}\,\epsilon^{\mu\nu\kappa\rho}(\theta\sigma_{\rho}\eta)(\bar{\chi}\bar{\sigma}^{\lambda}\varphi) - \mathrm{i}\,\epsilon^{\mu\nu\lambda\rho}(\theta\sigma_{\rho}\eta)(\bar{\chi}\bar{\sigma}^{\kappa}\varphi) \\
&- \mathrm{i}\,\epsilon^{\kappa\lambda\mu\rho}(\theta\sigma^{\nu}\eta)(\bar{\chi}\bar{\sigma}_{\rho}\varphi) + \mathrm{i}\,\epsilon^{\kappa\lambda\nu\rho}(\theta\sigma^{\mu}\eta)(\bar{\chi}\bar{\sigma}_{\rho}\varphi)]. \qquad \text{(A13)}
\end{aligned}
$$

We leave the verification of these as an exercise in which the following identities may also prove useful:

$$\sigma^{\mu\nu}\sigma^{\lambda} = -\tfrac{1}{2}[\eta^{\mu\lambda}\sigma^{\nu} - \eta^{\nu\lambda}\sigma^{\mu} + \mathrm{i}\,\epsilon^{\mu\nu\lambda\rho}\sigma_{\rho}] \qquad \text{(A14)}$$

$$\bar{\sigma}^{\mu}\sigma^{\nu\lambda} = \tfrac{1}{2}[\eta^{\mu\nu}\bar{\sigma}^{\lambda} - \eta^{\mu\lambda}\bar{\sigma}^{\nu} + \mathrm{i}\,\epsilon^{\mu\nu\lambda\rho}\bar{\sigma}_{\rho}] \qquad \text{(A15)}$$

$$
\begin{aligned}
\sigma^{\mu\nu}\sigma^{\kappa\lambda} = -\tfrac{1}{4}[&\eta^{\mu\kappa}\eta^{\nu\lambda} - \eta^{\mu\lambda}\eta^{\nu\kappa} + \mathrm{i}\,\epsilon^{\mu\nu\kappa\lambda} \\
&+ 2(\eta^{\mu\kappa}\sigma^{\nu\lambda} + \eta^{\nu\lambda}\sigma^{\mu\kappa} - \eta^{\mu\lambda}\sigma^{\nu\kappa} - \eta^{\nu\kappa}\sigma^{\mu\lambda})]. \qquad \text{(A16)}
\end{aligned}
$$

APPENDIX B

CYCLIC SYMMETRY OF THE OPEN-STRING SCATTERING AMPLITUDE

The proof of this requires the use of the residual conformal invariance that preserves the conformal gauge choice made in (7.13), (7.14). This residual symmetry is in fact SL(2, R): the group of 2×2 real matrices of unit determinant. It is the group of all (1, 1) analytic mappings $z \to z'$ of the upper half complex plane into itself:

$$z \to z' = \frac{az + b}{cz + d} \tag{B1a}$$

with

$$a, b, c, d \qquad \text{real}$$
$$ad - bc = 1. \tag{B1b}$$

Such mappings are called 'Möbius transformations'.

It is easy to see that this symmetry is a (finite-dimensional) subalgebra, generated by L_1, L_0, L_{-1}, of the infinite-dimensional Virasoro algebra. The generator L_n is associated with the infinitesimal transformation

$$z \to z' = z - \epsilon z^{n+1} \tag{B2}$$

as given in (12.16). Thus the finite operator $e^{\lambda L_n}$ generates the transformation obeying

$$\frac{dz}{dt} = -z^{n+1} \tag{B3a}$$

with

$$z(0) = z \qquad z(\lambda) = z'. \tag{B3b}$$

Hence

$$e^{\lambda L_1} z\, e^{-\lambda L_1} = z' = \frac{z}{1 + \lambda z} \tag{B4a}$$

$$e^{\lambda L_0} z\, e^{-\lambda L_0} = z' = e^{-\lambda} z \tag{B4b}$$

$$e^{\lambda L_{-1}} z\, e^{-\lambda L_{-1}} = z' = z - \lambda \tag{B4c}$$

and we see that the general transformation is given by (B1a). The restriction

$ad - bc = 1$ is convenient, since there are only three independent parameters.

The zero-momentum state $|0; 0\rangle$ which appears in (12.59) (uniquely) has the property that it is annihilated by all three generators of the SL(2, R) algebra

$$L_i|0, 0\rangle = 0 \qquad i = 0, \pm 1 \tag{B5}$$

with L_i given in (12.24). We leave it as an exercise to verify this statement, but note in passing that, since it is annihilated by L_0, and *not* by $L_0 - a$, $|0; 0\rangle$ is *not* a physical state. We note also that the SL(2, R) subalgebra generated by L_i ($i = \pm 1, 0$) is unaffected by the central extension term in (7.96), so

$$[L_1, L_{-1}] = 2L_0 \tag{B6}$$

just as for the classical algebra.

To prove the cyclic property we consider a general SL(2, R) transformation

$$\Lambda(T) = \exp(\lambda_{-1}L_{-1} + \lambda_0 L_0 + \lambda_1 L_1). \tag{B7}$$

Then it follows from (B5) that the state $|0; 0\rangle$ satisfies

$$\Lambda(T)|0; 0\rangle = |0; 0\rangle. \tag{B8}$$

Also, the transformation $\Lambda(T)$ generates the conformal transformation (B1), as we have seen, and since the vertex operators $V_\Lambda(k, y)$ that appear in (12.64) have conformal dimension $J = 1$, as we argued after (12.13), it follows from (12.19) that

$$\Lambda(T)\left[V_\Lambda(k, y)\frac{\mathrm{d}y}{y}\right]\Lambda(T)^{-1} \equiv V_\Lambda(k, y')\frac{\mathrm{d}y'}{y'} = V_\Lambda(k, y)\frac{\mathrm{d}y}{y}. \tag{B9}$$

Thus each of the factors $V_{\Lambda_i}(k_i, y_i)\,\mathrm{d}y_i/y_i$ ($i = 3, \ldots, N-1$) appearing in (12.64) can be replaced by the SL(2, R)-transformed factor, if we choose. Further, it is easy to see that the transformation (B1) preserves the order of the events (12.55) on the boundary of the world sheet.

This would be sufficient to demonstrate the invariance of the integral in (12.64) *if* we were integrating over all of the variables y_i ($i = 1, 2, \ldots, N$), whereas in actuality $y_N = 0$, $y_2 = 1$ and $y_1 \to \infty$ are *fixed*. If we were to allow integrations over all of the y_i, including $i = 1, 2, N$, this would overcount because we would be including configurations that can be mapped into each other by the use of the Möbius transformations given in (B1). In fact the overcounting would be by an infinite amount, equal to the volume of the non-compact group.

This is analogous to the problem encountered when quantizing gauge theories using the path integral technique, as discussed in Bailin and Love[(1)], Chapter 10, for example. The naive first guess is to perform a functional integral over all gauge configurations $A^\mu_a(x)$, including those that are gauge transformations of each other. Then because the Yang–Mills Lagrangian is gauge invariant, the integrand is constant over the infinite surface in gauge field space obtained from a given $A^\mu_a(x)$ by applying all possible gauge transformations.

Similarly, in the present case integrating over *all* of the y_i overcounts each inequivalent set of y_i by the volume of SL(2, R). As explained in §10.4 of Bailin and Love[(1)], the correct procedure is to impose delta-function constraints on *three* of the integration parameters

$$y_i = y_{i0} \qquad (i = 1, 2, N) \tag{B10}$$

and to include in the integrand the Fadeev–Popov determinant arising from the (infinitesimal) SL(2, R) transformations (B4)

$$y_{i0} \rightarrow y'_{i0} = y_i - (\lambda_{-1} + \lambda_0 y_{i0} + \lambda_1 y_{i0}^2). \tag{B11}$$

The required determinant is then the Jacobian

$$\begin{aligned}
\det\left(\frac{\partial y'_{i0}}{\partial \lambda_j}\right) &= \begin{vmatrix} 1 & y_{10} & y_{10}^2 \\ 1 & y_{20} & y_{20}^2 \\ 1 & y_{N0} & y_{N0}^2 \end{vmatrix} \qquad (i = 1, 2, N;\ j = -1, 0, 1) \\
&= (y_{10} - y_{20})(y_{20} - y_{N0})(y_{N0} - y_{10}) \\
&\rightarrow -y_{10}^2 \qquad \text{as } y_{10} \rightarrow \infty,\ y_{20} = 1,\ y_{N0} = 0
\end{aligned} \tag{B12}$$

and this is precisely what is required to convert the factor $(y_{10}y_{20}y_{N0})^{-1}$ associated with V_{Λ_1}, V_{Λ_2} and V_{Λ_N} into $y_{10}y_{N0}^{-1}$ as appears in (12.64) when k_1 and k_N are tachyons. Thus the amplitude (12.64) *is* invariant under the (Möbius) transformations, and can be written in the more symmetric form

$$\begin{aligned}
A_N = g^{N-2} \int \prod_{i=1}^{N} \frac{\mathrm{d}y_i}{y_i} \prod_{j=2}^{N} \theta(y_{j-1} - y_j)\delta(y_a - y_{a0})\delta(y_b - y_{b0})\delta(y_c - y_{c0}) \\
\times (y_a - y_b)(y_b - y_c)(y_c - y_a)\langle 0; 0| V_0(k_1, y_1) V_0(k_2, y_2) \cdots \\
\times V_0(k_N, y_N)|0; 0\rangle
\end{aligned} \tag{B13}$$

and we previously selected $(a, b, c = 1, 2, N)$ with the fixed values

$$(y_{10}, y_{20}, y_{N0}) = (\infty, 1, 0). \tag{B14}$$

We now use this invariance to demonstrate the cyclic invariance of the amplitude (12.64). For simplicity we consider only the case when all N particles are tachyons. The general Möbius transformation preserves the cyclic order of the coordinates y_i. In fact, since

$$\frac{\mathrm{d}y'}{\mathrm{d}y} = (cy + d)^{-2} \tag{B15}$$

the transformation preserves the actual order of those events for which $cy_i + d > 0$, and the actual order of those with $cy_i + d < 0$. The cyclic aspect operates when we have coordinates of both types. We consider just such a transformation so that points y_i satisfying

$$y_N < y_{N-1} < y_{N-2} < \cdots < y_3 < y_2 < y_1 \tag{B16}$$

are mapped into y_i' satisfying

$$y_{N-1}' < y_{N-2}' < \cdots < y_2' < y_1' < y_N' \tag{B17}$$

for example. To verify the cyclic property we need to be able to commute the vertex factor $V_0(k_N, y_N')$ past the preceding vertex factors so that the resulting expression is just the cyclic permutation applied to (B13). In fact commuting $V_0(k_N, y_N')$ past *all* of the preceding vertex operators gives an overall factor of unity. To see this we note that commuting $V_0(k_N, y_N')$ past any one of the tachyon vertex operators gives a phase factor

$$V_0(k_i, y_i')V_0(k_N, y_N') = V_0(k_N, y_N')V_0(k_i, y_i')\, e^{i\pi k_i . k_N}. \tag{B18}$$

This may be proved by using

$$e^A\, e^B = e^B\, e^A\, e^{[A, B]} \tag{B19}$$

when $[A, B]$ is a c-number. In our case the required commutator is

$$[i\, k_i . X(y_i'), i\, k_N . X(y_N')] = i\,\pi k_i . k_N \tag{B20}$$

since $y_i' < y_N'$. Then taking $V_0(k_N, y_N')$ past all of the $V_0(k_i, y_i')$ gives an overall phase factor

$$\exp\left[i\,\pi k_N . \sum_{i=1}^{N-1} k_i\right] = e^{-i\pi k_N^2} = 1. \tag{B21}$$

Thus the SL(2, R) invariance allows us to prove that the amplitude (B13) is equal to a similar expression in which the cyclic transformation (12.57) has been applied.

Reference

1 Bailin D and Love A 1993 *Introduction to Gauge Field Theory* (Bristol: Institute of Physics Publishing)

APPENDIX C

COHERENT STATE METHODS

In order to evaluate quantities such as (12.78), it suffices to consider a single oscillator mode with annihilation operator α and creation operator $\alpha^\dagger$ satisfying

$$[\alpha, \alpha^\dagger] = 1. \tag{C1}$$

This is because different modes commute, and we can always normalize the operators so that (C1) is satisfied. In order to establish (12.79), therefore, we need only consider operators A, B with

$$A = a_1\alpha^\dagger + a_2\alpha \tag{C2a}$$

$$B = b_1\alpha^\dagger + b_2\alpha. \tag{C2b}$$

Then

$$:\mathrm{e}^A: = \mathrm{e}^{a_1\alpha^\dagger}\,\mathrm{e}^{a_2\alpha} \tag{C3a}$$

$$:\mathrm{e}^B: = \mathrm{e}^{b_1\alpha^\dagger}\,\mathrm{e}^{b_2\alpha} \tag{C3b}$$

and

$$\langle 0|:\mathrm{e}^A::\mathrm{e}^B:|0\rangle = \langle 0|\mathrm{e}^{a_2\alpha}\,\mathrm{e}^{b_1\alpha^\dagger}|0\rangle. \tag{C4}$$

The 'coherent' state $|b_1\rangle$ is defined by

$$|b_1\rangle \equiv \mathrm{e}^{b_1\alpha^\dagger} = \sum_{n=0}^{\infty} \frac{b_1^n}{\sqrt{n!}}|n\rangle \tag{C5a}$$

where

$$|n\rangle \equiv \frac{1}{\sqrt{n!}}(\alpha^\dagger)^n|0\rangle \tag{C5b}$$

is the standard (normalized) number operator eigenstate satisfying

$$N|n\rangle = n|n\rangle \tag{C6a}$$

with

$$N \equiv \alpha^\dagger\alpha \tag{C6b}$$

and

$$\langle m|n\rangle = \delta_{mn}. \tag{C6c}$$

So the right-hand side of (C4) is the scalar product of two coherent states

$$\langle a_2|b_1\rangle = \sum_{m,n} \frac{a_2^{*m} b_1^n}{\sqrt{m!}\sqrt{n!}} \langle m|n\rangle = e^{a_2^* b_1} = e^{\langle 0|AB|0\rangle}. \tag{C7}$$

Thus we have shown that

$$\langle 0|{:}e^A{:}{:}e^B{:}|0\rangle = e^{\langle 0|AB|0\rangle} \tag{C8}$$

which is (12.79).

This may be generalized to

$$\langle 0|{:}e^{A_1}{:}{:}e^{A_2}{:} \cdots {:}e^{A_n}{:}|0\rangle = \exp\left(\sum_{i<j} \langle 0|A_i A_j|0\rangle\right) \tag{C9}$$

which arises when we calculate the N-tachyon scattering amplitude, using (12.64) for example. Using (12.80) we see that the contribution of the non-zero modes to the integrand is

$$\prod_{i<j} (1 - y_j/y_i)^{-k_i \cdot k_j}. \tag{C10}$$

The contribution from the zero-mode pieces is easily determined using (12.61). We get

$$\langle 0; 0|Z_0(k_1, y_1) Z_0(k_2, y_2) \cdots Z_0(k_n, y_n)|0; 0\rangle = \left(\prod_i y_i\right)\left(\prod_{i<j} y_i^{-k_i \cdot k_j}\right) \tag{C11}$$

so the integrand of (12.64) is

$$\langle 0; 0| \prod_i (V_0(k_i, y_i)/y_i)|0; 0\rangle = \prod_{i<j} (y_i - y_j)^{-k_i \cdot k_j}. \tag{C12}$$

Using the symmetric form (B13) of (12.64) established in Appendix B we see that the N-tachyon scattering amplitude is given by

$$A_N = g^{N-2} \int \prod_{\substack{i=1 \\ i<j}}^{N} dy_i \prod_{j=2}^{N} \theta(y_{j-1} - y_j) \delta(y_a - y_{a0}) \delta(y_b - y_{b0}) \delta(y_c - y_{c0})$$
$$\times (y_a - y_b)(y_b - y_c)(y_c - y_a) \prod_{k<l} (y_l - y_m)^{-k_l \cdot k_m} \tag{C13}$$

which is the Koba–Nielsen formula[1].

Reference

1 Koba Z J and Nielsen H B 1969 *Nucl. Phys.* B **12** 517

APPENDIX D

CLOSED-STRING INTEGRALS

The integral (12.109), which is required in order to evaluate the four-tachyon amplitude in the closed-bosonic-string theory, is a special case of the more general integral

$$I(\alpha, n; \beta, m) = \int \frac{d^2z}{\pi} |z|^\alpha |1 - z|^\beta z^n (1 - z)^m \tag{D1}$$

which is convergent for

$$\begin{aligned} &\mathrm{Re}(\alpha + \beta + m + n + 2) < 0 \\ &\mathrm{Re}(\alpha + n + 2) > 0 \\ &\mathrm{Re}(\beta + m + 2) > 0. \end{aligned} \tag{D2}$$

We substitute an integral representation of $|z|^\alpha$

$$|z|^\alpha = \frac{1}{\Gamma(-\alpha/2)} \int_0^\infty ds\, s^{-\alpha/2 - 1} e^{-s|z|^2} \tag{D3}$$

which follows from the standard definition of the Γ function

$$\Gamma(p) = \int_0^\infty du\, u^{p-1} e^{-u} \tag{D4}$$

by changing the integration variable to $u = s|z|^2$. Similarly for $|1 - z|^\beta$. Then

$$I = \frac{1}{\Gamma(-\alpha/2)\Gamma(-\beta/2)} \int_0^\infty ds\, s^{-\alpha/2 - 1} \int_0^\infty dt\, t^{-\beta/2 - 1} \times \int \frac{d^2z}{\pi} z^n (1 - z)^m e^{-s|z|^2 - t|1 - z|^2}. \tag{D5}$$

Next we do the z-integral. This is easily done using the generating integral

$$J(\lambda, \mu) \equiv \int \frac{d^2z}{\pi} \exp[-s|z|^2 - t|1 - z|^2 + \lambda z + \mu(1 - z)]. \tag{D6}$$

This integral separates into standard (Gaussian) integrals when Cartesian coordinates are used for $z = x + \mathrm{i}\, y$. Then

$$J(\lambda, \mu) = \frac{1}{s + t} \exp\left(\frac{\lambda t + \mu s - st}{s + t}\right) \tag{D7}$$

so the z-integral in (D5) is obtained by differentiating with respect to λ, μ:

$$\int \frac{d^2 z}{\pi} z^n (1-z)^m e^{-s|z|^2 - t|1-z|^2} = \frac{t^n s^m}{(s+t)^{n+m+1}} e^{-st/(s+t)}. \quad \text{(D8)}$$

Of course this result follows only when n, m are integral, but we can continue to the case when they are non-integral. Next we change integration variables to u and x where

$$u = \frac{st}{s+t} \quad \text{(D9}a\text{)}$$

$$x = \frac{t}{s+t}. \quad \text{(D9}b\text{)}$$

The domain of integration is

$$0 \leqslant u < \infty \qquad 0 \leqslant x \leqslant 1. \quad \text{(D10)}$$

Then the required integral (D5) is[1]

$$\begin{aligned} I &= \frac{1}{\Gamma(-\alpha/2)\Gamma(-\beta/2)} \int_0^1 dx\, x^{n+\alpha/2}(1-x)^{m+\beta/2} \int_0^\infty du\, u^{-\alpha/2-\beta/2-2} e^{-u} \\ &= \frac{\Gamma(1+n+\alpha/2)\Gamma(1+m+\beta/2)\Gamma(-\alpha/2-\beta/2-1)}{\Gamma(-\alpha/2)\Gamma(-\beta/2)\Gamma(2+m+n+\alpha/2+\beta/2)}. \end{aligned} \quad \text{(D11)}$$

Reference

1 Gross D J, Harvey J A, Martinec E and Rohm R 1986 *Nucl. Phys.* B **267** 75

INDEX

9 780750 302678

For Product Safety Concerns and Information please contact our EU representative GPSR@taylorandfrancis.com Taylor & Francis Verlag GmbH, Kaufingerstraße 24, 80331 München, Germany

Printed and bound by CPI Group (UK) Ltd, Croydon, CR0 4YY
01/05/2025
01858571-0001